Breaking History

JARED KUSHNER

Breaking History

A White House Memoir

BROADSIDE BOOKS

HarperCollins books may be purchased for educational, business, or sales promotional use. For information, please email the Special Markets Department at SPsales@harpercollins.com.

Broadside Books™ and the Broadside logo are trademarks of HarperCollins Publishers.

FIRST EDITION

Library of Congress Cataloging-in-Publication Data has been applied for.

ISBN 978-0-06-322148-2

22 23 24 25 26   LSC   10 9 8 7 6 5 4 3 2 1

To Arabella, Joseph, and Theodore—
may you always chase your dreams and use your
God-given potential to improve the lives of others.

If you can fill the unforgiving minute
    With sixty seconds' worth of distance run,
Yours is the Earth and everything that's in it,
    And—which is more—you'll be a Man, my son!
                           —Rudyard Kipling

Contents

Preface

I never planned to write a book, but then again I never planned to work in the White House.

As my time in government was coming to an end, several friends encouraged me to record my memories while they were still fresh. After years of nonstop action, I paused long enough to see the panorama of all I had experienced inside one of the most consequential presidencies. While I thought this chapter of my life was closing, I realized that my service would not be complete until I captured this history.

The story that follows is not your typical White House memoir, because mine was not a typical Washington experience. My untraditional role as senior adviser to a unique president made for a journey that would be hard for a writer to script if it wasn't true.

When Donald J. Trump announced his candidacy, I had no intention of getting involved in his campaign. Before long, however, I met men and women across the country who felt like Trump was finally giving them a voice, and they inspired me to play a bigger role than I had ever expected. After the 2016 election, Ivanka and I left behind our lives in New York and moved to Washington with our three young children. We knew we would face challenges, but we had no idea of the intensity of the storm that awaited us. It was probably better that we didn't.

Nothing could have prepared us for the ferociousness of Washington—the attacks, the investigations, the false and salacious media reports, and perhaps worst of all, the backstabbing within the West Wing itself. On several occasions I wondered if Ivanka and I had made the wrong decision about working in government. Yet we had been given this unexpected chance to serve, and it was up to us to make it count.

Each day was a race against our limited time in office. In an environment of maximum pressure, I learned to ignore the noise and distractions and instead to push for results that would improve lives. Across four years, I helped renegotiate the largest trade deal in history, pass bipartisan criminal justice reform, and launch Operation Warp Speed to deliver a safe and effective COVID-19 vaccine in record time. Humbled by the complexity of the task, I orchestrated some of the most significant breakthroughs in diplomacy in the last fifty years. In what has become known as the Abraham Accords, five Muslim-majority countries—the United Arab Emirates, Bahrain, Kosovo, Morocco, and Sudan—signed peace agreements with Israel. And Saudi Arabia and other members of the Gulf Cooperation Council resolved a bitter diplomatic and economic rift with Qatar, paving the way for additional peace deals in the future.

The Abraham Accords were a true turning point in history. If nurtured, they have the potential to bring about the complete end of the Arab-Israeli conflict that has existed ever since the founding of the State of Israel, seventy-five years ago. Already, hundreds of thousands of Arabs can make pilgrimages to the holy sites in Jerusalem. Israeli and Arab innovators, scientists, and business leaders are forging partnerships to create jobs, build infrastructure, and improve the lives of people throughout the Middle East and around the world.

As we advanced our strategy in the Middle East, we couldn't publicly discuss our approach or the positive signs we were seeing from Arab leaders. Our negotiations progressed on a knife's edge. A single untimely leak could have prompted traditionalists in the region to oppose Arab leaders who were bravely breaking with the past to make peace with Israel. Experts initially dismissed our goals as impossible, and critics delighted in my every stumble. Yet I pursued what I believed was the most logical pathway forward. Since I left government, people have often asked me how we reached these breakthroughs. I have done my best in this book to chronicle the surprising events that made them possible.

Throughout the Trump presidency, the media relied on leaks by officials who often had personal agendas. Until I saw high-stakes politics from the inside, I didn't realize how much goes on that the press fails to capture. The gap between the media's portrayal of events and the reality is far wider than I ever imagined. I eventually came to see that staff in the White House can spend their time trying to shape public perception, or they can spend it getting things done. Every administration wrestles with this challenge. It is the ticking clock in the background of every story in this book.

Many authors—including former senior administration officials—have tried to explain Trump through a conventional lens. Most of these accounts fail to convey how Trump thinks, why he acts the way he does, and what really happened in the Oval Office. The truth was often hiding in plain sight. Through his untraditional style, Trump delivered results that were previously unimaginable: five major trade deals, tax cuts for working families, massive deregulation, the lowest unemployment in fifty years, criminal justice reform, a COVID-19 vaccine in less than a year, confronting China, defeating ISIS, no new wars, and peace deals in the Middle East. In this book, I don't try to speak for Trump, but I do share a lot of previously undisclosed personal interactions that will hopefully give readers a deeper understanding of Trump's personality and management style.

During my four years in the White House, I learned countless lessons that changed my perspective about how the government—and the world—really works. Three stand out.

The first is that it's easy to make promises, but it's hard to achieve results. Trump came into office without an army of experienced bureaucrats and Washington insiders. Finding people who both believed in his agenda and knew how to operate the levers of power proved to be an ongoing challenge. At every turn, people within the government tried to prevent the president from keeping his promises to move the American embassy to Jerusalem, withdraw from the Iran deal, build the wall on the nation's southern border, and renegotiate NAFTA, among many bold actions. I met hundreds of smart, competent, and patriotic people who worked tirelessly behind the scenes to get things done. Yet it takes only one bureaucratic barrier, congressional complication, or powerful individual to stop progress. Washington is programmed to resist change, even though change is what voters say they want most.

I remember one meeting that typified the resistance Trump faced in Washington from both Republicans and Democrats. A veteran of the George W. Bush administration came to see me to discuss US-China trade policy. While he fully agreed with our aims on China, he thought that using tariffs was a grave mistake. When I asked him what he would recommend instead, he suggested more rounds of talks. I said the first thing that came to mind: "So you want us to accomplish something you couldn't by doing it the same way you did it?" For the Washington establishment, the answer to that question was a resounding yes. Many Beltway insiders are experts at pointing out problems, but they're even better at shutting down solutions. When confronted with the potential risks of change, they play it safe for fear that any disruption to the current system will jeopardize their political careers. This explains why even some of Trump's own cabinet members clashed with him and those of us who believed that it was time to take calibrated risks and deliver more opportunities for the American people. Instead of spending endless energy diagnosing the problem, I focused on clearly defining the optimal solution and then worked backwards to reach the best possible outcome.

Second, I learned that our political differences are not always as insurmountable as we think. Ordinarily, the Washington game revolves around the party out of power trying to stop the party in power from accomplishing its priorities. While initially I found this frustrating, I learned to keep moving ahead and to focus on the long game. Almost all of the greatest accomplishments of the administration involved former adversaries coming together to make the lives of normal people better. Rather than starting from two different sides of the table on any given issue—from criminal justice reform to peace deals in the Middle East—I tried to bring everyone to the same side of the table to agree on shared goals and search for win-win solutions. I wasn't always successful, but it is the responsibility of those in power to try. We can't solve problems by talking only to those who agree with us. For anyone who's looking to advance bipartisanship, I hope this book provides insight into how it's possible—and why it often fails.

Finally, we all have the ability to make a difference in the lives of others, whether it's in our own families, communities, states, or on a national scale. In each case, the way to find solutions is by engaging with one another—not by criticizing each other or virtue signaling. If we try to understand the perspectives of others, and work to find common ground, we can move beyond the stalemates of the past and forge a new path forward. No problem is too big to solve.

As George Orwell once wrote, "It is difficult to be certain about anything except what you have seen with your own eyes." On these pages, I recount my personal story. I do not detail every action of the president or the administration, of which there are enough to fill volumes. While this book is primarily about my time as senior adviser, I open with a few defining moments from my life that shaped and prepared me for this unexpected opportunity to serve my country. Many of the quotes in the book are drawn from published records, such as transcripts, but others come from private conversations. In these cases, I've relied on my memory and extensive interviews with colleagues and counterparts. In some instances, I recreated dialogue to help readers experience what it was like to be in the room.

In Washington, history books were often my best survival manuals. They helped me realize that my predecessors had confronted similar problems. I learned to contextualize my situation, shift my approach, and navigate complex challenges. I hope that through this story, other leaders, dreamers, and risk takers—from all backgrounds, political persuasions, and industries—will be inspired to go beyond what's comfortable and chase the impossible.

My journey is a mostly unknown part of history. Now I am ready to share it in hopes that it enhances our shared journey.

Breaking History

Sentenced

I'm going to be arrested."

As my father told me the startling news over the phone, I was walking the block from my apartment in Lower Manhattan to the subway station on Astor Place. It was a muggy July morning in 2004. I had just completed my first year of law school at New York University, and I was on my way to my internship at the office of Robert Morgenthau, the legendary New York district attorney. I'd been working long days, carefully reviewing wiretap transcripts and helping to secure warrants for brave cops who had gone undercover to infiltrate a drug ring.

Across the Hudson River in New Jersey, my father was having a very different experience with a US attorney. He was ensnared in an investigation led by a brash, ambitious, and hard-charging federal prosecutor named Chris Christie.

The focus of Christie's investigation was a private family feud that had boiled over into public view as my father battled with his brother Murray and brother-in-law Billy, who were attempting to dismantle his control of the company he had spent his life building. They coordinated with an accountant in my father's company to surreptitiously access documents. Then they turned them over to the government and the media, alleging mismanagement and illegal avoidance of taxes.

It was an astonishing betrayal. In building his business into a billion-dollar enterprise, my father had made his siblings fabulously wealthy. The lawsuit and investigation had placed a heavy burden on him, and he reacted in anger. Billy's infidelity was an open secret around the office, and to show his sister Esther what kind of man she had married, my father hired a prostitute who seduced Billy. He had their resulting tryst recorded and sent the tape to Esther, who turned it over to the Feds. Unbeknownst to my father, Esther was cooperating as a witness in their investigation. My father was arrested and charged with witness tampering and violating the Mann Act, a century-old statute against transporting a prostitute across state lines. He had gone too far in seeking revenge, and now he was paying dearly.

After hanging up with my father, I rushed down the stairs into the subway station and waited for a few minutes on the platform before entering the 6 train and riding to my stop on Canal Street. When I emerged from the subway, I walked my normal path to the DA's office building and tried to turn my attention to the files on my desk. But my mind was racing. How could this really be happening to my dad? He had worked his whole life to build a great company and provide good-paying jobs to his employees. He had given generously of his time and money to serve the community. I also worried about my mother and what it would mean for her.

I stared at my computer screen for twenty minutes, but for the first time in my life, I couldn't push myself to keep working. I wanted to be there for my dad, just as he had always been there for me. I left the office, drove to New Jersey, and picked him up after his arraignment. During the ninety-minute trip home, he looked out the window and didn't utter a single word. It was the longest drive of my life. That afternoon he paced on the patio, adjusting his stride to account for his ankle tracker. I didn't know what to say or do, so I walked with him in silence, trying to support him simply by being at his side. After what seemed like an eternity, my father paused, turned to me, and said, "In life, sometimes we get so powerful that we start to think we're the dealers of our own fate. We are not the dealers. God is the dealer. Sometimes we have to be brought back down to earth to get perspective on what is really important."

Two days later, I arrived back at my apartment on Mercer Street in the NoHo neighborhood of Manhattan. The moment I opened the door, the weight of reality hit me. I'd been strong for my father and my family, but now I sat alone on the floor, with my back against the wall. For the first time since I was a kid, I put my face in my hands and cried.

I tried to make sense of my emotions. I was angry at my uncles and aunt. I was angry at my father. I was angry at my father's lawyers, who had known about his revenge plot and assured him that there was nothing illegal about it. I was angry at Chris Christie, who knew my father had been a major backer of his Democratic rivals in New Jersey.

When I woke up the next morning, I felt like I had a concrete block in my stomach. As I laid in bed staring at the popcorn ceiling of my apartment, I realized that my anger wasn't going to lead to anything productive. I was at a critical crossroads and had to make a choice. I could choose to be angry about things I could not control, or I could choose to help. I knew the answer immediately. I had to help my father, who had been through a lot and who was about to suffer more. I had to help my mother, who was the kindest person I knew and didn't deserve to have her husband of thirty years taken from her. I had to help my two sisters, Dara and Nikki, and my brother, Josh, who was about to begin his freshman year of college.

Despite my resolve, that first day back in the DA's office was agonizing. That night, I boarded the subway to go home, but when I got to my stop, my legs froze. I couldn't muster the strength to get up. I skipped my stop and rode the 6 train all the way to the end of the line in the Bronx and back downtown. For the next few hours, I watched New Yorkers get on and off the train—workers heading to the night shift, homeless people looking for their next meal, teenagers causing mischief, senior citizens trying to shuffle out of the train car before the doors shut on them. I studied their faces and saw, maybe for the first time, how much was weighing on everyone around me. Perhaps this woman had just lost her job, or that man couldn't feed his family. Maybe the person sitting across from me had just received a diagnosis of cancer.

It made me realize a simple truth: everyone has difficulties, but it's up to each of us to choose whether we are going to focus on ourselves or on helping those we love. I decided not to look back, but to look forward.

Improbable Existence

My family's mere existence is improbable. I'm here today only because my grandparents survived the Holocaust and later came to America. They taught me one of the most important things that I've ever learned: life is a gift that can be taken from us in an instant.

My grandmother, Rae Kushner, was sixteen when the Germans invaded Poland in 1939. Her family of six lived in Novogrudok, a quiet town located in eastern Poland, now part of Belarus. In 1941 the Germans seized control of the area and relocated about thirty thousand Jewish people to a ghetto. Over the next two years, the Nazis systematically exterminated the occupants of the ghetto, including Rae's mother and sister. In one round of killings, the Germans brought the remaining educated Jews—about 150 doctors, lawyers, professors, and teachers—down to the town square. While an orchestra played and my grandmother and the other occupants of the ghetto looked on, the Germans shot them in the head, one by one. The Nazis then forced fifty young Jewish girls, including my grandmother, to clean up the blood and stack the bodies on wagons to be hauled off to a mass grave. All the while, the Germans were dancing in the square. The music continued to play as the young women washed the blood off the stones.

By 1943, only a few hundred of the original thirty thousand Jews were left. Risking death, my grandmother and the remnant secretly dug a six-hundred-foot-long tunnel and waited patiently for a nighttime thunderstorm to cover their escape. About 250 people crawled through the narrow tunnel. The younger people went first, because they could move more quickly through the tunnel and had the best chance of escaping, but Rae chose to wait toward the back with her father. In a twist of fate, this decision likely saved her life. Her brother emerged from the tunnel with the rest of the young people only to be shot and killed by the Nazis. Of the 250 people who entered the tunnel, only 170 escaped into the nearby forest. Rae, her father, and her younger sister were among the survivors. They fled deep into the woods and found refuge with the partisans—a group of freedom fighters who created hidden camps deep in the forest and carried out daring acts of resistance against the Nazis. Among the partisans, Rae reconnected with a young man from a neighboring town, Joseph Berkowitz, the youngest of eight children born into an impoverished tailor's family.

When the war came to an end, Rae and Joseph fled to Hungary, where they quickly married. The day after their wedding, they trekked through the Austrian Alps and snuck across the border into an Italian displaced persons camp. They applied to come to America, using my grandmother's last name, Kushner, since my grandfather had accrued a rap sheet from smuggling cigarettes into the camp to provide for his family. As my grandmother recalled years later, "We would go anywhere where we could live in freedom, but nobody wanted us."

They waited three and a half years in that refugee camp to come to America. Like so many others during that time, they knew they had finally made it when they spotted the Statue of Liberty in New York Harbor. Two days after arriving, my grandfather showed up early at a construction site in Brooklyn, willing to work hard, with one limitation: he was afraid of heights. The foreman told him that he should consider going to New Jersey, where the buildings were not as tall, so he began commuting two hours from their tiny Brooklyn apartment to jobsites in New Jersey. He worked seven days a week, sleeping at jobsites to maximize work and spare the daily bus fare. Only on major Jewish holidays would he go home. He earned the nickname Hatchet Joe by using the dull end of a hatchet—which required fewer, though much heavier blows—to hammer nails.

My grandfather was a simple, quiet man who had no formal school-
ing. But he spoke six languages, and he lived the American dream, starting a successful construction company that built thousands of homes. A
lifelong smoker of Camel cigarettes, he died in 1985 from a stroke at the
age of sixty-three. I was just four years old, so much of what I know about
him is through my father and grandmother's recollections. She was proud
of their survival story.

In many ways "Bubby Rae," as we called her, was a typical European
immigrant, full of life, sharp in wit, and overflowing with love. When I
was a young kid, I'd go over to her house on Saturday evenings and sit in
her lap as she played gin rummy with her friends, placing five-cent bets.
She promised us she had given up smoking, but the bathroom always
smelled like smoke after she used it. When we confronted her about the
smell, she retorted: "Your dog really needs to stop smoking." She doted
on her grandchildren, slipping us quarters to play games at the arcade or
a piece of candy while my parents were looking in the other direction.

My dad met my mother, Seryl, when they were both eighteen. On
their first Shabbat together, my mom still wasn't old enough to buy wine.
They were married by the time they were twenty. My parents raised us
in Livingston, New Jersey, a middle-class suburb forty-five minutes west
of Manhattan. My mom is an incredibly selfless and caring person, who
taught us to treat others with respect and take responsibility for our actions. She never made excuses for me. When I got in trouble, she always
sided with my teachers and told me that it was my responsibility to figure
out how to get along and make things work.

Like my grandfather before him, my father worked all the time. After
he briefly practiced law, he started a company with my grandfather. My
dad purchased, financed, and managed the properties, and my grandfather ran construction of the new buildings. My dad had no experience
in construction, and when my grandfather died unexpectedly, he had to
find a way to finish a project that was in process. My grandfather's close
friend Eddie Mossberg, also a Holocaust survivor, sent workers from his
own jobs to help my father complete the project. To this day, my father
still recounts this act of kindness, and it has inspired him to help many

others who face hardship. My father's company grew quickly, and he began outcompeting the same companies that had employed my grandfather a decade earlier when he was Hatchet Joe.

On Sundays, my dad would take me to the office so we could spend time together. On the days we toured properties, we'd stop for a treat at a local farm stand and buy fresh bread, butter, and famous Jersey corn. Right on the side of the highway, we'd tear off the husks and eat the corn raw off the cob. My dad always treated me like an adult, asking what I thought about a potential deal or what I noticed about jobsites—which one was nicer, what the manager could be doing better, or why one commanded higher rent than another. I worked every summer once I turned thirteen. My first job was on a construction site, beginning at six o'clock in the morning. I worked under the scorching sun alongside carpenters, plumbers, and electricians, who taught me how to hammer, saw, wire, and clean. When I got home each night, I was so filthy my mom would hose me off before letting me into the house. Each summer I gained more knowledge and responsibility, eventually helping my father manage rental properties and creating financial models for projects.

During my senior year of high school, I woke up at 4:30 each morning to train with my dad for the New York City Marathon. I will never forget what he told me as we ran up the big hill at the north end of Central Park: "Running is like life. When there's a big hill at the end, don't look up, keep your head down and watch your feet. Don't think about the top of the hill, just think about your next step. Before you know it, you will achieve your goal and be at the top of the hill."

In 1999 I was thrilled to learn that I had been accepted into Harvard. Like most students on campus, I was initially nervous about how I would perform against the world's top students, but I quickly learned that while many kids had high IQs, some didn't work hard or have common sense.

I met my best friend while I was in the laundry room, switching loads. Nitin Saigal was from India and quipped that because I wasn't taking economics professor Marty Feldstein's legendary Economics 10 class my freshman year, one day I'd be working for him. We hit it off immediately and roomed together for nearly a decade, until I married Ivanka. Today,

Nitin remains one of my closest friends. He manages a successful hedge fund and is one of the hardest workers I know.

My sophomore year, an acquaintance tried to sell me an apartment in Cambridge. I told her that I liked living on campus, but I asked a few questions and learned that apartments in Cambridge cost 30 percent more than apartments just across the street in Somerville. I saw an opportunity. The Somerville apartments were just as close to campus and, once retrofitted, could be listed very near to Cambridge prices. I called my dad and pitched him on purchasing a number of older apartments in Somerville. He agreed to put up half the capital if I could raise the rest. I began slipping off campus after class to show bankers potential investment sites. A few months later, I had posted my share of the financing. At the age of nineteen, I bought my first building. From that point on, I would go to class, then to the jobsites, where I would check on the progress, issue work orders to the contractors, and make deals with tenants.

I made plenty of mistakes. On one purchase—an historic apartment building at 82 Monroe Street—I took the seller at face value when he quoted the number of units in the building. But after I purchased it, I discovered that many of the apartment units were illegally constructed. The lower number of rentable units dramatically reduced the projected revenue and eliminated much of the return that I had told investors we would make. After looking at several scenarios, I concluded that to salvage the project, I had to convert the building into condominium units, a far more involved and extensive construction project than I was anticipating. It took us longer, but we ultimately made a nice profit. The experience taught me the importance of conducting due diligence on every detail of a business deal, even those typically taken for granted. Growing up so close to my dad's business, I had been immersed in real estate, but I learned that nothing could replace the experience of being responsible for an entire project, where I had to answer to investors, manage contractors, and keep tenants happy. I graduated from Harvard with honors, while making millions of dollars from my real estate investments.

During my college years, I interned in New York each summer. The night before one interview, my dad asked me what time I planned to leave our house in the morning for an appointment at nine o'clock. I planned to leave at eight. "What if there is traffic?" he asked. I had accounted for that. "But what if there is an accident in the tunnel?" That seemed unlikely, but I would leave earlier just in case. "The only excuse for being late is that you didn't leave early enough," my father said. I left at six o'clock, breezed into the city, and waited in a Starbucks for two hours. I got the internship.

My most valuable experience was working at SL Green Realty Corporation, where I met Marc Holliday, who ran the company and was widely viewed as an up-and-coming star in the real estate business. One evening, he asked me to run a complex analysis for his negotiation the next day. I stayed up all night to get it done. When he reviewed it the next morning, he thought I had done a good job, but added that if I wanted to be great, I needed to internalize concepts around eight principles of real estate. He offered to extend my internship by several weeks and spend an hour on Fridays walking me through each principle. This education was better than any I received in school.

After interning at Goldman Sachs and Morgan Stanley, I realized that I did not want to go into investment banking. So I applied and was accepted to New York University's dual JD/MBA program. During my first year, I was inspired by the public policy focus of the law program and wanted to start my career in public service as a prosecutor. After my father's arrest, however, as I watched a prosecutor inflict havoc and hardship on our family, I began to have second thoughts. I didn't think I could do that to others.

My father ultimately decided not to fight his case in court. He recognized that he had let his emotions get the better of him and felt that he had sinned before God and was ready to take responsibility for his actions. He knew that fighting the charges would be a painful five-year ordeal for our family and diminish morale at his company. He pleaded guilty and was willing to accept the consequences, which the judge decided would be two years in federal prison.

* * *

In April 2005, during my second year of graduate school, I traveled with my parents to the federal prison in Montgomery, Alabama. My mom and I gave my father one last hug before he walked inside. I later learned that as he entered, a prison guard smirked and whispered in his ear, "Welcome. They love to fuck billionaires in here."

The prison tightly controlled his calls, and we had to split the time between my mom and the four kids. I got about three minutes a week, ninety seconds at a time. The timing was unpredictable, and if I missed the call, that was it. I kept my phone with me all the time, even when I showered.

I offered to drop out of grad school to help manage the company full-time, but my dad pleaded with me not to make that sacrifice. We compromised that I would stay enrolled, but spend the bulk of my time helping with the business. We were fortunate that my dad's close friend and mentor Alan Hammer, a lawyer and experienced real estate executive, generously offered to run the company in his absence.

Every weekend I flew with my mom to Alabama for a six-hour visitation with my dad. The first time I saw him lined up with all of the other inmates in his green prison uniform, it was hard not to cry. We were always the first to arrive and the last to leave, and we spent countless hours sitting in the waiting room with the other families, eating popcorn and Pop-Tarts from the vending machine. For years after, I couldn't stand the smell of either. We often became so engrossed in our conversations that we would forget we were inside a prison—until a siren rang, calling for my father to line up against a wall for the regular count of all the prisoners.

Prison is a great equalizer, and my dad's fellow inmates grew to love him because he is down-to-earth. He spent time reading, exercising, and working in the cafeteria. At night, he sat in the library and doled out advice. One visitation day, we were surprised to see two mothers smother him with hugs. He explained that he was teaching their sons how to interview for a job.

On another trip, we were sitting on benches outside, soaking in the heat from the sun, when an inmate yelled across the yard, "Hey, it's Charles the Great!" My dad turned to me and quipped, "Maybe I don't want to leave here—no one in my company ever called me that."

During this difficult period, Chris Christie sought to punish my father in a way that would hurt the most: by putting other Kushner Companies executives in jail, bankrupting the family business, and shutting it down for good. I often played the office psychologist to employees at every level of the company, who came to me worried that the company would collapse, and that they would lose their jobs. Every day felt like a kick in the gut. At the lowest points, I would tell myself that at least my dad wasn't gone forever. I had to learn how to absorb bad news, put on a strong face, and keep moving forward. I couldn't have known this at the time, but being thrust unexpectedly into a role leading our company prepared me for an equally unexpected, but much more consequential, role in the federal government.

Eleven and a half months after entering prison, my father was released to house arrest. It was the happiest day for our entire family. But it almost didn't happen. Christie tried to invalidate my father's earned time credits and block his release. Thanks to the brilliance of Washington lawyer Miguel Estrada, Christie's cruel and punitive effort failed.

My father's time in prison was the most humbling, difficult, and formative experience of my life. It had a way of uncluttering my thinking. I learned to separate the fleeting—money, power, and prestige—from the enduring: the way we react to difficult situations, the faith we hold on to, and the people we love. I had now seen for myself the truth of my grandparents' maxim: life really can change in an instant.

{ 3 }

Making It in Manhattan

Shortly after my father's release from prison, we finished the biggest real estate deal in our company's history, with what at the time was the highest price ever paid for a single real estate asset in the United States. For $1.8 billion, we bought a midcentury skyscraper located at 666 Fifth Avenue. Maybe the bad-luck street number should have given us pause: the purchase closed in early 2007, right before the market collapsed at the onset of the Great Recession. Twenty months later the major investment firm Lehman Brothers filed for bankruptcy, and office vacancy rates in midtown Manhattan tripled overnight.

We thought 666 Fifth Avenue could be worth $2.5 billion, a valuation driven in large part by the building's pristine commercial space and prime storefronts on New York's iconic Fifth Avenue. In the lead-up to the crisis, the building was collecting rents of about $120 per square foot—a rate that soon dropped precipitously. I remember Steve Roth, founder of Vornado Realty Trust and one of the smartest real estate moguls in New York, remarking as the crisis hit, "I'm getting sixty-dollar rents now in my best buildings. Do you know why I'm not getting fifty? Because the tenants aren't asking for it." We had counted on the revenue from renters to service our debt payments, and we found that we were falling short of the amount we needed. Titans of finance and real estate began circling our investment like vultures. Plenty of people told me that there was no way to recover. I saw it differently. There was no way I was going to let the investment fail.

I had very little leverage, so I was willing to talk to anybody. To salvage the purchase, I restructured the debt to prevent foreclosure and raised more than $500 million by selling a 49 percent interest in the retail space to the Chera family and the Carlyle Group. I brought in a real estate investment firm to co-own the building, and modernized the retail and commercial space to attract more lucrative tenants. I gradually convinced Brooks Brothers to sell their lease, which we rented to Uniqlo for a record $300 million. For several years, I tried unsuccessfully to convince National Basketball Association commissioner David Stern to give up his prized lease for the NBA store, which was located in the ground-floor retail space. Then I met rising NBA executive Adam Silver, and enlisted his help to negotiate a deal. Stern used to call and rib me: "Leave Adam alone! We are never leaving the store!" Silver explained to me that Stern's money-losing push for the NBA to open a retail store had initially been used by his antagonists at the owners meetings to embarrass him. After Stern dug into the operations and had the store turning a profit, he proudly opened every owners meeting, where the league announced billion-dollar deals, with an update about the couple hundred thousand dollars of profit generated by his beloved Fifth Avenue store. Silver and I ended up becoming close friends. I tried for three years to get them to sell the lease—Silver drove a hard bargain. Eventually, Stern made a good deal to give it up. We later sold the space to Inditex, the parent company of Zara, for $324 million—a record price per square foot.

Navigating the fallout of 666 was the biggest challenge and learning experience of my business career. Being thrust into complex, high-stakes negotiations at a young age gave me unique training. I forged relationships with many of the titans in the industry, which proved invaluable moving forward. I did not win every negotiation, but I gained credibility by being honest about our difficult situation, offering constructive solutions, and seeking successful outcomes for all parties. My goal was to increase the size of the pie rather than eliminate slices from it. Two of my creditors, with whom I developed close personal relationships, told me flat out over lunch meetings that friendship was separate from business and that they were going to do everything in their power to make sure I lost the building. Fortunately, others were more magnanimous and went out of their way to help find a win-win outcome. At one point I flew to California to meet with Tom Barrack, a real estate giant whose firm was one of our creditors. I expected him to be hostile and jockeying for the kill, but after our meeting, he became an ally. "Most people in your position are looking to take advantage of their lenders," he said. "I appreciate your pragmatism and I'll work with you to figure this out."

After salvaging our investment in 666, I didn't fear failure in business. I learned how to focus on important decisions and ignore petty distractions. I got better at mitigating potential downsides, taking calculated risks, identifying market trends, and developing in up-and-coming neighborhoods.

My first successful deal in New York City was the purchase of a building on 200 Lafayette Street from John Zaccaro. No one thought he would ever sell. I met with him and offered to put down money immediately and sign whatever contract he put in front of me. While the building was in terrible shape, I knew that if I achieved my business plan, I would make a substantial profit. At the time, I had been helping my brother start and build his venture investing business, Thrive Capital, and I saw that start-ups like his wanted more modern offices spaces that didn't yet exist in New York. I thought this building could serve a new niche. After convincing Zaccaro to sell me the building for $50 million, I went looking for a partner. I found Avi Shemesh, an Israeli immigrant who started as a gardener and built a multibillion-dollar real estate firm. As Avi and I stood on the roof of 200 Lafayette Street, he asked, "How large are the floors?" "Seventeen thousand rentable square feet," I replied. He inquired if this was the right floor size for the tenant I wanted. "It's what we've got." He liked my honesty and enthusiasm for the project. "Jared, I'm making this investment, but not because of the building. I'm betting on you." After twenty months of executing my plan, we sold the building for nearly $150 million.

After that success, I went on a major buying spree, acquiring more than twelve thousand apartments across the country and completing $14 billion of transactions in roughly ten years. One of the best deals I made was purchasing the Jehovah's Witnesses headquarters in Brooklyn. When I heard that they were selling, I called their representative, Dan Rice, and asked him to let me participate in their auction. Located on the river next to the Brooklyn Bridge, the properties were unbelievable. They were the best-run buildings I had ever toured—they were so clean you could eat off the floors. I went to the representative's office that day and asked how much he wanted. He quoted $325 to $350 million. I told him I'd pay $375 million if he promised not to have an auction. He called his board, got approval, and shook my hand. The next day, a competitor offered him a higher number, but he said, "Nope, we're Jehovah's Witnesses; we honor our word." He sold the property to me, and after renovations and rebranding, it is now worth close to a billion dollars.

With every new purchase, I focused not on the last dollar but on the next deal. I saw the potential in buildings that most people overlooked and learned how to make that vision into a reality through building consensus, motivating hundreds of people, making quick decisions, and solving problems as they arose. Before long, many of the big players started following me to the changing neighborhoods in which I was investing.

People found that they could make money by working with me, which led to many incredible opportunities. I never forgot what Greg Cuneo, a consultant who became a friend and mentor, advised while we negotiated with subcontractors on the 200 Lafayette Street project: in his thick Italian accent he urged, "Tutti mangia"—loosely translated, "Everyone has to eat." He added, "If you make too good of a deal, they will cut corners and not perform."

In addition to building a reputation through real estate deals, I also met New York's top business leaders through another investment I had made in 2006. That July, I visited Arthur Carter, the owner and publisher of the *New York Observer*, a weekly newspaper read by New York's elite. I told him that I wanted to buy the paper. He said that Robert De Niro and Jane Rosenthal were far along in negotiations but were raising new issues at the last minute. I put a check for $5 million on the table. He said if I closed by Monday, it was mine. I worked all weekend on the due diligence to finalize the deal. In the *Observer*, I saw an opportunity to bring a sophisticated paper into the digital age, while making helpful business connections in the process. I soon learned that, particularly in journalism, change is like heaven: everyone wants to go there, but nobody wants to die.

One of the real estate giants who noticed the paper was Donald J. Trump. I will never forget receiving a letter in the mail from him: upset about his placement on the *Observer*'s annual Power List, he asked to be removed. "Interestingly, the name Trump is used prominently in your title and mentioned in the snippet along with the person ranked #1. I guess you're trying to get people to read the article." It ended, "P.S. Please stop sending me your paper, so I don't have to read bullshit like this anymore!" I'm sure I wasn't the first to receive a message from Trump regarding a press article, and I certainly wouldn't be the last.

* * *

Around the time of my 666 Fifth Avenue purchase, Donald Trump suggested to his daughter Ivanka that she talk to the guy who was actively buying buildings to see if I was interested in purchasing any of their properties. In the spring of 2007, we had lunch. We spoke about business, but the conversation soon turned to NASCAR, New Jersey diners, and other unlikely interests that we had in common. That led to a second lunch at my favorite Indian restaurant, the Tamarind Flatiron & Tea Room on Twenty-Second Street, where we talked for three hours. We both had to keep calling our assistants to reschedule our other meetings for the day.

Ivanka was not what I had expected. In addition to being arrestingly beautiful, which I knew before we met, she was warm, funny, and brilliant. She has a big heart and a tremendous zest for exploring new things. Soon I was taking Ivanka to parts of the city she had never seen before, using our dates to check out neighborhoods where I was looking to purchase property. We walked the streets, observed the people, and debated which neighborhoods would evolve over the next few years. On Sunday mornings we would take our backgammon board to a new restaurant and sit there for hours as we played games, read the papers, and sipped coffee.

I loved how she always treated everyone with charm and respect, whether they were business leaders, waiters, or cabdrivers. She made everything fun. We also seemed to have a great deal in common. Both of us worked with our fathers in the family business, but we also had started our own companies. We were both driven and ambitious, with a healthy appetite for adventure.

When I realized that I was falling in love with Ivanka, I grew concerned about our different religions. As hard and painful as it was, I broke up with her. Ivanka told me it was the worst decision of my life. She was right. Several months later, our mutual friend Wendi Murdoch invited me away for a weekend with her and her husband, Rupert Murdoch—the owner of News Corp, then the parent company of Fox News and the *Wall Street Journal*—on their boat *Rosehearty*. I had first met Wendi and Rupert through my work with the *Observer*, and they had become good friends. To my surprise, Ivanka was there. She was equally shocked, but it wasn't long before we got back together.

That same weekend, Rupert made the final offer to the Bancroft family to purchase the coveted Dow Jones Company. He shared with me a letter he had just sent to board members informing them that if they didn't accept his offer by Monday, he was going to pull the offer, and the stock would fall. I was amazed by his negotiating style. Rupert struck me as an intellectual, in addition to being a brilliant businessman. When we spent time together, he started his days by reading every line of his company's newspapers, as well as the competition's. He devoured books and gave me his favorites. On that Sunday, we were having lunch at Bono's house in the town of Eze on the French Riviera, when Rupert stepped out to take a call. He came back and whispered in my ear, "They blinked, they agreed to our terms, we have the *Wall Street Journal*." After lunch, Billy Joel, who had also been with us on the boat, played the piano while Bono sang with the Irish singer-songwriter Bob Geldof. Rupert joked to me that we were clearly the least talented people there.

As the months went on, Ivanka told me that she was open to exploring the possibility of converting to Judaism. We began meeting with a rabbi and studying and practicing Shabbat together. I saw that Ivanka was enjoying these rituals. After a few Friday evenings eating takeout from 2nd Ave Deli—my favorite New York deli—Ivanka decided she wanted to learn how to cook to make our Friday nights together more special. She loved it and quickly became an excellent chef.

As our relationship turned more serious, Ivanka suggested that I should try to get to know her father, so I called Trump and asked if I could see him. He suggested lunch the next day in the grill at Trump Tower—an unusual offer, as he rarely met people for lunch. As we sat down, I could feel my voice shake as I managed to say that Ivanka and I were getting more serious and that she was in the process of converting.

"Well, let me ask you a question," he said. "Why does she have to convert? Why can't you convert?"

I replied that it was a fair question, but Ivanka had made the decision on her own, and we were both comfortable with it.

"That's great," he said. "Most people think I'm Jewish anyway. Most of my friends are Jewish. I have all these awards from the synagogues. They love me in Israel." Then he added, "I just hope you're serious because Ivanka is in an amazing place in her life right now. You know, Tom Brady is a good friend of mine and had been trying to take Ivanka out . . ."

Before he got any further, I quipped, "If I were Ivanka, I'd go with Tom Brady." He looked at me with complete seriousness. "Yeah, I know," he sighed.

A few months later, I made a clandestine trip to Trump Tower to ask for Ivanka's hand in marriage, and I mentioned that I had planned a surprise engagement. Later, I learned that right after I left, Trump picked up the intercom and alerted Ivanka that she should expect an imminent proposal. That night, I took her to see *Wicked* on Broadway. I had asked my brother Josh to scatter rose petals across my apartment and light candles right before we came home. But the show started late and ran long, which rarely happens. The engagement ring was in my pocket the entire time, and I was anxious that the candle wax would be melting all over the place. In the hallway outside my apartment door, I nervously pulled out the ring and proposed to Ivanka. Fortunately, she said yes.

We got married at Trump National Golf Club in Bedminster, New

Jersey—a majestic and serene getaway with lush trees and rolling hills an hour from Manhattan that remains one of our favorite places. Trump drove out each week leading up to the wedding to check on the construction of the tents. He was respectful of our Jewish traditions, and before he walked Ivanka down the aisle, he asked for a yarmulke to wear. We were happy to share the day with so many friends, but I did not forget one person's advice: Never let go of your wife's hand on your wedding day.

I had planned an African safari for our honeymoon, but when we got to Amsterdam, bad weather delayed our connecting flight until the next day. The airline wouldn't release our luggage, so we stopped at a gift shop in the airport, bought the cheapest coats, scarves, and hats we could find, and went to explore the city together. I scrambled and reserved a room at the Dylan, and we finally made it there after running around the city in the snow for several hours. We were drenched from head to foot, and didn't have a spare change of clothes. Some brides would have had a meltdown as their dream honeymoon was thrown off track, but Ivanka smiled and improvised. We donned hotel bathrobes and slippers and went down to dinner at the hotel's Michelin star restaurant. It was a blast and the perfect start to our marriage—the first of many unexpected adventures.

During our early years of marriage, both Ivanka and I were busy growing our respective companies and building relationships with members of New York society, but most often we preferred to have dinner just the two of us. We took turns planning date nights exploring the city. We'd go rock climbing at Brooklyn Boulders, trapeze at the South Street Seaport, take cooking lessons at a local restaurant, or play shuffleboard at a new bar in a trendy neighborhood. Soon, our first child, Arabella, arrived and added more joy to our lives. I would sit in her room for hours and watch her sleep as I worked on my laptop. Joseph followed two years later. We tried to raise them both with as normal an upbringing as possible, teaching them good values, spending quality time with them, and observing Jewish traditions together. Life was full. We had no idea that our world was about to turn upside down.

4

"Everything Will Be Different"

After I announce this week, everything will be different."
Ivanka and I were gathered with her family for lunch in the clubhouse at Trump National Golf Club in Bedminster to celebrate Trump's sixty-ninth birthday. Typically, he's totally focused on the present moment, especially when surrounded by his family. But this day, his mind was on the future. Trump interrupted the typical banter and ribbing to utter his prediction about what would happen after he declared his intention to run for the presidency. We had no idea where this would go or how it would change our lives. We just knew that with Trump, there is always something going on, and it's never boring.

Trump asked Ivanka to introduce him for his campaign announcement on that coming Tuesday, June 16, 2015. She told him that she would do it, but only if he was serious this time. He had explored a presidential bid in years past but ultimately had decided not to run. Preparing to introduce her father was a new challenge for Ivanka: she was not political and had never given a nationally televised speech. As we worked on her remarks, I tried to reassure her. "Don't worry," I said. "It's just the introduction. No one will notice it unless you screw something up." That didn't help.

At the time, I was serving jury duty. I asked my supervisor for Tuesday afternoon off to attend a family event. A driver picked me up in an SUV and sped toward Trump Tower. In the back seat, I changed into a suit. As I arrived and waited for the elevator up to Trump's office, where Melania, Eric, and Don Jr. had assembled, I could hear the melancholic echo of the song "Memory" from the 1981 musical *Cats*—the tune that Trump had chosen as Ivanka's walk-out music. While Ivanka delivered her introductory remarks, the rest of us took the elevator down to the atrium. Right before he descended the iconic gold escalator, Trump turned to Don Jr., Eric, and me. "Okay, kids," he said, "now we find out who our real friends are."

Trump took the makeshift stage, framed by eight American flags. His campaign manager, Corey Lewandowski, released the text of his prepared speech to the press, but Trump didn't use a word of the script. Instead, he delivered a forty-five-minute off-the-cuff speech that was a thunderclap above the Republican political landscape. He spoke as an outsider confronting a corrupt and feckless political establishment that had traded away American manufacturing jobs, failed to secure our borders, upended our health-care system, and plunged the country into two endless wars costing trillions.

I thought the speech was vintage Trump: raw, authentic, and effective. In other words, nothing like a politician. As a novice to politics, I didn't realize that one line would become a flashpoint: "When Mexico sends its people, they're not sending their best. . . . They're sending people that have lots of problems, and they're bringing those problems with us. They're bringing drugs. They're bringing crime. They're rapists. And some, I assume, are good people." The press immediately seized on the "rapists" comment. I later learned that Trump was inspired to use the controversial line by a Customs and Border Protection officer who had come to his office to enroll him in Global Entry. During the screening, Trump asked the officer how things were looking on the border. The officer told Trump that things were a mess, that they were sending busloads of people back to Mexico every day, but they kept coming back faster than we could return them. Trump asked him what kind of people they were—were they families or young children? "No," the officer said. "The people we're sending back on buses mostly have criminal records, even including some rapists and murderers."

Trump has a habit of seeking information and opinions from people whose views are often overlooked. As a builder, he would visit construction sites and ask the frontline workers for their input on serious design questions. When Ivanka was leading the renovation of the Old Post Office in Washington, DC, a general contractor gave him a complicated blueprint for the heating and air-conditioning system. Trump turned to one of the hard-hat workers and asked what he thought about the schematic.

"It's stupid," the worker responded. "You're putting all of the ductwork in these precise locations to keep temperatures stable during a once-in-a-hundred-year hot or cold day. Just blow air up through the middle, and you'll save on the cost of installing all of the ductwork. Guests walking the thirty seconds from the elevator to their room might be two degrees warmer if there are extreme weather conditions, but you'll save millions." Trump called for the plans to be redrawn immediately.

Dealing with the crises confronting the Trump Organization fell on the shoulders of Ivanka and her brothers. She drafted an op-ed for her father to clarify his position on immigration—that he was for legal immigration and against human trafficking, drug smuggling, and ungoverned borders. To help her draft the op-ed, Ivanka called upon Hope Hicks, a communications ace who joined the Trump Organization in 2014 and quickly earned both Ivanka's and Trump's trust. As early as January 2015, Trump told Hope, "I'm thinking about running for president, and you're going to be my press secretary." A brilliant communicator who remained poised under pressure, Hope helped craft the campaign's message, fielded hundreds of press inquiries each week, and designed creative events that brought out the best in Trump.

When Ivanka and Hope brought the op-ed to Trump, however, he refused to back down. "I haven't said anything wrong, and the media knows that I haven't said anything wrong," he insisted. "I don't plan to follow their rules, and they just want me to apologize for entering this race. There is no way I am doing that." Despite the fact that he had lived a glamorous life in a gilded, three-story penthouse on Fifth Avenue, Trump understood intuitively what other politicians had long ignored: citizens across the country were feeling the effects of globalist trade and immigration policies that jeopardized their jobs, safety, and very way of life. And they were angry.

This first of many media crises taught me what I later called the "three rules of Trump." Number one, controversy elevates message. Number two, when you're right, you fight. And number three, never apologize. Most politicians follow polls, but Trump changed the polls. Before he entered the political arena, immigration wasn't a hot issue. Suddenly, people were talking about the very real immigration crisis on our southern border—a problem that other candidates had desperately tried to avoid. The debate was playing out on Trump's terms.

Shortly after the president's campaign launch, Rupert Murdoch tweeted, "When is Donald Trump going to stop embarrassing his friends, let alone the whole country?" A week later, on July 21, the *New York Times* published a tabloidesque story that described Rupert's disparaging views of Trump and his chances as a candidate.

Trump called me. He'd clearly had enough. "This guy's no good. And I'm going to tweet it."

"Please, you're in a Republican primary," I said, hoping he wasn't about to post a negative tweet targeted at the most powerful man in conservative media. "You don't need to get on the wrong side of Rupert. Give me a couple of hours to fix it."

I called Rupert and told him I had to see him.

"Rupert, I think he could win," I said, as we sat in his office. "You guys agree on a lot of the issues. You want smaller government. You want lower taxes. You want stronger borders."

Rupert listened quizzically, like he couldn't imagine that Trump was actually serious about running. The next day, he called me and said, "I've looked at this and maybe I was misjudging it. He actually does have a real following. It does seem like he's very popular, like he can really be a kingmaker in the Republican primary with the way he is playing it. What does Donald want?"

"He wants to be president," I responded.

"No, what does he really want?" he asked again.

"Look, he doesn't need a nicer plane," I said. "He's got a beautiful plane. He doesn't need a nicer house. He doesn't need anything. He's tired of watching politicians screw up the country, and he thinks he could do a better job."

"Interesting," Rupert said.

We had a truce, for the time being.

Within four weeks of entering the race, Trump skyrocketed to first place in the polls. At the first debate, in August, Fox News anchor Megyn Kelly brought up provocative comments Trump had made when he was in the entertainment business. From that point forward, he was locked in a brutal battle with Fox News. He got a call from the network's CEO, Roger Ailes. "Donald, in your business, your assets are buildings," said Ailes. "In my business, my anchors are everything I have. If you attack my anchors, I'm going to have to come after you with the full force of the network. We need to find a way to deescalate this thing." Trump was undeterred by the threat.

The more irreverent Trump was toward the media and the political establishment, the more my friends in New York thought he was on his way out, but he kept climbing in the polls. I was glad to see Trump's growing momentum, but I had no plans to get directly involved in his campaign. Our company was on a hot streak, and I was focused on growing our portfolio.

* * *

On a November morning in 2015, Trump called and asked if I wanted to come to a rally that evening in Springfield, Illinois. Like any smart son-in-law, I said yes. I knew this was an opportunity to see how the billionaire developer from New York City was playing in America's heartland.

We were greeted by a crowd of fans waiting at the airport and lining the road to the venue. The Prairie Capital Convention Center was packed. We felt the pulse of energy from backstage. The event manager greeted Trump at the entrance. "Congratulations, sir," he said. "You just broke the attendance record for this arena, previously held by Elton John."[1] Trump quickly joked, "See, Jared, and I don't even have a piano. Imagine if I played the guitar."

As we went to a reception area to meet with local officials and volunteers, I was surprised by Trump's willingness to shake every hand and pose for a picture with everyone who asked, even though he was a germophobe. This was a big sign to me about his total devotion to winning the race. When he took the stage, more than ten thousand people erupted in cheers and applause. I walked around the arena and watched in amazement as my father-in-law interacted with the enthusiastic crowd. He riffed for an hour, occasionally looking at the few notes he had jotted down on the plane. In contrast to media reports that described his rallies as a breeding ground for lunatics and neo-Nazis, I saw normal people: hardworking moms and dads as well as students and grandparents. People of different ages, races, and backgrounds believed someone was finally speaking for them. His message about illegal immigration, unfair trade deals, and endless foreign wars resonated. When Trump promised to end Common Core, the crowd went wild.

I couldn't believe it. Weeks before, I had attended a dinner hosted by the Robin Hood Foundation, one of the largest philanthropic groups in New York. The group's chairman, a finance billionaire, had given a speech hailing Common Core as the savior of American education and urging participants to call their contacts in Washington to support it. When I heard the crowd's reaction that night in Springfield, it reminded me of a book that Rupert Murdoch had given me months earlier: Charles Murray's *Coming Apart*, which makes a case that over the last fifty years America has divided into upper and lower classes that live apart from each other, geographically and culturally.[2] They attend different schools, consume different foods, and seek different forms of entertainment. They share so little, and have such minimal contact, that they no longer understand each other. Now, as I stood among my father-in-law's supporters, I was beginning to understand why Trump's message resonated with so many Americans. Washington's upper-class elites were out of touch with the lower- and middle-class citizens they supposedly represented, leaving their constituents feeling forgotten and disenfranchised. While these decisions did not hurt people like me in New York, they were stripping opportunity from many families and communities throughout Middle America. The rally was a wakeup call for me.

On the flight home, as we chowed down on McDonald's Big Macs, Filet-o-Fish sandwiches, and fries, I told Trump how much I enjoyed watching him connect with the crowd. I was moved by the patriotism that so many of his supporters had expressed. I mentioned that he could do more with his Facebook page to engage many energetic supporters like those in the Springfield arena. Trump was an early adopter of Twitter and had already revolutionized politics with his viral turns of phrase. He suggested that I talk to Dan Scavino, who was managing content for his Twitter and his other social media accounts. Scavino had started working for Trump as a golf caddy when he was sixteen, and over the course of a decade, he had climbed the ranks and proven to be an indispensable executive at the organization.

As we sat in Scavino's campaign office in Trump Tower, a small concrete, windowless room with a plastic card table and folding chairs, we tested Facebook's ad options. Soon we asked Corey Lewandowski to give us a budget so that we could experiment with tactics to boost content on the platform. As manager of the nimble campaign team, Lewandowski had many good qualities. He worked around the clock, was staunchly loyal to Trump, and had a good sense of how to connect with the Republican primary base. But he couldn't see the strategic significance of getting into Facebook. He gave us $500 per month, which seemed more like a ransom payment to get us off his back than a calculated investment, but it was enough to start trying out different tactics in Iowa and New Hampshire.

Scavino and I soon learned that if we targeted a message to the appropriate demographic, it would catch like wildfire, spreading across the social-media platform and receiving tens of thousands of likes and shares for little cost. As Trump's strongest admirers revealed their support for him online, their friends also started to publicly acknowledge support.

Because so much of this was happening organically, Scavino and I struggled to spend our tiny budget. When we went to purchase an ad, Trump's message had already reached and fully saturated the demographics we had planned to hit. By 2017, Facebook changed its algorithms, making it more difficult to get as much free, organic exposure as we did during the campaign.

Back in the summer, I was walking through Trump's corporate office in Trump Tower when I passed by the desk of Amanda Miller, head of marketing and communications for the Trump Organization. I noticed a wide-brimmed, old-school red baseball hat with four words in bold white lettering: MAKE AMERICA GREAT AGAIN. "You've got to be kidding me," I said, laughing. Amanda said that Trump had called her to his office and designed the hat himself and asked her to order a thousand. She'd ordered a hundred, thinking he'd never know the difference. Soon after, Trump wore the hat on his visit to the southern border, and it became the hottest thing on the internet. It even appeared on the front page of the *New York Times* style section in an article by Ashley Parker entitled "Trump's Campaign Hat Becomes an Ironic Summer Accessory." The demand was so incredible that I worked with Amanda to create an online store, where we started selling roughly $8,000 in hats per day. By December, when I attended a rally in Iowa, red hats blanketed the crowd. When I looked closely, I saw that there were twelve knock-off hats for every official one. We could sell a lot more of our authentic hats if we scaled marketing. Amanda introduced me to Brad Parscale, the vendor of the campaign's website, and we worked on a plan to start spending $10,000 a day on Facebook ads to sell the hats, bypassing Lewandowski's budget restrictions and correctly guessing that by the time he noticed the large expense, we would have positive results to share. Soon we increased online hat sales tenfold from $8,000 to $80,000 per day, which funded most of the campaign's overhead costs.

At the suggestion of former Speaker of the House Newt Gingrich, I began filming daily Facebook videos of Trump riffing on trending topics. The videos went viral, picking up traditional news coverage and reaching more than seventy-four million viewers before the Iowa caucuses. For this project, I was given a budget of $400,000, but only spent $160,000 because supporters shared the videos faster than we could spend the money.

Just as the campaign's lack of structure and experience created room for innovation, it could also lead to colossal mistakes. On January 13, two weeks before the Iowa caucuses, the *New York Times* reported that the campaign had virtually no ground game in the first primary state: "Mr. Trump . . . may well win the caucuses, now less than three weeks away. But if he does, it will probably be in spite of his organizing team, which after months of scattershot efforts led by a paid staff of more than a dozen people, still seems amateurish and halting." By that time Trump and I were talking more frequently, and he asked what I thought about the news reports. "These articles make me look incompetent," he said. "I'm running as a businessman—if I can't run a campaign, how can I run the country?" Lewandowski reassured me that we had a great operation. But there were signs that trouble was brewing. Parscale and I built a mobile tool to help many likely first-time caucus-goers find the closest location. We asked Lewandowski for our Iowa voter list, which he had previously told me included eighty thousand emails. When Parscale got the list, he called me, alarmed: the data file had roughly twenty thousand names, and the file quality was garbage.

A week before the caucuses, Trump announced that he intended to skip the Fox News primary debate in Des Moines because Megyn Kelly was scheduled to be a moderator. Ailes struck back with a sarcastic statement: "We learned from a secret back channel that the Ayatollah and Putin both intend to treat Donald Trump unfairly when they meet with him if he becomes President." But when Trump doubled down, calling Megyn Kelly a "third-rate reporter who is frankly not good at what she does," Ailes grew nervous about the bad ratings that would result from a Trump boycott, and he called Trump to negotiate. It played out like two old friends looking for an off-ramp from a situation neither wanted to escalate further. Trump had planned to host a rally to raise money for veterans in lieu of the Fox News debate. Ailes agreed to donate $5 million to a veterans' organization of Trump's choice in exchange for his participation in the debate.

Ailes took this agreement to Rupert Murdoch, who told him, "No way!" Trump asked me to speak with Rupert and get him to approve the deal. I called Rupert and suggested that it would be a win-win-win: the vets would get $5 million, Fox News would receive a huge ratings bump, and Trump could declare victory. "Are you crazy?" Rupert exclaimed. "Once I start paying one person, I have to pay everyone to show up to debates. No. The answer's no," he said, abruptly ending the call. That night as we landed in Iowa, Trump skipped the debate, raising millions of dollars for veterans and stealing the thunder from the Fox News debate.

The morning of the caucuses, the *Des Moines Register* poll, the gold standard in Iowa, reported that Trump was five points ahead of Texas senator Ted Cruz. Trump had asked Ivanka, who was six months pregnant with our son Theodore, to speak at one of the largest caucus sites in the state, alongside Cruz and Kentucky senator Rand Paul, while he spoke at another caucus location. When we arrived at the DoubleTree convention center in Cedar Rapids, where more than 2,500 caucus voters from dozens of precincts had converged, all the other campaigns had large booths manned by packs of volunteers. They displayed slick posters and gave away loads of swag. Ivanka and I could not find a single Trump campaign staffer on site. I called Lewandowski, who promised that a team was on its way, but I could hear in his deflated voice that he wasn't sure how he was going to make that happen. I asked the campaign aide who was accompanying us, a woman from Arizona named Stephanie Grisham, to grab the other side of an empty card table. We carried it to the entrance and set up a makeshift display. While Ivanka shook hands and took photos with supporters, Grisham and I frantically looked up site-specific caucus instructions, the information most people requested when they came to our table. This lack of professionalism at the most important test to date was not a reassuring sign of Lewandowski's management.

As we boarded a small plane to Des Moines to meet up with the rest of the family, the initial results showed Trump stuck several points behind Cruz. We were silent on the forty-minute flight as the race slipped away. Despite leading by an average of seven points in ten polls in the days before the primary, Trump lost Iowa by more than three points.

Lewandowski knew he was on shaky ground, and rather than bringing in the talent he needed to help our campaign succeed, he seemed to become more insecure and territorial. I tried to help him by recruiting Bill Stepien, an experienced campaign operative recommended by one of my few Republican friends, Ken Kurson, the editor of the *Observer*. Stepien was the first political person I'd encountered who made any sense. He explained his straightforward approach to running a campaign: first, determine how many people you think are going to vote, and work backward to find blocks of voters that add up to 51 percent of that number. Then, do whatever it takes to get them to the polls. Since Lewandowski was constantly traveling with Trump, the campaign desperately needed someone like Stepien to organize the headquarters and field operations. I pitched Trump and Lewandowski on hiring him. After Trump initially agreed to bring him on, Lewandowski claimed that Stepien would be too high-profile and would cause problems.

Knowing that New Hampshire could be decisive for him, Trump spent the week barnstorming the state. On election eve, I got a call from Ailes, who told me that the Fox News exit polls were showing that Trump was going to win the state by more than ten points. We were staying with Trump and the rest of the campaign staff in an outdated hotel that had hot tubs in the middle of the bedrooms. Ivanka and I went up to Trump's room, which we knew was right above ours because the sound of his blaring TV had woken us up at four o'clock that morning. I relayed the news from Ailes. Trump was elated. New Hampshire had validated his conviction that he could win.

Less than two weeks later, Trump won big again in South Carolina. But the race was far from over. A consortium of establishment Republican politicians, donors, and media influencers began to mount a full-throttle campaign to prevent the outsider candidate from winning the Republican nomination.

5

An Unlikely Upset

B y March, the primary had effectively narrowed to a two-man battle between Trump and Cruz. Trump was driving virtually every
news cycle and honing his populist message, but we knew that
if he was going to become the front-runner, he needed to show skeptical
Republicans that he was going to offer concrete plans and serious policy
solutions.

I reached out to Howard Kohr, executive director of the American
Israel Public Affairs Committee (AIPAC), a well-established advocacy
group, and offered to have Trump participate in a question-and-answer
session at their upcoming convention in Washington, DC. Four days
before the event, Trump called and said he wanted to give a big policy
speech instead. To compose the speech, which would be the first scripted
policy speech of the campaign, I worked with Ken Kurson. The first draft
sounded nothing like the candidate, but after reading through it, Trump
gave us extensive edits that made it his own.

The day before the event, Lewandowski called me from Mar-a-Lago,
my father-in-law's palatial beachfront estate in Palm Beach, Florida.
"You have to call him ASAP," Lewandowski said. "He wants to cancel
the speech." Trump had seen news reports indicating that protesters
were now coming to AIPAC. I immediately got on the phone with him.
"These protesters are not going to be like the ones at your recent rallies,"
I said. "If anything, it will look like seventy people getting up to buy a
hot dog or use the bathroom in a stadium of twenty thousand. Canceling

at the last minute will look weak and will isolate your pro-Israel voters."
Trump was also reluctant to use a teleprompter. He had poked fun at the
politicians who used them. "You can use it as notes," I suggested. "We
have a teleprompter set up in the ballroom with the speech loaded. Try
it for fifteen minutes and see what you think. If you don't like it, you
never have to see one again." He practiced for more than an hour, and
the teleprompter operator commented that he was a natural. AIPAC was
back on.

As Trump took the podium the next day, I paced back and forth behind the stage. To my surprise, Trump mostly stuck to the script, with one
exception. He read the line, "With President Obama in his final year,"
and then added one exclamation: "Yay." The delighted crowd erupted in
applause. Their frustration toward President Obama had been building
since he signed the Iran deal a year earlier. As Trump walked offstage, he
gave me a rare compliment: "Good job."

Even his critics praised the address. Charles Krauthammer hailed the
speech as presidential. The only negative call I received was from Kohr,
who said Trump's playful comment about Obama's impending departure
had elicited backlash from the White House. Kohr was going to put out
a statement. I was shocked. "You're making a big mistake," I warned.
"Trump just made AIPAC hotter than ever, and he now has a one-in-two
shot of winning the nomination. Why would you alienate someone who
has that much potential to be president of the United States?" The statement went out the next day. Trump didn't forget it. During his four years
as president, he never returned to address the AIPAC conference, despite
being a hero to its attendees.

One month later, at the Mayflower Hotel in Washington, DC, Trump
again delivered a substantive, scripted speech that we entitled "America
First." It called out decades of rudderless, dangerous, and wasteful foreign
policy perpetrated by the leaders of both political parties. And it proposed a new vision that departed from the previous thirty years of failure
in Washington. But the event at the Mayflower was also significant for a
different reason. During a small cocktail reception, Jeff Sessions, the US
senator from Alabama, and I were introduced to roughly forty guests,

including Sergey Kislyak, Russia's ambassador to the United States. We shook hands, exchanged niceties, and moved along. At the time, I thought nothing of it; these sorts of functions are always bustling with foreign dignitaries. Little did I know that our benign encounter would become central to an enormous, convoluted, and ultimately pointless investigation.

On March 27 Ivanka and I welcomed our third child. Because he was a political baby, conceived and born during a presidential run, we named him after the twenty-sixth president of the United States, Theodore Roosevelt, a courageous and transformative president who, like Trump, was energetic and irreverent. A week after Ivanka gave birth, her father called and asked if she would come to a rally on Long Island that evening and introduce him. He was coming off a loss in the Wisconsin primary, and Ivanka always lifted his spirits. She agreed, and her appearance at the rally—along with the momentum the campaign was gaining—reinvigorated Trump. From that point forward, he won every single remaining primary.

Ivanka and I were with her father in Trump Tower as the Indiana primary results came in on May 3. Fox News flashed a breaking headline: Cruz was dropping out of the race. Against all odds, Trump had achieved a victory never before reached by a Republican candidate without a political or military background. He had put everything on the line, fought for what he believed in, and defeated sixteen candidates to earn the Republican nomination. Ecstatic, Trump turned to Ivanka and me: "Can you believe we pulled this thing off?"

The more time I spent with Trump on the campaign trail, the more I began to understand why millions of people felt like the American dream was becoming harder and harder to achieve. I agreed with my father-in-law that the status quo was no longer working. Washington and its ruling class needed to be disrupted. I didn't want to look back in twenty years and regret not having gone all-in on an insurgent effort to change America for the better. So as the campaign entered the general election phase, I became more involved than I had anticipated, including on personnel and finance decisions, our digital advertising strategy, and the president's travel schedule. I learned that when Trump worried about details, he grew frustrated and distracted. When his team was running operations well, he was more focused on the strategy and message. A happy Trump was a winning Trump.

I began working with Brad Parscale to ramp up our digital operation. I was connected to the founder of a tech company who had purchased record amounts of digital advertising, and he agreed to fly down to San Antonio to meet with Parscale and his team. He arrived a few minutes early and asked for Parscale, who was in the conference room having a team meeting. Not one to waste time, the tech company founder walked right into the conference room. "Who are your top three advertising designers?" he asked Parscale. Without waiting for an answer, he began barking orders to everyone else in the room as if he were a drill sergeant. He sent one person to buy air mattresses so that the staff could start working in shifts, twenty-four hours a day. "You only have a hundred and fifty days until the end of the campaign, you are far behind the competition, and you are going to need every second between now and then to make up ground," he said.

We set up a trading-floor-style operation, where the advertising teams competed against each other to drive engagement and raise donations for the campaign. To run the operation, we tapped Gary Coby, the Republican National Committee's most impressive digital expert and one of the few people in the Republican establishment who wanted Trump to win. Each day, the team that achieved the best return on investment received additional money to buy a larger share of ads the following day. The teams tested everything, down to whether the "Donate Here" button should be green or red, or whether an ad performed better with an eagle, an American flag, or a picture of Trump. This highly competitive environment produced staggering results. Under Parscale and Coby's leadership, the advertising teams tested more than a hundred thousand ad combinations each day, gathering real-time public opinion data that preceded polling by several days and informed the campaign about which messages resonated most with voters. We quickly saw that ads did better when they were focused on Trump's pro-America policies, like building the wall, rather than on attacking other candidates. In total, our campaign produced about six million ad variants, far surpassing the Clinton campaign's sixty-six thousand.[3]

In the last four months of the campaign, our digital operation raised more than $250 million in small-dollar donations—an unprecedented number—and persuaded millions of voters to support Trump in the process. Andrew Bosworth, the Facebook executive who oversaw the company's advertising during the 2016 election, later wrote a memo in which he argued that Trump was elected not because of Russia or "misinformation" but "because he ran the single best digital ad campaign I've ever seen from any advertiser. Period."[4]

* * *

Around the same time that we stood up our digital operation, Trump decided to expand his campaign's leadership. He wanted to elevate Paul Manafort—a seasoned campaign consultant who had joined our campaign back in March—to the position of comanager of the campaign. Trump asked me to break the news. When I met with Lewandowski, I explained that it wasn't personal. The campaign was growing; the stakes were increasing.

"Take this as a sign of your success," I reassured him. "Trump won the primary, and you're doing a great job."

Lewandowski started to whimper and walked away, but he pulled himself together for a meeting with Manafort and the campaign leadership. Early that evening, he called me, sobbing. "I can't do this anymore," he said. "I've given up my whole life for this."

I asked Lewandowski what would make him happy, and he suggested that Manafort could be the chairman instead of comanager of the campaign. I hung up, thinking we had reached an amenable solution, but Lewandowski called me two more times that evening, sounding incoherent and threatening to quit. Exasperated, I updated Trump on the situation. I was afraid that Lewandowski was cracking under pressure. "I'll handle it tomorrow," Trump told me.

The next morning, Trump called Lewandowski and said, "Look, you're a very good campaign manager, but there are also sixteen other really good campaign managers who are sitting at home now because they didn't have me as their client. Jared is not your psychiatrist. I am not your psychiatrist. You either get your act together or go home."

6

"I Am Your Voice"

The clash between Lewandowski and Manafort didn't take long to manifest. Both wanted to lead the search for a vice presidential running mate. I asked Trump how he wanted to proceed, and he replied, "I will run the search myself."

Manafort suggested that a vice presidential candidate should be able to deliver on three fronts: a clean rollout that excited supporters, a winning debate performance, and the ability to not steal the spotlight. After going through half a dozen names, Trump narrowed the vice presidential search to three: Mike Pence, Newt Gingrich, and Chris Christie. Christie had endorsed Trump in February, becoming one of the first major figures in the Republican establishment to do so. Trump well understood the tension between him and my family. When Christie offered his endorsement, in fact, Trump called my father and asked if he was comfortable with him accepting the endorsement. Trump and my father had become close friends after Ivanka and I married, and my father appreciated the sincere gesture. He told Trump that he was happy with his current life and encouraged him to do whatever was best for the campaign. Suddenly, my family's old nemesis was a political ally.

After we read the vetting files of the three candidates, I joked that Christie's file read like a John Grisham thriller, Gingrich's read like a Danielle Steel romance novel, and Pence's read like the Bible. I thought Pence was the perfect choice. A midwestern governor with experience in Washington as a congressman, he was respected by evangelicals, and his steady nature counterbalanced Trump's enthusiasm. I suggested that Trump invite the Indiana governor to Bedminster for a round of golf so they could get to know each other. I had no clue how painful this informal interview would be for Pence, who was not an avid golfer and probably would have preferred a CIA interrogation. Trump gave him three strokes per hole, and the round took four hours—more than double the time Trump usually takes to play eighteen holes. At the end of their round Trump good-naturedly poked fun at Mike for notching a hole-in-zero on a par three, when he shot an actual par on the hole.

Having run a family business for decades, Trump was accustomed to consulting his adult children on big decisions, so he wanted them to meet Pence before he made the announcement. Trump was campaigning in Indiana and planned to bring Pence back to New York with him. But his plane, Trump Force One, busted a tire and was grounded for the night. Eric called and said we needed to get to Indiana right away. The next morning, the media was surprised when Trump and his family walked through the front door of the governor's mansion. The Pence family showed us around their home. During the brief tour, Karen Pence pointed out that the furniture was all made by prison inmates through a program she supported, and Pence gave Trump a book called *The Forgotten Man*, a history of the Great Depression.[5] Inside, Pence inscribed a note: "To Donald Trump, with great admiration for the way you have given voice to the Forgotten Men and Women of America." Since the visit was last minute, Karen displayed flowers she had picked from her garden that morning and served breakfast in aluminum takeout trays from a local restaurant. Pence opened with a simple prayer, asking the Lord to watch over our family as we fought for the country.

On Friday, July 15, Trump announced Pence as his running mate. Over the next five years, I kept waiting for Pence to break character—to do what most politicians do behind the scenes and criticize others, complain about situations, and push back on requests to travel to events—but he never did.

Manafort and Lewandowski's coleadership of the campaign was short-lived. In both style and strengths, they were polar opposites. While Lewandowski was quick, visceral, and instinctual, Manafort was measured, methodical, and analytical. It didn't help that they were viciously sabotaging each other, each claiming the other was leaking to the press. By the middle of June, Lewandowski was out. As Manafort took the helm, Trump asked me to handle the campaign's finances. I brought in Jeff DeWit, treasurer of the state of Arizona, Sean Dollman, and Steven Mnuchin, the campaign's national fundraising chairman, to help manage the cash flow and track expenses. We had learned from watching our primary opponents like Scott Walker and Jeb Bush that expenses can quickly balloon out of control as a campaign grows. We wanted to avoid their mistakes.

* * *

Leading up to the Republican National Convention in July, Manafort had suggested that Trump's acceptance speech should be packed with poll-tested slogans and themes. We later learned that Manafort had spent roughly $300,000 to have a pollster named Tony Fabrizio craft the message. To draft the speech, Manafort called upon Stephen Miller, a former top aide to Jeff Sessions, who I had installed as the campaign's primary speechwriter and policy coordinator months earlier.

When Trump reviewed the draft, he hated it. For three hours, he dictated a new speech to Miller. Trump wanted to focus on the recent horrific attacks on police. On July 7, a deranged gunman had shot and killed five Dallas police officers. Days later, in Baton Rouge, Louisiana, another murderer attacked six police officers, killing three of them and badly wounding the others. Trump fumed about the evil and injustice. He felt that President Obama had stoked hatred toward law enforcement, putting police officers everywhere in jeopardy. When they finished the draft, Trump said, "I like the speech just like this—don't change a thing."

It was Sunday, July 17, just four days before Trump's most important speech to date. Stephen called me in a full panic: "The reading did NOT go well. He gave me an entirely new speech that will make his past controversial comments seem tepid by comparison."

I had planned to depart for Cleveland the next day with Ivanka.

Trump had asked her to introduce him, and I wanted to support her as she prepared for her big moment. After initially receiving a stilted draft of her introduction from Manafort, she scrapped it and wrote her own remarks. But Stephen Miller asked me to stay in New York and help him finalize the speech with Trump. Knowing the stakes, I asked Ivanka if that was okay with her. As usual, she was prepared for her moment and felt it was more important for me to stay back.

Stephen and I printed out the speech that Trump had dictated and laid the twenty-two pages out on a large conference-room table in Trump Tower. Trump has a near-photographic memory, so we knew he would notice if so much as a comma was out of place. If we tried too hard to change his words, he would double down. So we reorganized the paragraphs for logical flow and tweaked the lines that we thought would cause too much backlash. The next day, we nervously handed the new draft to Trump. As he read it, he frequently paused and asked: "Why did you change this line?" or "Why did you move this?"

By the time we got through two and a half hours of edits, Trump was both exasperated and satisfied. "Now, please don't touch it this time for real."

On July 21, Trump delivered his convention address to a packed Quicken Loans Arena in Cleveland and thirty-five million viewers across the nation. It was a smashing success. The next morning, the front page of the *New York Times* ran the headline, "Trump, as Nominee, Vows: 'I Am Your Voice.' " Ivanka crushed her speech as well. She delivered her message with ease and grace, highlighting many of the same issues she had championed in her company, including supporting mothers in the workforce and making childcare affordable and accessible. They weren't traditionally Republican issues, but she knew her father would endorse them.

Manafort executed a highly successful convention that was authentic to Trump and ensured that he secured the delegates for the nomination. A few weeks later, however, reports began percolating about Manafort's business dealings with the Kremlin-backed political party in Ukraine. At the time, Manafort was struggling to develop chemistry with my father-in-law. He spoke slowly and muffled his words. Trump would brush him off, and Manafort never modified his approach. It didn't help that Lewandowski regularly criticized Manafort on CNN and called Trump to point out all of the ways in which Manafort was failing the campaign. Behind the scenes, Manafort was doing an excellent job building out the infrastructure of the campaign, but publicly he was taking on so much water that his position was becoming untenable.

In the lead-up to Trump's decision to fire Manafort, I had been working twenty-hour days for weeks on end, splitting time between overseeing my business and helping to build and run various parts of the campaign. Ivanka had assumed a disproportionate share of the parenting duties while also helping lead her father's business and running her own company. In August we took a weekend trip to recharge before entering the final campaign sprint. Less than twenty-four hours after we left, I received a call from Trump. "It's time for Paul to go," he said. "I like him, but he doesn't have the energy we need." He mentioned Steve Bannon and asked me to bring him on board.

Bannon was the executive chairman of Breitbart News Network, a news website with strong ties to Trump's antiestablishment conservative base. A former naval officer, Harvard Business School alum, and Goldman Sachs banker, he came highly recommended by Republican donor Rebekah Mercer. At sixty-three, Bannon cut an unorthodox political profile. He was gruff and unkempt, with a perpetual five o'clock shadow, and he had never led a campaign. When I called Bannon to pitch him on joining the team, he responded with his trademark bluntness: "I don't want to join a sinking ship. You have an undisciplined candidate. You have no operation."

I pushed back: "This is a much better opportunity than you think. The RNC has a good ground game that we are integrating with our field operation. I just hired Jason Miller, a communications pro with extensive campaign experience, to manage our messaging and build up our press team, and we have a state-of-the-art digital data operation that I built like a start-up—you just haven't heard about it because the people running it aren't political."

I reported back to Trump that I had made a deal with Bannon that he be campaign CEO, and I recommended that he promote Kellyanne Conway to campaign manager. I had hired Conway about a month earlier as a polling consultant to help with our messaging. While Trump was initially hesitant to hire her given all the negative things she had said about him when she worked for Ted Cruz, he grew to appreciate her skill at defending his campaign on television. She would make history as the first female Republican presidential campaign manager. Trump signed off on the plan, but Manafort was still technically the campaign chairman.

Early in the morning of August 19, Trump called me and wanted to sever Manafort's involvement with the campaign. Another story had broken that alleged shady dealings with Ukraine. I met Manafort for breakfast at eight o'clock that morning at Cipriani on Fifth Avenue, a wood-paneled Italian restaurant across the street from Central Park.

"I hate doing this because I have grown very fond of working together and appreciate the amazing contributions you have made, but it's time to make a change," I told Manafort.

He was shocked. "Okay, I understand, let me have a week to figure this out."

"I wish I could give you a week," I said, "but this needs to be done today. I have a draft statement thanking you for your service. Trump lands in Louisiana at ten thirty this morning, and he wants to have the news out beforehand."

Manafort was angry, but he took it like a gentleman. We went back to the campaign headquarters, where he signed off on the statement, and we released it just before Trump landed. Manafort packed up and left. With just eighty-one days separating us from Election Day, the sprint to the finish had begun.

7

"We're Going to Win"

Trump's promise to make Mexico pay for a new wall on the United States' southern border had become a live-wire issue on the campaign, equally controversial in the United States and Mexico. Back in the spring of 2016, a friend reached out to me to relay a message: Luis Videgaray Caso, President Enrique Peña Nieto's finance minister, wanted to make contact with the Trump campaign. I figured this was a joke, but she insisted, "This is a very serious and important reach out, and he is a very serious person." I had no idea how important Luis would become to me in the years ahead.

In a dingy hotel cafeteria in a Maryland suburb of Washington, DC, Don Jr. and I met Luis for coffee. During the discussion, we found more in common than one would have thought. Luis, who has a PhD in economics from MIT, was cerebral and brilliant at politics. He looked past the media's spin on Trump's statements and saw an opportunity. American presidential candidates rarely paid attention to Mexico. He felt that the United States and Mexico could improve their relationship through modernizing the North American Free Trade Agreement, or NAFTA, the trade deal between the United States, Mexico, and Canada that Trump routinely condemned. He also believed that reforming immigration, and stopping the flow of illegal guns and cash, would be mutually beneficial. Most surprising of all, Luis was sure that Trump was going to win the election, and he wanted to establish a relationship immediately.

After a couple of false starts, I was able to arrange a breakfast between

Trump and Luis in Bedminster. During the breakfast, Trump floated the idea of traveling to Mexico City to meet with President Peña Nieto.

"If we invite you," Luis explained, "we would also invite Hillary."

Trump laughed. "That's okay, she'll never come." He was right.

I knew a trip to Mexico would be a big risk, but I also knew that Trump was at his best when he was doing the unexpected. At the time, Trump was down by thirteen points, his polls reeling from the most recent campaign turmoil and shakeup. We needed to play big to stay in the game. A trip to Mexico would catch everyone by surprise. It would show that Trump could conduct himself presidentially on the world stage, which would counter the media narrative. It would also show that he wasn't against the Mexican people; he was against the unimpeded flow of illegal immigration. I asked Bannon for his thoughts, and he agreed that the trip was worth the risk, so we began to plan the logistics.

Each detail of the visit had to be meticulously scripted and flawlessly executed. We needed to keep our plans a secret. If word leaked in advance, it could put pressure on President Peña Nieto to withdraw the invitation. There were also massive security implications; Trump's comments on illegal immigration had supposedly drawn the ire of notorious drug kingpin El Chapo, who reportedly had placed a $100 million bounty on Trump.[6] The Friday before we were scheduled to depart, Trump called off the trip. Apparently, his campaign fundraising chairman Steven Mnuchin had told him that one small misstep could turn the whole trip into a humiliating debacle. At the time, there were signs that Trump was making a comeback in the polls, and he didn't want to push his luck.

"It's too risky," he said. "What if we travel all the way down to Mexico, and he stands next to me at the podium and lectures me, saying, 'I'm not paying for your stupid wall.' It would be a disaster, and the campaign will be over." I tried to explain that I trusted Luis and thought I could mitigate, though not fully eliminate, these risks. But deep down, I was a bit relieved; I knew that if anything went wrong, I had full responsibility.

I called Bannon and asked what he thought we should do. "This trip is too good to let pass," he replied. The two of us met with Trump in Bedminster and addressed his concerns before he decided to proceed. The night before departure, the news broke that we were heading to Mexico. People were shocked.

As we boarded the unmarked plane for Mexico City, I made sure that a campaign staffer loaded an important delivery that I had commissioned as a gift to the Mexican president: red hats, embroidered with five words, "Make Mexico Great Again Also."

We arrived in Mexico City on August 31. Trump and Peña Nieto met in private before emerging for a press conference with Mexican reporters as well as a few American journalists who had jumped on a plane as soon as they found out about the trip. Both politicians delivered statements, holding their ground on their key issues but showing that the United States and Mexico had many overlapping interests.

"Even though we may not agree on everything," the Mexican president stated, "I trust that together, we will be able to find better prosperity and security."

"A strong, prosperous, and vibrant Mexico is in the best interest of the United States and will keep and help keep, for a long, long period of time, America together," Trump said.

This could not be going better, I thought. Then just as Trump was about to conclude, ABC's Jonathan Karl shouted a question, asking whether they had discussed Trump's plan to make Mexico pay for the border wall. We had agreed with the Mexicans that Trump and Peña Nieto would not take questions from reporters. But when Karl shouted the million-dollar question, Trump answered, "We did discuss the wall; we didn't discuss payment of the wall." I looked at Luis, who hurried to get someone to cut off the public address system. As the reporters began to yell questions, Mexican music started blaring from the loudspeakers, and the politicians walked offstage. But the damage was done. For the Mexican public, it was unthinkable that their president could have discussed the wall without raising opposition to Trump's payment proposal. It was an insult they couldn't bear, and it made Peña Nieto look weak and potentially complicit.

The press conference triggered a political nightmare for the Mexican leadership—especially for Luis, who took the blame for his role in planning the trip. He resigned the next week. I felt terrible. I called Luis and told him I was sorry he had resigned. I relayed the news to Trump, and he put out a tweet: "Mexico has lost a brilliant finance minister, and wonderful man who I know is highly respected by President Peña Nieto. With Luis, Mexico and the United States would have made wonderful deals together—where both Mexico and the US would have benefitted." Luis was incredibly honorable in the way he conducted the visit, and we trusted him implicitly each step of the way. When he took the fall, he did so gracefully and without bitterness, believing that he did the right thing for his country.

For Trump, the trip was a massive success. It showed voters that he could go into the lion's den and fight for American interests. Afterward, the campaign settled into a positive groove. To maximize the schedule for the final stretch of the campaign, I consulted Newt Gingrich, a political mastermind and former Speaker of the House. He knew how to coordinate political travel to highlight messaging that would reach voters in swing districts. David Bossie came on as deputy campaign manager. He was a tremendous help in leading and executing the operations, as was Bill Stepien, who I was finally able to bring on board. Eric Trump did an amazing job organizing the campaign's ground game, and Don Jr., Ivanka, and Lara traveled around the country, drawing large crowds and increasing our campaign presence in all the swing states.

As we entered the last two months, I learned that our ad teams were not getting timely internal feedback on their video scripts. When campaign political consultant Larry Weitzner explained the problem to me, I told him to skip the approval process and just spend the money making the ads; I would show them to Trump and get his approval. I called Roger Ailes, who had recently resigned from Fox News, and asked if he would oversee Weitzner and edit our ad scripts. He agreed, and Weitzner worked closely with him to make some of our most effective ads in the final stretch. Trump was mostly staying on message, crowds continued to swell at rallies in swing states, and we were raising tens of millions of dollars from small individual donors online—a solid signal that our message was connecting. A few of our internal polls even had Trump pulling ahead. We had survived many controversies that would have sunk any traditional politician. The biggest controversy, however, was yet to come.

On Friday, October 7, as Trump was preparing for the second debate with Hillary, Hope Hicks received an inquiry from the *Washington Post*. They'd found a video of Trump having a vulgar conversation with Billy Bush during his time on *Access Hollywood*. I stayed late at Trump Tower that evening to help Trump prepare an apology, which he recorded and released as a video message that night. It was the first time since Ivanka and I were married that I broke from observing the Sabbath. I regretted my father-in-law's words, but as I had learned from my own family, forgiveness means not defining people based solely on their past transgressions.

The next day, Republican National Committee chairman Reince Priebus came up from Washington, DC. "You have two choices: withdraw now or lose in the worst landslide in the history of presidential elections," he told Trump. The rest of us looked at each other in bewilderment. Anyone who knew Trump knew there was zero chance he was going to withdraw. Meanwhile, we noticed that hundreds of people had gathered in front of Trump Tower to show their support. Trump insisted on going out to thank them. Secret Service rushed into action, and ten minutes later, Trump went down and spoke off the cuff. The visual of Trump surrounded by adoring fans on Fifth Avenue was just the image we needed to hold off the calls for him to bow out and get to the debate the next day. Trump went on to deliver an amazing performance under fire. We were back in the game.

Election Day was approaching, and the campaign was entering its final push. I asked our political directors in the swing states how much money they needed to win. The total amount they quoted exceeded $25 million. When I showed the numbers to Bill Stepien, he took one look at them and said, "Only $1.25 million of this will make a difference. The state directors are padding their requested budgets, so if they lose, they can tell their future clients they would have won if they were given enough money." I went with Bill's recommendation, knowing that if we overspent and still lost the election, Kellyanne Conway and Steve Bannon would be long gone, and I would have to be the one to ask Trump to write another check. I wasn't going to put myself, or him, in that position unless I was convinced that every extra dollar would push us closer to victory.

Trump was like a gladiator in the arena, delivering speech after speech in the closing days. Knowing that his previous undisciplined tweets and off-the-cuff comments had hurt his chances with some voters—and wanting to win badly—he focused on keeping his message tight. He joked at a rally in Pensacola, Florida, with thousands of fans, "We've gotta be nice and cool . . . no side-tracks, Donald, nice and easy, nice and easy."

During the final night on the campaign trail, in front of a packed arena in New Hampshire, Trump thanked his family. "I've been reading about all these surrogates going all over for Hillary Clinton, but I had my family. I had the best surrogates of all."

*   *   *

On the morning of November 8—Election Day—I was working in my office at 666 Fifth Avenue. I had just traversed the country and watched Trump perform at 10 rallies in the final 48 hours. I received a message from Savannah Guthrie, host of NBC's *Today Show*. I had not met her— and I rarely talked to reporters—but I was curious what the media was thinking about the election, so I returned the call. After hearing that her colleagues thought Trump was going to lose in a landslide, I predicted with cautious optimism that he was going to win. I had studied our data. In 2012, Republican presidential candidate Mitt Romney had failed to turn out enough voters in smaller, rural counties across America. Trump was significantly outperforming Romney in these areas, and he was motivating people who didn't typically vote. I also knew that the election could easily swing the other way if Hillary even slightly outperformed him in key suburbs. After months of nonstop action, it was unsettling to wait for results with nothing more to do.

At five o'clock, shortly before the first polls closed, deputy campaign manager Dave Bossie called me with exit polling data. He warned me that it looked like a nightmare. I had promised Trump that I would update him, so I nervously called up to his apartment. "The exit polls aren't great," I told him before I walked him through the numbers. "They show us behind, but Stepien thinks their methodology is flawed, and our voters are working Americans, so they will likely be heading to the polls late. Let's see what happens." I will never forget his response: "We left it all on the field. I worked my butt off, and there is nothing more we could have done. I am proud of what I've done. I am proud of the team. I am proud of you. Win or lose, let's have some fun tonight as a family." I was blown away. He couldn't have been calmer or more at peace.

As the night progressed, things started to look more positive: just after ten thirty, the Associated Press called Ohio for Trump. Parscale came over to me and whispered in my ear, "Our data science team says the optimistic models are playing out as we expected based on the actual turnout data. It's a rust-belt Brexit. We are going to win." They called Florida roughly fifteen minutes later, when the massive vote margin in the panhandle dropped. Preliminary results from Pennsylvania were in line with our optimistic data-modeling scenario. Phone calls and texts from well-wishers began pouring in. I asked Stephen Miller if we had a victory speech. The answer was yes, but it was a very rough draft that spent more time gloating than bringing the country together. At 11:00 p.m., Trump reviewed the speech. After seeing a television clip of despondent Hillary supporters at the Javits Center, he knew it wasn't the right tone. He wanted to be gracious.

While we waited for the results to trickle in, Ivanka, Stephen, and I huddled in the dining room of Trump's apartment to rewrite the speech as Trump dictated what he wanted to say. As we wrote, we looked out at the Peninsula Hotel to the south, where the Clinton campaign had reserved the rooftop bar for a victory party to taunt our group in Trump Tower. The press was holding back on actually calling the race, so we went over to the New York Hilton in midtown around 2:00 a.m.

Trump hadn't wanted to spend millions of dollars on an election-night party like 2012 presidential candidate Mitt Romney had done. In fact, he had told me that he didn't want to spend a single dollar on a party. He suggested that if he won, he'd just send out a tweet, and supporters would spontaneously gather. "If we lose, I'm going right to my beautiful 757 plane and heading to Scotland to play golf for a few months," he joked. When I insisted that we needed to have some venue, he told me to get the cheapest ballroom I could find, and that's what we did. At two thirty in the morning, as we were discussing what to do next, Kellyanne Conway got a call and brought Trump the phone. Hillary spoke to him for less than one minute. She offered her congratulations on a hard-fought campaign and conceded. President-elect Donald Trump walked out onstage before an elated audience and delivered a fifteen-minute victory speech. Trump became the first true outsider to be elected to the presidency. His victory changed the course of history.

Before the night was over, I rang Luis Videgaray, who seemed surprised to hear from me on such an historic night. "You bet correctly," I said. "I want to thank you. Now we have a chance to fix the US-Mexico relationship."

Intellectually, I always believed Trump could win. But emotionally, I never let myself think about what would happen if he did. As messages started to flood my in-box from new and old friends all over the world, I began to realize that Ivanka and I had a major life decision to make. The hard part was about to begin.

8

"I'll Never Get Used to This"

As a New Yorker, I was used to sitting in traffic, not causing it. Yet thirty hours after Trump's victory, I found myself in a Secret Service motorcade rolling through the streets of Manhattan toward LaGuardia Airport, flanked by a counterassault team with semiautomatic rifles and night-vision goggles. Through the bulletproof windows of the armored Chevy Suburban, I watched the blocks pass by as a phalanx of NYPD officers held back the cross traffic and pedestrians. Even FDR Drive was closed off to Manhattan traffic. For the first time, I began to appreciate how much my life was going to change.

I asked Trump if he thought he'd ever get used to it.

"I grew up as a kid from Queens," he said. "I'll never get used to this. This will always be cool to me."

As we caravanned onto the tarmac at LaGuardia Airport, Port Authority fire trucks blasted their water cannons fifty feet into the sky, forming an arch over Trump's 757 aircraft, which he had parked there for decades. The Port Authority officers loved Trump. He always greeted and thanked them when he arrived at the airport. Their salute that day was the ultimate sign of respect.

We were on our way to see President Obama, who had invited Trump and Melania to visit the White House. As we drove up the long circle drive that forms a ring around the South Lawn of the White House, where the president's Marine One helicopter lands, we were greeted by two Marines in formal dress, standing at attention. President Obama met us outside and graciously ushered us into the Diplomatic Reception Room, the elegant oval room often used to greet foreign leaders during state visits. Dan Scavino was filming Trump and Melania's entrance, but a White House protocol official told him to turn off the camera as First Lady Michelle Obama greeted Melania. As the two went upstairs to the Executive Residence for tea, President Obama led Trump and the rest of us across the famous colonnade that connects the residence to the West Wing. I had seen the passageway on television, but this was the first time I'd walked along the storied corridor past the Rose Garden, and I tried to take it all in during the forty-five-second walk.

As someone who always paid attention to real estate, I was shocked by the limited square footage of the West Wing. Desks lined the perimeter of cramped, windowless rooms where administrative assistants were stacked on top of each other. Senior staff offices were scattered throughout the three-story structure. This was the exact opposite of the open workspaces that I had found conducive to collaboration in my companies. It was beautiful, but it didn't seem designed for running the free world in the modern era. As soon as we stepped into the Oval Office, however, I understood why the place enamored people. It was breathtaking. The eighteen-foot ceiling decorated by an oversize plaster presidential seal; stunning views of the Rose Garden and South Lawn with the Washington Monument towering in the distance; the custom oval rug covering the hundred-foot circumference of the oak and walnut floor; the ornate carvings of the timbers salvaged from the British vessel HMS *Resolute* to make the iconic desk. The Oval Office is the greatest home-court advantage in the world. I would later watch heads of state, business titans, and powerful lawmakers become so awestruck by the grandeur of the room that they stumbled over their words, trying to deliver their carefully prepared remarks during their precious few minutes with the president of the United States.

Obama and Trump met privately for about an hour and a half. Afterward, Trump described Obama as a candid politician who was cordial— warm even. He recalled that Obama kicked off their meeting with a backhanded compliment: "I've been watching your speeches for the past years, and I must say you are an amazing politician. On so many issues, I still can't figure out where you stand. Are you for guns, are you against guns? Are you pro-life? Are you pro-choice? You have this amazing ability to be on every side of an issue." Obama warned Trump not to hire General Michael Flynn as national security adviser and said that he believed North Korea was America's greatest threat.

After their meeting, Obama motioned to an aide to bring in the press. Hope Hicks, Dan Scavino, and I were invited to stand in the back of the room. The serenity of the Oval Office was shattered as reporters rushed in and began yelling one question after another as cameras clicked twenty times per second. It was unlike anything I had experienced, even on the campaign. Just as quickly as the stampede had begun, it was over, and Obama offered Trump a piece of advice: If you don't answer their questions, they will stop asking. It was a good suggestion, but I couldn't imagine my father-in-law ever adopting it.

On our way out to the motorcade, as we passed through the colonnade again, Obama turned back toward me and asked, "Have you and Ivanka decided if you are coming to Washington?" I said that we had not. "You definitely should," he encouraged. "You could do a lot of good here."

On the trip back to New York City, Melania mentioned that the White House living quarters were dated and were going to need work before they moved in. Trump turned to her and said, "Honey, don't do too much. It's the White House—it's perfect. If it was good enough for Honest Abe Lincoln, it's good enough for us."

Trump was particularly reflective. He felt the gravity of the responsibility entrusted to him. He genuinely wanted to help the country unite. He asked Ivanka to call Chelsea Clinton, who we knew socially, to convey that Trump had no intention of looking backward and hoped to have a cordial relationship with Hillary to unite the country. He even told Ivanka to invite Hillary and Bill for dinner in the coming weeks. Ivanka did call Chelsea, but days later Hillary backed Jill Stein's challenge to the election, and Trump ended his outreach.

While Trump was intent on building bridges, the Clinton campaign was busy hatching plans to cripple the Trump presidency before it started. As Jonathan Allen and Amie Parnes reported in *Shattered*, less than twenty-four hours after their loss, at the Clinton headquarters, campaign heavyweights John Podesta and Robby Mook came up with the idea of blaming Hillary's loss on Russian interference.[7] When news reports started percolating, I thought the claims were absurd and would never be taken seriously. I had no idea that the fabricated story would loom over Trump's administration for years.

In the days immediately following Trump's improbable victory, he seemed optimistic about resetting his relationship with the media. He asked Hope to invite the editorial staff of Condé Nast—the publisher of the *New Yorker*, *Vanity Fair*, *GQ*, and *Vogue*—to his office for a meeting and worked hard to charm them. He did the same with the *New York Times*.

Afterward, the *New York Times* published one of the most unfair stories in its history. The heads of Obama's intelligence agencies—CIA director John Brennan, director of national intelligence James Clapper, and FBI director Jim Comey—had come to Trump Tower to give the president-elect his first intelligence briefing. As the meeting wrapped up, Comey pulled Trump aside and told him about the existence of the notorious Christopher Steele dossier, a salacious and patently false file that we later learned had been funded by the DNC. Many journalists had seen it, but they couldn't confirm the unfounded rumors. The FBI knew it was unverified, but Comey decided to brief Trump on the dossier. The briefing itself was newsworthy, so the *New York Times* could now justify reporting about it.

The rest of the press obsessed over the dossier, the Clinton campaign amplified it, television talking heads said it was the tip of the iceberg, and a narrative about collusion between the Trump campaign and Russia took root. Watching this unfold, I mostly dismissed the claims because I knew they were baseless. A credible media would realize that, I thought.

* * *

Setting up a new administration is a monumental undertaking. Back in May of 2016, Trump appointed Chris Christie to head the transition. I was in the meeting.

"What should we do about Chris?" Trump started, looking in the governor's direction. Christie explained that he really wanted to lead the transition and could do a good job with it. "Well, what about the Charlie issue?" Trump asked, referring to my father's complicated relationship with Christie.

"I spoke to my father this morning after your request," I said. "And he holds no grudge and thinks you should do whatever you think is best for you and the country."

"So, we are good then?" Trump asked.

"My father is good," I responded. "Between Chris and me, if we're going to work together, I should express that I felt the way you handled my father's case was overzealous, and it brought serious hurt to me and my family."

Christie explained that he had just been doing his job, that my dad had committed a crime, and that it wasn't personal.

"Well, respectfully, I have a different point of view on that," I said. "If it wasn't personal for you, then how come you challenged my father's release date after he had served his sentence? I hope you can understand how brutal it is for a family to have a loved one in prison. The only solace is having a date when your nightmare will end. When the prosecutor comes back and challenges the release date, and it gets delayed indefinitely, that's devastating to a family. So don't tell me it wasn't personal, because if it wasn't personal, you would have let him come home on time."

"You know, the crime your father committed was terrible," Christie started to say.

Then Trump interjected: "Chris, it was a family dispute."

"Look, at the end of the day, it doesn't matter," I said. "Donald wants you to do a job. You have my word: I've put that in the past, and I'll do everything I can to support you in that effort. I'm here to help. This is about an opportunity to help the country. This is about service. And we can take the personal situation that has happened and put it aside, because that's not relevant right now to what we're doing."

"Okay, let's do that," Christie agreed.

Six months later, a few days before the election, Steve Bannon called me in a panic. "Christie is trying to get on the plane," Bannon said. "We've got to keep him off. He wants to talk Donald into letting him be chief of staff. We can't let that happen—the transition effort is a train wreck. He's angling to slot his closest political cronies—including anti-Trump establishment types—into the most important appointments, regardless of their qualifications." Both Bannon and Steven Mnuchin reviewed Christie's transition materials and believed they failed to meet proper vetting, research, and professional standards for hiring key personnel. They told Trump and me that Christie was unprepared for the task. "Plus," Bannon added, "Chris is politically radioactive. He has an eighteen percent approval rating and is enmeshed in the Bridgegate scandal. Trump shouldn't have to carry his baggage."

Trump didn't want his incoming administration tainted by the legal mess in New Jersey. "I want a nice, clean Mike Pence administration—not a corrupt New Jersey administration," he told us.

In fairness to Christie, during the summer months, Trump had made clear that he did not want to focus on the transition before the election. At one point, when I told Trump that I had attended a three-hour transition meeting, and assured him that Christie and I were working well together, my father-in-law gave clear instructions: "Don't spend another minute on the transition. Romney spent all this time on the transition, with his binders of women, and he lost the election. Spend all your time on the campaign, and if we win, we will figure it out." So Christie had been working for months without much support from the campaign leadership.

To get the transition on track, I quietly reached out to Chris Liddell, a former chief financial officer for both Microsoft and General Motors who had been executive director of the Romney Readiness Project. Liddell volunteered to help immediately. He arrived less than twenty-four hours later, the day before Thanksgiving, and worked through the holiday. Liddell became a trusted friend and confidant and was one of the few people who served in the White House for all four years of the presidency.

As we raced to set up the administration, business executives, politicians, and military brass came to Trump Tower and Bedminster to inter-view for cabinet positions. Romney called me and pitched himself for secretary of state, pledging to be loyal to Trump. While Trump flirted with the idea, he decided not to take the risk of hiring Romney, who had criticized him throughout the campaign. The revolving door of Trump Tower was buzzing with high-profile candidates coming in and out con-stantly. When Trump interviewed James Mattis for defense secretary, he asked the four-star Marine general about his thoughts on torture.

"I don't believe in it," Mattis said with conviction.

"What do you mean?" asked Trump.

"I can get these guys to talk with a cup of coffee and a cigarette better than I can by waterboarding."

We were all impressed. Given the serious threats from Iran and ISIS, Trump wanted a general who knew the situation on the ground, was already up to speed, and could quickly build morale with the troops. Mattis seemed like a natural fit. He had a storied military career and a reputation for being beloved by frontline service members. While his nickname in the military was "Chaos," Trump thought "Mad Dog" was better and started using it. Before the interview was over, Trump offered Mattis the job.

During the campaign, Trump had asked me to be the point of contact for the representatives of foreign countries who occasionally contacted us. I agreed, assuming it would be a minor responsibility among my growing list of duties, but it became far more intense during the transition as we began receiving hundreds of meeting requests from dignitaries. Foreign governments hadn't planned on a Trump presidency and were scrambling to establish contact with a bunch of Washington outsiders.

Months earlier, I had met Henry Kissinger, the historic former sec-retary of state and national security adviser under presidents Richard Nixon and Gerald Ford. The advice he had given me then rang even truer now: "Trump is talking about a lot of critical issues that have been ignored, which is making foreign leaders nervous. Don't reassure them. Right now, they're all doing reassessments. They are taking inventory of their relationship with America and determining what they have that they don't want to lose and what they are willing to give up to keep it. That puts the United States in a better negotiating position if you win." He further warned me to "be careful who you interact with in every country. A relationship with you is valuable currency in a capital city. Select carefully whom you want to give that power to."

In one of Trump's many congratulatory calls with foreign leaders, he spoke with Prime Minister Shinzō Abe of Japan. Abe told Trump that he would love to meet. Honored, Trump invited him to visit Trump Tower. Soon after, I received a call from Obama's chief of staff, Denis McDonough, who explained that typically a president-elect declines meetings with foreign leaders out of deference to the current commander in chief. "One president at a time," he told me. I relayed the message to Trump. "Forget protocols," he said. "It's not a big deal. It's just a meeting. If the leader of Japan wants to travel halfway across the world to see me, I am happy to meet with him."

Trump also took a call with Tsai Ing-wen, the president of Taiwan. The call broke a diplomatic norm and the Chinese interpreted it as a challenge to their One China policy, which claimed that Taiwan was a Chinese province rather than an independent nation. Outraged, the Chinese sent over one of their highest-ranking diplomats to meet with our team: Yang Jiechi, the director of the Central Foreign Affairs Commission. To avoid the crowds and barrage of cameras outside Trump Tower, the national security transition team advised that the Chinese should come to my office at 666 Fifth Avenue. Before the Chinese arrived, Peter Navarro, an eccentric former professor whom I hired as the campaign's trade adviser after reading his book *Death by China*, insisted that I refrain from greeting them at their car, which I had offered to do as a courtesy.[8] Navarro was adamant that such a gesture would be interpreted by the Chinese as weakness. When they arrived, Yang read from a script, while a second Chinese official looked intently at me to gauge my reaction. A third Chinese official sat nearby, taking notes. When Yang got to the talking point on Taiwan, he looked up sternly and drew a hard line: "The territorial integrity of China is nonnegotiable." After the meeting, Navarro demanded that the Secret Service sweep the office for bugs.

Hess Corporation chief executive John Hess and Blackstone Group founder Steve Schwarzman asked me to meet with several high-ranking Saudi officials, who were eager to strengthen their relationship with the United States after the disaster in the Middle East with the previous administration. When Flynn, Bannon, and I met in New York with a small Saudi delegation, led by Dr. Fahad bin Abdullah Toonsi, they explained that they had a fraught dynamic with Obama over his positions on a number of challenges in the region—including Syria, ISIS, Iran, and Yemen—and were anxious to begin a new and hopefully more productive relationship with our administration. Bannon and I were tough. We told the Saudis that they needed to stop funding terror, improve their record on women's rights, pay for their own military, and begin taking steps toward working with Israel. We weren't interested in building the relationship if they weren't committed to making real progress on these goals. Fahad assured us that change was underway and that we would be very surprised by the reforms that they planned to make. The kingdom had a new young leader, Mohammed bin Salman, colloquially known as MBS, who wanted to transform Saudi Arabia. They would come back with a plan to show how we could make progress together.

As I interacted with dozens of foreign officials, from the United Kingdom's newly appointed foreign minister Boris Johnson to the Norwegian foreign minister Børge Brende to the Russian ambassador, I learned diplomacy on the fly. There was no rulebook for success or protocol officer guiding our interactions. When I developed new relationships in business, I would spend time listening and learning before showing my cards, and I took a similar approach here. After Trump tapped Exxon oil executive Rex Tillerson for secretary of state, I handed off most of the files and turned my attention to other pressing domestic issues, which was a relief.

* * *

On a cold, quiet afternoon in December, Ivanka and I were walking with our kids in Bedminster and taking time to think about what we were going to do next. Before election night, we hadn't let ourselves focus on what a victory would mean for us. But as we thought about it, we realized that we couldn't imagine looking back one day, knowing that we had walked away from an opportunity to help solve some of the greatest challenges facing our nation and the world. Through the campaign, we had seen firsthand how the president-elect's message resonated with millions of forgotten men and women. Their stories led the two of us—longtime Manhattanites with limited exposure to national politics—to believe in the core principles Trump was fighting for. Hundreds of Republican power players, who had done nothing to help him on the campaign and in many cases actively opposed and undermined him, were vying for top positions. We believed that he needed people like us—family members who understood him and were committed to helping him succeed, without any hidden agenda. We knew that this opportunity to serve would come at a steep cost. It would mean giving up our thriving businesses, leaving our lives in New York, and coming to Washington amid claims of nepotism.

As we mulled it over, I received a call from Risa Heller, who handled public relations for our companies. "The *New York Times* just asked me for comment. They have been told that you are moving to DC and leading the Middle East peace effort."

"That's bullshit," I replied. "We haven't made a decision yet. Who's their source?"

"Their source is your father-in-law," Risa said.

Trump's announcement of my appointment to the *New York Times* was his way of offering me the job. Ivanka and I decided together to take this once-in-a-lifetime chance to serve the country we love. When we told Trump that we were coming to Washington, he was happy, but warned us that we had to be very careful: "You're too young, too skinny, too rich, and too good looking. They'll be gunning for you."

As I prepared to exit Kushner Companies, I was glad that I had recruited my younger sister Nikki to come join the family business the year before. At the time, I never imagined that I would be leaving to enter government service, but now Nikki, who had spent ten years as an executive at Ralph Lauren, would be able to assume some of my responsibilities and help my father and longtime partner Laurent Morali lead the company. I knew that our family business would inevitably come under scrutiny because of my government service, but fortunately we ran an extremely professional and aboveboard business and I was confident the company would sail through the scrutiny. Because of my father's situation, we had learned that money was not the most important thing. My family was prepared to prioritize my government service over the company's profit. For that, and for their constant love and support, I will always be deeply grateful.

A few days after Ivanka and I had made the decision to move to Washington, I took our daughter Arabella, who was five at the time, on a dinner date. Just as our pizza arrived, my phone rang. It was Israeli ambassador Ron Dermer, right-hand man of Prime Minister Bibi Netanyahu. Born and raised in Miami Beach, Florida, Dermer studied at both Wharton and Oxford before renouncing his American citizenship and diving into Israeli politics. He was appointed as Israeli ambassador to the United States in 2013, and we maintained regular contact throughout the campaign.

I had a rule not to let work interrupt our father-daughter time, but I took the call. Dermer told me that twenty blocks south, at the United Nations headquarters, several countries, led by Egypt, were preparing to introduce a resolution to denounce Israeli jurisdictional claims in the West Bank as having "no legal validity" and as being "a flagrant violation of international law." Dermer was hearing that the Obama administration intended to abstain. If the United States abstained, it would be an unprecedented abandonment of Israel. It would also threaten our future efforts to forge peace by tilting negotiations toward the Palestinians and discouraging them from negotiating directly with the Israelis.

As we ordered ice cream, my phone rang again. It was Mike Pence. I looked helplessly at Arabella as it began to dawn on me that working in government would be far more time-sensitive and consequential than my old job. Pence had heard similar rumors about the resolution. After hanging up, I dialed Denis McDonough, who had given me his number when we visited the White House and told me to reach out if I ever needed anything. He said that he had no knowledge of the resolution, but would keep me updated. That was the last I heard from him.

Unsure what the Obama administration would do, I thought it was important for Trump to make clear that he opposed the resolution. Though it is rare for a president-elect to comment on a policy of an outgoing president, Trump agreed that it was worth breaking protocol for an issue this important. Working with David Friedman and Jason Greenblatt, our campaign's top liaisons to the pro-Israel and Jewish community, we drafted a statement, which Trump modified and pushed out on Twitter and Facebook: "Peace between the Israelis and Palestinians will only come through direct negotiations between the parties and not through the imposition of terms by the United Nations. This puts Israel in a very poor negotiating position and is extremely unfair to all Israelis."

The next day, President Fattah el-Sisi of Egypt called to let us know that his team had not been working under his direction and that Egypt was going to rescind the resolution. For a moment, it looked like we had succeeded and were already making an impact.

Two days later, Malaysia, New Zealand, Senegal, and Venezuela resubmitted the resolution. Flynn, Bannon, and I stayed up late into the night calling dozens of ambassadors, pressing them to oppose or abstain from voting on the resolution. We were rookies, we didn't know the key players in the countries on the UN Security Council, but we weren't going to let the resolution pass without doing everything we could to stop it. On many of these calls, we were introducing ourselves for the first time. At one point, I asked Dermer if he had any influential contacts in Russia whom Flynn could call—other than the Russian ambassador. After our first meeting with the Russian ambassador during the transition, both Flynn and I had determined that he didn't have any sway in Moscow. Dermer later reminded me of this conversation as proof that we had not colluded with Russia.

On December 23, UN Security Council Resolution 2334 passed 14 to 0. In a move that many suspected was punishment for Bibi denouncing the Iran deal in a 2015 address to Congress, the Obama administration abstained. Despite our efforts, we had not flipped a single vote. After forty-eight hours of working the phones, I was exhausted and deflated by the result. I called Bannon to commiserate on our first failure at the United Nations. Without missing a beat, Bannon said, "Welcome to the NFL, partner."

9

Learning on the Job

I vanka and I stood on the inaugural platform in the cold drizzle as Donald J. Trump took the oath of office and became the forty-fifth president of the United States. As we looked out at a sea of people who had come to witness the historic moment, newly sworn-in President Trump delivered a message that rang across America and around the world: "January twentieth, 2017, will be remembered as the day the people became the rulers of this nation again. The forgotten men and women of our country will be forgotten no longer."

Trump invited us to stay the night in the Lincoln Bedroom, a stately room located on the second floor of the White House where President Abraham Lincoln had hosted his legendary cabinet meetings. Before we left for the balls, we lit Shabbat candelabras and prayed. A member of the White House residence staff told us that it was the first time Shabbat candles had been lit in the private residence. At one ball, Trump—who's not known to be a dancer—asked the vice president and Second Lady, as well as Ivanka and me and her siblings, to follow shortly behind him and Melania, so that they didn't have to dance alone for a full three-minute song.

The next day, Ivanka and I walked through the West Wing with Trump as he went from room to room, selecting artwork to hang in the Oval Office and in his private study. A team of White House workers followed him around, moving furniture and paintings to the places he directed. For the Oval Office, he requested a portrait of Andrew Jackson, America's first outsider president. At the end, Trump thanked the team and its supervisor, a gentleman named David Jagdahne, and asked if he could tip the guys a couple hundred dollars because they did an amazing job. David laughed and told him they were federal employees and it was their greatest pleasure to work for the president of the United States. As we walked out, David asked me, "Is he always like this? He is treating us all like equals. I have spent more time with him in his first twenty-four hours than I did with President Obama in eight years."

My office was located on the first floor of the West Wing, next door to Bannon's and two down from Chief of Staff Reince Priebus. Unlike my New York office, a steel-and-glass skyscraper with modern furniture, this office was cramped, narrow, and dark, illuminated by light falling through a single ground-floor window that looked out on a few shrubs. Walking in my first day, I discovered an old desk, a few built-in shelves, and a worn-out couch with nylon upholstery. The only modern element was the two phones—a black one for general use and a yellow one for secure communications only. I hadn't yet gotten around to changing anything, other than putting a picture of my grandparents on my desk, hanging a mezuzah from my rabbi on the doorpost, and placing an HP-12C calculator that Marc Holliday had given me in my drawer. Despite the office's modest size, it had one highly coveted feature: it was the closest to the Oval Office. I was told that its former occupants had included George Stephanopoulos and John Podesta.

As Ivanka and I left the White House to take the kids to our new home for the first time, two gentlemen introduced themselves and told me that they would be my Secret Service agents. Up to that point, I did not know that I would have a Secret Service detail, but this was just one of the many changes in our life. My detail assigned me the Secret Service code name "mechanic," because they had observed me quietly and methodically fixing problems behind the scenes during the presidential campaign. We quickly came to see our Secret Service detail as an extension of the family. The kids would frequently run out the front door, where the agents faithfully stood guard, and throw the football or color the sidewalk with chalk as they talked to the agents. Years later, in January of 2021, the press wrongly reported that we would not allow the Secret Service to use a bathroom in our home. This was one of many false reports. When I offered to set up a way for the agents to access our home and use the restroom, the leader of our detail declined. They were looking for a larger space that could double as a command post. We set up a pantry inside our home for the agents, and for the next four years we kept it stocked with snacks, coffee, and other drinks. Whenever we ordered meal deliveries, we ordered extra for them, and on Sundays our kids loved baking cookies and sharing them with the agents.

During the first few days on the job, every hour felt like a race. The policy team rushed to draft dozens of executive orders so that the president could follow through on his campaign promises. The press team cycled through an onslaught of inquiries on everything from the inauguration crowd size to turf wars within the West Wing. I tried to navigate the unfamiliar realm of government, which seemed to be filled with endless processes and obstacles designed to prevent anything from getting done. Foremost on my list of priorities was finding a workable solution to the North American Free Trade Agreement, or NAFTA, negotiated by George H. W. Bush and signed by Bill Clinton in the 1990s. In the decades since, NAFTA had sent tens of thousands of US manufacturing jobs to Mexico, shutting down steel mills and factories in Midwest cities and towns, where generations of workers had made lifelong careers in good-paying jobs, earning a stable living for their families.

Trump's promise to tear up NAFTA animated the campaign and broke from Republican free-trade orthodoxy, reflecting his long-held belief that the deal hurt American workers. When their factories closed, some found new work, but many did not, and drug use and crime now plagued these once-thriving working-class communities. While Trump had agreed to let me try to renegotiate the agreement, I knew his patience was somewhere between thin and nonexistent. At any moment, he could act on his desire to terminate the $1.3 trillion deal completely, which would create tremendous uncertainty for American businesses that traded with Canada and Mexico—our two largest export markets, covering about 40 percent of America's annual exports. This uncertainty would give us a weaker hand to play in our looming trade negotiations with China.

To hammer out the details of a new deal, I invited Mexico's freshly minted secretary of foreign affairs, Luis Videgaray Caso, to Washington. Following our election-night call, President Peña Nieto had asked Luis to return to government as Mexico's top diplomat and primary interlocutor with Washington. He was reluctant at first, having just settled into a calmer lifestyle with his family, but he accepted because he knew our established trust uniquely positioned him to help Mexico navigate the complicated road ahead.

Luis came to my office early on the morning of January 26. As we strategized, Bannon heard that a senior diplomat was with me, so he joined as well. The three of us discussed the trade talks that we planned to announce in the next few weeks at a White House event with President Peña Nieto and Canadian prime minister Justin Trudeau. To finalize our plans, we called Trudeau's chief of staff, Katie Telford, a talented operator whom I had met during the transition.

"Are the meetings still on?" Telford asked.

"What do you mean?" I said.

"Didn't you see his tweets this morning?"

We had missed the president's tweets because Luis and I had placed our cell phones in a secure, soundproof box, which was White House protocol when discussing sensitive national security matters. Our phones were surprisingly susceptible to foreign infiltration: hackers could turn on our microphones and cameras to record conversations—even when the phones were powered off. I grabbed my phone to pull up the tweets: "The U.S. has a 60-billion-dollar trade deficit with Mexico. It has been a one-sided deal from the beginning of NAFTA with massive numbers . . . of jobs and companies lost. If Mexico is unwilling to pay for the badly needed wall, then it would be better to cancel the upcoming meeting."

I hung up, left Luis and Bannon in my office, and headed to the Executive Residence. As I raced along the same colonnade we had crossed with President Obama in November, I thought about all the White House aides before me who must have run down this walkway when the cameras were gone, as I was doing now. Until that moment, I had always thought of the colonnade as a majestic and dignified walkway. But

I started to wonder if, for people who worked inside the White House, it was actually a panic corridor.

Because Melania had not yet moved to Washington, I walked straight into Trump's bedroom, where he was reading documents with the news blaring on the television. The previous day, he had signed an executive order for the secretary of Homeland Security to direct all available resources toward constructing the border wall. In the hours following the signing, anonymous sources within the Mexican government had told the *New York Times* that President Peña Nieto was considering canceling his impending visit to the White House. Seeing these reports, Trump, who doesn't like it when people cancel on him, had decided to go on offense.

I confronted him about the tweet: "Luis is in my office right now. He assures me they are willing to make some major changes, and we have a plan to announce the NAFTA renegotiations during Peña Nieto's visit. This could derail the whole thing."

Trump skeptically asked if I thought we could actually make a deal with Mexico, and I urged him at least to let me try. Realizing that he might have fired off his tweet prematurely, he responded half jokingly, "I can't make this too easy for you."

I had put Trump in a difficult situation. Mindful of the numerous priorities he was juggling, I had not yet updated him on the positive indications we had received from the Mexicans and Canadians. I couldn't expect him to know what I didn't tell him. I was used to running my own business, but I was no longer the boss. I was a staffer, and my approach needed to change. From then on, I provided more frequent updates. In this case, however, it may not have made a difference. A natural negotiator, Trump was establishing his opening posture with Mexico and all foreign nations. Projecting weakness and predictability now would put him at a strategic disadvantage. As I stepped out of the president's bedroom, I wondered why I was taking on this impossible problem. Holding NAFTA together until we could negotiate a better deal cut against the instincts of a president who was inclined to tear it up and deal with the fallout.

Meanwhile, back in my office, Luis sat awkwardly with Bannon, who advised him to embrace the conflict and stand up to Trump. "It will make you a hero," he said. "Your poll numbers will go up immediately." Luis remained silent. When he later told me about the conversation, it struck me as odd. Bannon had previously agreed that renegotiating a deal could bring back jobs and benefit American workers. I couldn't figure out why he was now trying to blow it up.

* * *

On March 2, Trump traveled to Newport News, Virginia, to commission the USS *Gerald R. Ford*, a spectacular new nuclear-powered aircraft carrier that was about two years late and $2.8 billion over budget. Trump was in a great mood. He was clearly having fun as he toured the massive ship. It had a familiar feel to him, like visiting one of his hotel construction sites. As he inspected the ship's new electronic catapult system, he told the crew he thought it was too expensive and complex. The old steam-powered system had worked perfectly fine for decades. A similar scene played out when he inspected the new magnetic elevators, which, he noted, would malfunction if they got wet.

As Trump made his way back to Washington, Attorney General Jeff Sessions announced that he had recused himself from any investigation into accusations that the campaign had colluded with Russia. The recusal shocked the president, who had told the press during his tour of the ship that he had "total confidence" in Sessions. Just as I had, Sessions had shaken hands with Russian ambassador Sergey Kislyak during the cocktail reception at the Mayflower Hotel in 2016. It was an entirely innocuous interaction, one that Sessions would repeat scores of times with various ambassadors throughout the campaign and transition. When news broke that he had failed to disclose this exchange in his clearance paperwork and during his Senate confirmation hearing, Democrats immediately called for his resignation, and the press drummed up the story to ridiculous proportions: "Why Would Jeff Sessions Hide His Talks with Sergey Kislyak?" questioned the *New Yorker*. "Sessions Discussed Trump Campaign-Related Matters with Russian Ambassador, U.S. Intelligence

Intercepts Show," read the *Washington Post*. Sessions's recusal proved to Democrats that their baseless attacks would yield political rewards. His decision ultimately led to the appointment of a special counsel with virtually unlimited power and resources to investigate the phony claims of Russian collusion.

The next day Trump summoned Reince Priebus, Steve Bannon, and Don McGahn, his White House counsel, into the Oval Office and reamed them out. "Has Sessions come to his senses? If he doesn't want to oversee the Department of Justice, then he should just resign," Trump declared. He couldn't understand why Sessions would recuse, or how the White House team had allowed the attorney general to do it. The collusion narrative was pure politics compounded by media hysteria, and Sessions surrendered to it. Neither he nor the Trump campaign had colluded with Russia. Half the time, we couldn't even collude with Trump; his team often contradicted him in the press. The gap between the facts I knew to be true and the media's reporting could not have been wider.

As Trump continued dressing down his leadership team, Marine One landed a few dozen yards away, at the center of the South Lawn. It was ready to take Trump to Florida for an event at a charter school, which Ivanka had arranged, before heading to Mar-a-Lago for the weekend. Like most members of the senior staff, Bannon and Priebus felt like they needed to be in the room with the president at all times, and they had made sure they were included on the flight manifest to Florida that afternoon. But Trump had different plans. "Why are you coming to Florida for the day? I don't need you there. Stay here. There was no reason for Sessions to recuse himself, and this is going to unleash a disaster." An adept student of American political history, Trump had watched previous special counsels dismantle past administrations. By nature of their appointment, special counsels seemed to think that their investigations needed to find a smoking gun to rationalize their existence, regardless of the merits of the case.

That evening, Priebus called and said that Maggie Haberman was writing a story about how Priebus and Bannon were in trouble with the president, who had kicked them off the trip to Mar-a-Lago. Someone had leaked to Haberman, a White House correspondent for the *New York Times* who had covered Trump for more than twenty years. Priebus wanted to tamp down the story and asked if I would tell her it wasn't true. Up to that point, I had never called Haberman. Priebus gave me her number, and I tried to head off the story, telling her that it wasn't a big deal. The president had asked Priebus and Bannon to stay back and work on time-sensitive issues at the White House.

Minutes after I spoke to Haberman, Bannon called. "How fucking dare you leak on me? If you leak out on me, I can leak out on you twenty-eight ways from Sunday."

I pushed back hard. "Steve, are you fucking kidding me? This wasn't a leak. I spoke to Maggie because Reince asked me to call to defend you guys and dispel the story. I've been with you in the trenches. When have you ever seen me talk to a reporter? I don't talk to the press. I've never leaked on anyone. I wouldn't know how to leak. Don't accuse me of anything."

A few days later, Bannon apologized. I accepted it and asked him never to do that again. "It's not the game I play," I told him. "I'm a foxhole guy. If I have a problem with people, I tell them, but I don't air grievances through the press. It doesn't help the team, and it doesn't help the president."

Bannon's behavior became more erratic, which confirmed the warnings that a few of my friends from New York had conveyed: he had been a destabilizing presence in his previous organizations, where he always seemed to leave with an explosion. There was an obvious uptick in negative stories about Ivanka and me, which portrayed Bannon as the savior against our supposedly liberal crusade. Shortly after one of these stories appeared, a reporter rang Hope Hicks, saying, "I'm not your friend, but I'm being your friend." The reporter revealed that the leaks were coming at the direction of Bannon.

On March 27, the *New York Times* reported about a routine and unremarkable meeting I had during the transition with a Russian banking executive named Sergey Gorkov. Democrats heralded it as evidence that I had colluded with the Russians and sought business-related favors from Gorkov. In reality, I met with Gorkov at the specific request of the Russian ambassador, who implied that Gorkov was a direct line to the Russian leadership. We did not discuss business—it was simply a thirty-minute introductory meeting. I received an update on Russia's foreign policy priorities and communicated Trump's desire to form new relationships on shared objectives, such as reining in Iran and countering Islamic extremism. As a gift, Gorkov brought a bag of earth from Novogrudok, the town in Belarus where my grandparents had escaped. I turned it over to the transition office as a foreign gift. I never followed up or talked with Gorkov again. After being pressured by the media, Republican senator Richard Burr, chairman of the powerful Senate Select Committee on Intelligence, announced that he planned to question me about the meeting and any involvement the campaign might have had with Russia.

The same skills that made Bannon valuable as a fighter during the campaign made him toxic in the White House. Even though he was only with the campaign for the final eighty-eight days, he positioned himself as the keeper of the Trump flame. He wrote down all of the president's campaign promises on a whiteboard—most of which Trump made months before Bannon joined the campaign—and he made sure to display the list to the endless parade of reporters who filed through his office. I, too, believed in most of Trump's policies, but I realized that achieving them would often demand time, effort, and technical expertise. We were now playing with live ammunition, I explained to Bannon, and we needed to lay out the options and help the president execute them in a thoughtful and strategic way. I also believed that as staff, we should keep our heads low, get things done, take the blame for mistakes, and make sure the president received credit for any success.

Bannon's approach was on full display when Trump signed an executive order on his seventh day in office, resulting in a public-relations mess. The order blocked travel from countries that failed to meet commonsense standards for preventing terrorists from traveling to the United States. The seven countries covered by the policy were Muslim-majority, but nothing changed with dozens of other Muslim countries around the world that had better vetting standards and controls in place. At a time when ISIS remained strong, and national security experts were concerned about domestic terrorist attacks, this policy made sense. Yet Bannon bulldozed it through the approval process, keeping it hidden from me and most of the senior staff until the president had already signed the document. The lack of planning caused confusion about how and when federal agencies should implement the travel ban. It unleashed chaos at our airports and created an information vacuum about why Trump was taking this action. The Democrats framed the action as a "Muslim ban," which it was not. But the facts got lost in the chaos that flowed from Bannon's botched rollout.

Bannon tried to bolster his position by using me as a foil: the liberal New Yorker who was undermining Trump's agenda and was riddled with business conflicts. He was also demonstrating the truth of a warning Ivanka had received early on from a former senior aide to Nancy Pelosi: "In Washington, if you don't define yourself, your enemy will happily do it for you." I should have pushed back from the beginning. Instead, I took a page from the Howard Rubenstein public relations playbook: refrain from engaging with the press to avoid drawing more unwanted attention. This let Bannon and others define me. But there's no guarantee that the opposite strategy would have worked. We didn't know the players in the press or how to speak to them, and Bannon was a black belt in the dark arts of media manipulation.

I was far from the only person Bannon turned against. During the transition, Bannon expressed frustration that Kellyanne Conway inserted herself in discussions and leaked to the press to constantly overstate her role. Bannon bet that he could engineer her exit in the first three months. He was convinced that she wouldn't pass a White House drug test, and he didn't hide his disappointment when she did.

It quickly became obvious that the White House was very different from my experience in the private sector. Bureaucracy, egos, and people's obsession with holding on to power stifled collaboration and progress on policy goals. In one instance Gary Cohn, the former Goldman Sachs president appointed to lead the National Economic Council, pulled me aside.

"Bannon is leaking on me nonstop," he said. "I'm not going to take this. I know how to fight dirty."

In retrospect, I should have just told him to take it up with Bannon directly, but instead I pulled Bannon into the Cabinet Room. "Steve, you gotta stop leaking on Gary. We're trying to build a team here."

"Cohn's the one leaking on me," Bannon retorted. "Jared, right now, *you're* the one undermining the president's agenda," he continued, his eyes intense and voice escalating into a yell. "And if you go against me, I will break you in half. Don't fuck with me." Bannon had declared war, and I was woefully unprepared.

From the beginning, the West Wing was fractured by competing camps. There were the Trump originals like Hope Hicks and Dan Scavino, who lacked government experience but had no political motivations and were entirely focused on seeing Trump succeed. There were the Trump ideologues like Peter Navarro and Stephen Miller, who believed deeply in his pro-American policies. There were the experienced executives like Gary Cohn and Dina Powell, who believed the White House should be run more professionally. And there were the RNC establishment types, who were skeptical of Trump but loyal to Chief of Staff Reince Priebus. It was an impossible situation for any chief of staff.

Compounding the problem, Priebus didn't have an existing relationship with Trump and deferred to Speaker of the House Paul Ryan on setting the White House legislative agenda. Ryan made health-care reform the Republicans' number one legislative priority. Since Obamacare's enactment in 2010, Republicans had voted to repeal the law more than sixty times. We quickly learned that they didn't have a real replacement plan because they had assumed that Trump would lose the election. They scrambled to draft one, and the result was a catastrophe.

One afternoon, just as Trump was preparing to leave the Oval Office for an event, Ryan called the president. "Where is your health-care plan?" Trump demanded. "We are getting killed for not releasing one."

"It's ready, and I'm trying to get it out, but your team is holding it up," Ryan responded.

Surprised, Trump looked around the room. "Who on my team is doing that?"

I raised my hand. Trump demanded to know why I was holding it up. I was blunt: "Two reasons. First, it doesn't do what you campaigned on—providing health insurance to more people, lowering prices, and preventing people from dying in the streets. Second, Paul hasn't shown us that he even has the votes to pass his bill."

Despite these concerning facts, Ryan released the plan, the American Health Care Act of 2017, which some studies say would have increased the number of uninsured people by twenty-three million. Fortunately, the bill died on its own.

10

The World Is Watching

I climbed into a Black Hawk helicopter after landing in Baghdad. American service members placed belts of bullets around their necks and positioned their machine guns. "It's a nice day out," Chairman of the Joint Chiefs of Staff General Joseph Dunford said. "Let's fly with the doors open."

As the helicopter lifted off and headed to an American military base, the hot desert air rushed through the cabin and the twin engines drowned out our voices. I looked down below and watched an unexpected scene unfold. Amid the charred buildings, turned-up asphalt, and other scars of war, we saw new signs of vibrancy. Makeshift storefronts, farmers markets, and carnivals were springing up in the war-torn city. As we flew, General Dunford pointed down at the roof of one of Saddam Hussein's former palaces, where a bomb had exploded and left a gaping hole. To our left, I caught a glimpse of a V-22 Osprey. A soldier was standing on the back platform strapped into a cable, with his machine gun ready. I looked up at the Black Hawk's rotors, which somehow seemed to freeze in the air. Three months ago, I was making real estate deals in New York, I thought. Now I'm flying over Saddam Hussein's bombed-out palace in Iraq with the head of the Joint Chiefs. What the hell am I doing here?

It was Monday, April 3, 2017. I certainly hadn't planned to travel to Iraq in the first few months of the administration, but a few weeks earlier, at an intimate dinner with the president and several of his top military leaders, General Dunford pulled me aside and invited me to join him on the trip. I had been listening intently to the spirited debate about how to approach the ongoing wars in the Middle East, and he suggested that the visit would give me a real sense of our force structure and capabilities in the region and a firsthand account that I could bring back to the president. My father taught me that executives can't make business decisions from a glass tower; they need to be on the ground, getting dirt under their fingernails and interacting with and learning from the men and women on the front lines. I accepted the invitation.

Two days before we left, a White House doctor stepped into my office and asked for my blood type. I was taken aback, but he reminded me that I was going into a war zone—this was clearly a different kind of job. On the long flight to Iraq, Dunford invited me to sit next to him in his executive quarters on the military plane, a Boeing 757. Inside his cabin, which was furnished with a large bolted-down table, a couch, and two captain's chairs, we spent the next four hours discussing a range of topics related to the Middle East. I asked him what he would do differently in Iraq if he could start from scratch. How should we change our objectives and strategy to make them more forward-looking? Who were the most valuable regional partners? Where should we invest our resources so they would have maximum effect? Why hadn't we gained more ground over the past sixteen years? I was impressed by Dunford's reservoir of knowledge. When it was time to get a few hours of sleep, the general took off his jacket, flipped it over to use as a blanket, reclined his seat, slipped on a camouflage sleeping mask, and within minutes was out cold. It was like sitting next to G.I. Joe.

While we were in Baghdad, a loud air-raid siren blared, and we were whisked into a secure area. Totally unfazed, Dunford explained that this sort of thing happened all the time. Insurgents had apparently fired off a few mortars. No big deal. The next morning, helicopters dropped us off at an installation about ten miles from Mosul, which at the time was an active war zone. General Dunford had his commanders explain the force structure—Americans were training and arming Iraqi soldiers to do the fighting. It was an impressive operation, and the way our forces were leveraging the capabilities of the Iraqis made me optimistic about the future stability of the country.

Later I learned that I had made a mistake: dressed for our morning meetings, I wore my sport coat and sunglasses to the war zone, and when someone handed me a flak jacket, I threw it on top. The N-E-R in my last name was covered by the Velcro, so my name read K-U-S-H. I didn't know that photos would be taken, and I was used to being a behind-the-scenes guy in politics—not a principal who had to think about optics. When the resulting visual quickly became a meme on social media, I thought it was hilarious. I clearly had missed the memo on the dress code, and *Saturday Night Live* and Jimmy Fallon were not going to let me forget it. Even Ivanka still pokes fun at me for this one.

When we met with Iraqi prime minister Haider al-Abadi, he asked to see me one-on-one. The prime minister took seriously Trump's public statements that he wanted countries to pay a larger share of the defense cost. Al-Abadi said that he was willing to pay something for US protection, but essentially hoped for the "cheapest deal." We probably could have gotten 20 percent of Iraq's oil revenues in exchange for our military support, but Secretary of State Rex Tillerson and Secretary of Defense Jim Mattis thought Trump was crazy for suggesting such a proposition and stalled the discussion indefinitely.

In another instance, Mattis and his leadership team came to talk to the president about their budget and claimed that $603 billion—the largest request since 2012—wasn't enough to keep the country safe. They needed $609 billion to achieve "military readiness."

"So with one percent more, you are military ready, but with one percent less you're not?" Trump queried skeptically.

After the brass had filed out, Trump pulled me aside and remarked, "These guys may be the best at killing people, but they sure don't understand money."

*   *   *

One of the greatest challenges from the very beginning was finding the right personnel to staff Trump's White House. The president needed talented people who aligned with his agenda, could adjust to his style, and knew how to operate the levers of bureaucracy. It was possible to find people with one or two of these traits, but rarely all three.

Trump was initially impressed by his choice for secretary of state, Rex Tillerson. The Texas oil tycoon had run an iconic American company and possessed long-established relationships with world leaders. But the two men began to clash almost immediately. Tillerson talked slowly, often didn't return phone calls, and siloed himself off from most of the State Department. The president grew to dislike Tillerson's swaggering style. During one dispute with head of White House personnel Johnny De-Stefano, Tillerson said, "That's right, Johnny, you can talk back to me freely. I don't know about you, but I'm all man." But more fundamentally, he and the president had opposite views and approaches on many foreign policy questions. Tillerson was risk-averse and wanted to manage the world's problems. Trump, on the other hand, was a calculated risk-taker and dealmaker who wanted to disrupt the ways of the past and change the world.

Early on, I scheduled a weekly lunch for Tillerson to meet with the president in a casual setting to help the secretary bridge his two biggest challenges: his lack of a preexisting relationship with Trump and his lack of alignment with the president's policy goals. A secretary of state becomes useless the instant his foreign counterparts know he doesn't speak for the president or have influence in the decision-making process.

Tillerson and I worked together to set up the first official call with President Xi Jinping of China. The Chinese remained outraged by the call Trump had taken from the president of Taiwan and his refusal to endorse the One China policy. We both thought it was important to establish direct communication between the two most powerful world leaders. It took weeks to negotiate the terms of the call, but when the two leaders finally spoke, it was friendly. As we had negotiated in advance, Trump invited Xi to come to Mar-a-Lago for their first in-person meeting. Over the next several weeks, I worked with Ambassador Cui Tiankai, China's longtime top diplomat to Washington, to carefully orchestrate the details of the trip.

On April 6, President Xi arrived at Trump's Palm Beach estate. Knowing that his tough rhetoric had put the Chinese on edge, Trump wanted to begin the visit with something that would break the ice. He asked Ivanka and me if five-year-old Arabella could greet the Chinese leader in Mandarin. She had grown up learning Mandarin thanks to the encouragement of our good friend Wendi Murdoch and to XiXi, our beloved nanny and tutor, who has been with us since Arabella was an infant. At the leader's welcome tea, Arabella recited Tang poetry. Xi was so impressed that he asked to meet XiXi and complimented her on Arabella's perfect Beijing pronunciation and the selection of poems. The flattering gesture put him at ease and the video circulated like wildfire in China. It was a major sign of respect that the granddaughter of the president of the United States knew Mandarin.

Scheduled to last fifteen minutes, Trump and Xi's introductory tea continued for well over an hour, and the two leaders quickly developed a warm and respectful dynamic. Trump treated Xi like a regular person, and Xi responded in kind. In one meeting Xi started going into the history of China, stretching back to the Opium Wars and the signing of the "unequal treaties," and continuing through the so-called Century of Humiliation, which ended with Mao Zedong's rise to power. Xi's forty-five-minute performance was fascinating, and Trump was taken by how even the stoic leader of the Chinese Communist Party could not hide the motivation his country derived from the Century of Humiliation. Xi was certain that China had learned from their past and would rise again. When Trump asked Xi how much influence he had over North Korea's mercurial young leader, Kim Jong Un, who was testing long-range missiles and threatening America with North Korea's nuclear arsenal, the Chinese president was surprisingly candid with his response: he'd had a relationship with Kim Jong Il, the deceased former leader of North Korea, but didn't really know his son.

After four hours of meetings, the president was summoned to a secure room that had been converted into a sensitive compartmented information facility, or SCIF, so that he could receive classified information and military briefings while in Palm Beach. Two days earlier, Syrian dictator Bashar al-Assad had launched a chemical weapons attack on civilians, killing more than eighty people and injuring more than five hundred. Trump was horrified, as we all were, by the photos of mothers and children suffocating to death, and he was concerned by the incoming intelligence on the situation. He felt strongly that the United States must make clear that it would not tolerate the use of chemical weapons. During the campaign, he had excoriated Obama for drawing a "red line in the sand" on Syria and then meekly pulling back from confrontation. Obama's failure to enforce the "red line" had undermined America's influence around the globe. Trump was determined not to repeat this mistake.

This was Trump's first big test as commander in chief, and he was acutely aware that his decision could have consequences for American troops in the region. He had asked Mattis to present him with strike options, sparking a fierce internal debate about what to do. Bannon vehemently opposed a strike, warning the president that it could begin a war. The rest of us believed that the president needed to send a strong signal, but that the strike would have to be surgical. The Syrian military base we were targeting was co-populated with Russians—if we accidentally hit one, it could start World War III.

One of the most inspiring parts of watching my father-in-law as commander in chief was seeing how he responded in moments of military crisis when no cameras were present. That day, he went around the room and asked tough questions of the generals, sought out opinions, listened intently, and carefully weighed the implications before taking decisive action. I was grateful that the president was surrounded by so many experienced people at the table, particularly the new national security adviser, H.R. McMaster, who thoroughly briefed the president on his options. A legendary Desert Storm tank commander, McMaster was a bulldog of a man, possessing so much physical energy that he exercised twice a day. As an active-duty officer, he'd written a scathing critique of America's handling of the Vietnam War in a book called *Dereliction of Duty*.[9] The book earned McMaster a reputation as an iconoclast general, which slowed his career trajectory. He was passed over twice for a promotion from colonel to one-star general. But by the time he entered the Trump administration, he had earned his third star.

Earlier in the day, the president had asked Mattis if the strike plan was going to work without creating an international incident. "No problem, sir," said Mattis, cool as a cucumber. "You have the finest and most lethal military equipment in the history of the world. These missiles will do what they were intended to do—one hundred percent. You don't have to worry."

During dinner that night with President Xi, the national security team let the president know that the strike was going very well. Fifty-eight of fifty-nine missiles "severely degraded or destroyed their target," and no Russian soldiers had been harmed. It was still nighttime in Syria, and we didn't expect to have conclusive satellite images for another few hours. When an aide whispered an update into Trump's ear, he immediately told Xi the news. Xi couldn't hide his shock. He was clearly impressed that Trump was so relaxed in such a consequential moment, and I got the sense that he didn't know what to make of Trump. The Chinese had never dealt with anyone like him before. No one had.

Part of what ultimately made Trump successful in his foreign policy objectives was that leaders found him unpredictable. He built warm rapport with his counterparts and approached each situation with an open mind. He was willing to change course at any minute and take calculated risks. His opponents never could tell whether he was bluffing or making a serious threat.

That day, a story broke in the *New York Times*: "Kushner Omitted Meeting with Russians on Security Clearance Forms." When filling out my security forms for the White House, I was required to disclose foreign contacts and relationships that had occurred within the last seven years. The process of going back through my records and calendars to produce this list took weeks, and our incoming National Security Council (NSC) team advised me to submit the first part of my application immediately to get the initial security clearance process started, and then follow up at a later date with my full list of foreign contacts. I followed this counsel, submitted the initial form, and weeks later filed my list of foreign contacts. When the press got hold of the fact that I had submitted my foreign contacts after the initial form—a process that is supposed to be classified and had never before leaked to the press—they further connected me to the false Russia collusion narrative, noting that I had left Ambassador Sergey Kislyak and Sergey Gorkov off the initial form. They took something benign and made it sound nefarious, and I couldn't fathom why it had been so unfairly framed.

I had not included any foreign contacts on my initial form, exactly as I had been advised to do, so I couldn't believe that the *New York Times* singled out the Russian contacts. You've got to be kidding me, I thought. Had I submitted a completed form and included two hundred foreign contacts while omitting only Kislyak and Gorkov, perhaps that would have justifiably raised eyebrows, even though my contact with those two was minimal and harmless. But no one—not even my harshest critics— could have honestly believed I was trying to claim I had met zero foreign nationals and traveled to no foreign countries over the past seven years. It was obvious to any fair-minded observer that this was exactly as I'd stated: a submission of an incomplete form with a complete disclosure to come. Questioning how they could have gotten this so wrong, I suspected that Bannon leaked and framed the information. He was one of only a handful of people who had access to the form.

Around the same time, the media ran a series of stories on Ivanka, claiming that because her business had previously applied for trademarks in China, she was profiting from her government position. In reality, following the 2016 election, numerous companies in China had filed hundreds of trademark applications to exploit Ivanka's name and brand on products completely unrelated to her. On March 8, the headline of a *Washington Post* article read, "From Diet Pills to Underwear: Chinese Firms Scramble to Grab Ivanka Trump Trademark." The article went on to say that "an astounding 258 trademark applications were lodged under variations of Ivanka, Ivanka Trump and similar-sounding Chinese characters between Nov. 10 and the end of last year." Ivanka had a successful business and owned hundreds of trademarks globally before her father ran for public office, and in May and June of 2016, after Trump entered the race, she submitted a number of additional trademark applications in an effort to protect her name in countries where trademark theft was rampant. Ivanka's applications had been caught up in the Chinese bureaucracy for a full year. When several of the requests were approved around the time of Xi's visit, the media tried to make it sound nefarious, but Ivanka had no control over the timing and was merely doing her best to prevent Chinese companies from counterfeiting her brand and deceiving customers.

Nearly a hundred days into the administration, I wanted to focus on real policy wins, but negative stories kept hitting me. On the campaign, the press mostly ignored me—probably because they had no idea the role I was playing behind the scenes. In the White House, however, I had a target on my back. A consortium of senior staffers saw me as a threat to their power and influence. When my father-in-law confronted me about my negative press, I complained that people were leaking on me. "Jared, this is the White House," Trump said. "If you want to work here, you have to figure out how to get people not to leak on you."

On Wednesday, May 3, FBI director James Comey testified before Congress regarding his infamous decisions during the 2016 election cycle. Trump watched the hearing with great interest and commented afterward that "something was off." He found Comey's testimony to be erratic and inauthentic. A few days later, while we were in Bedminster for the weekend, the president invited Ivanka and me to join him and Melania for dinner. Midway through, he called for Stephen Miller and began dictating a letter firing Comey. I encouraged Trump to wait until he got back to the White House, where he could get input from his legal counsel and chief of staff, which he ultimately decided to do.

When we got back to Washington, Priebus and McGahn met with the president and handled the situation from there. They asked Trump to hold off while they coordinated the matter with the Department of Justice leadership, who had lost confidence in Comey. On Tuesday, Trump fired the FBI director. Steve Bannon, who was kept in the dark about the discussions to prevent him from leaking, was furious. "This is the end of the presidency," he said. Soon after, in what I suspected was a Bannon leak, the press reported that I advised the president to make the decision, which was false. Democrats framed Comey's firing as an attempt to obstruct the FBI's Russia investigation and began calling for the appointment of an independent counsel. Soon after, deputy attorney general Rod Rosenstein appointed Robert Mueller as special counsel to lead the investigation.

One weekend while the president and the usual entourage of senior staff were in Mar-a-Lago, I had the White House maintenance team seal off the internal doorway between my office and Bannon's.

11

Riyadh to Rome

Jared, read my lips: we're not going to Saudi Arabia. Take no for an answer!"

I was having dinner with Trump on a Saturday evening in April. Along with Ivanka, we were tucked in the corner of the restaurant at the Trump International Hotel on Pennsylvania Avenue, where the president would occasionally visit when he needed a change of scenery from the White House.

Our meal had started with Trump and Ivanka reminiscing about the hotel itself, which they had painstakingly worked on together for several years. Formerly the historic Old Post Office, the hotel had opened the previous October, just before the election. The conversation turned to how our kids were adjusting to their new schools in Washington, DC.

As Trump talked, he relaxed. Yet the business of the presidency was never far from his mind. Its challenges and opportunities dominated his days and nights, and nearly all of my conversations with him touched on his massive responsibilities. This one was no different and soon he shifted our discussion to one of my major projects: his first foreign trip as president. I had been planning the trip for months. His first stop would be Saudi Arabia, where he would deliver a major address to fifty-four Muslim and Arab leaders. Our departure was just a few weeks away.

"I know you've been working hard on this trip," Trump said, "but I need to stay here right now." He was reluctant to leave the country while a growing chorus of Washington lawmakers and pundits called for the appointment of a special counsel to conduct a Russia investigation.

"Plus," he added, "Rex doesn't believe the Saudis will come through."

I began to explain why I disagreed with Tillerson, but the president cut me off. That's when he told me to take no for an answer.

During the campaign, I had picked up on Trump's decision-making style: his first answer was often provisional, used as a method of drawing out different viewpoints from his team and seeing how strongly we believed in them. So I didn't interpret his words as a hard no.

"Let me come by tomorrow," I said. "I'll show you the package I've negotiated, and you can make the final decision."

To end the stalemate, he agreed.

Around 10:00 a.m. the next day, a Sunday, I met Trump in the residence and briefed him on my proposal. In addition to convening Muslim and Arab leaders, the Saudis had promised to execute up to $300 billion in job-creating business deals with American companies, take unprecedented measures to block the financing of terrorism, open a global center to combat extremism, have the king of Saudi Arabia denounce violence in the name of Islam, purchase a substantial amount of US military equipment, and provide more military support in the fight against ISIS. On top of all of that, they would roll out the red carpet for Trump and show America tremendous respect with military flyovers and a magnificent state dinner, a noticeable contrast from President Obama's visit a year earlier, when they refused to greet him on the tarmac. The trip would give Trump an opportunity to forge stronger ties with the Arab world, issue a tough call to action against terrorism in the region, and lay the groundwork for normalizing relations between the Arabs and Israel. Plus, he would bury the media's false narrative that he was Islamophobic. All of this would be accomplished in just forty-eight hours.

"Let's give it a shot," Trump decided. "Tell Rex the trip is on, but I want everything in writing."

Back in January, during Trump's first call with King Salman bin Abdulaziz, the Saudi ruler told us to coordinate the potential trip with his son Mohammed bin Salman, the charismatic thirty-one-year-old deputy crown prince and minister of defense, known as MBS. Trump said that I would be his point person. When I got back to my desk, I already had an email from MBS asking to set up a call. In March, while MBS was in town to negotiate the details, a blizzard hit the Northeast. Chancellor Angela Merkel of Germany, who was scheduled to have lunch with the president, canceled at the last minute because her plane could not take off from Germany. I asked Trump if he would have lunch with MBS, since the deputy crown prince was already in town. Trump thought it was a great idea, despite White House National Security Council staff insisting that presidents don't have lunch with foreign officials who are not the head of state. Trump dismissed that bureaucratic protocol and decided to explore a potential partnership that could advance America's interests in the Middle East.

Because MBS was technically Saudi Arabia's third-ranking official, the National Security Council staff wouldn't let him skip the security checkpoint and drive right up to the West Wing, as they allowed for heads of state. So my deputy and only staffer at the time, Avi Berkowitz, waited in the snow outside the security checkpoint to meet MBS. When the deputy crown prince arrived, there was a paperwork issue and the Secret Service denied him entry. I ran to the gate and convinced them to let him through.

Despite the rough start to the visit, the lunch was a success. Trump told MBS directly that he wanted stronger cooperation in combating terrorism, countering extremism, and ending terrorism finance. He also expected Saudi Arabia to take on more of the defense burden in the region. America was not going to keep spending precious blood and trillions of dollars on endless foreign wars. It was a tough message, and Trump did not shy away. In response, MBS unveiled an ambitious and thorough antiterrorism plan. This ad hoc meeting reinforced my instinct that we should take a risk on Saudi Arabia for the president's first foreign trip. Trump gave me the green light to continue planning.

As I coordinated the trip, I found a talented and effective partner in the president's deputy national security adviser, Dina Powell, a veteran of the Bush White House and State Department. Ivanka had recruited Powell to the administration from Goldman Sachs. Powell's guidance and support helped me navigate the stiff internal resistance I encountered in planning the trip. I also found an ally in NSC director for the Middle East Derek Harvey, a former US Army colonel who believed that the trip was critical to strengthening America's relationships with Arab countries in our efforts to confront Iran's aggression.

In one Situation Room meeting, portrayed in Bob Woodward's book *Fear*, I argued forcefully for the trip against Tillerson, Mattis, and Mc-Master.[10] "I know you have a lot more experience than I do with Saudi Arabia," I told Tillerson. "But this is the way I view it: we can't allow the broken promises of the past to determine the future."

The secretary of state countered: he had engaged in serious business negotiations with the Saudis during his thirty-plus years as an oil ex-ecutive, and he didn't think we could trust them. "They never come through," he said. "They won't deliver on their promises."

I pushed back. "I may not have diplomatic experience, but I have done hundreds of hard transactions, and I can tell when people want to do deals and when they don't want to do deals. MBS is not just saying he wants to do it. His top negotiator is literally sitting in the Four Seasons down the street, ready to come over and put the finishing touches on these documents."

Tillerson and the others in the room waffled. They thought we were trying to do too much on a tight timeline. So I put it more bluntly: "If you don't like this idea, what's your idea? If you have an alternative, let's hear it. But no one has put forward any alternative. We didn't come to government to sit outside a cigar shop and talk about how the world should be. We're the ones in charge, and we need to get things done. I think the downside of my proposal is super low and the upside is super high. If it is a failure, I'll take responsibility."

After the meeting, I dialed MBS: "Everyone here is telling me that I'm a fool for trusting you," I said. "They are saying the trip is a terrible idea. If I get to Saudi Arabia, and it's just a bunch of sand and camels, I'm a dead man."

He laughed and assured me that he was also facing internal skepticism, but would not let us down: this was going to be a massive success for the president, the Saudis would deliver on their promises, and we would see changes in Saudi Arabia beyond our imagination.

* * *

Few times in my life have I felt as nervous as I was boarding Air Force One to depart for Riyadh, Saudi Arabia, on May 19. The whole world was watching the trip, and we needed to nail it. Short on sleep and running on adrenaline after months of working around the clock, I couldn't help thinking that it would have been so much easier to follow the pattern of previous presidents and travel to Mexico or Canada for a ribbon-cutting-type event.

As Air Force One began to accelerate on the tarmac of Joint Base Andrews, Chief of Staff Reince Priebus appeared in an absolute panic. He read me a tweet from the Associated Press: "Exclusive: Draft of Trump speech abandons the harsh anti-Muslim rhetoric of his presidential campaign." The staff secretary's office had circulated a draft of the president's keynote speech for the Arab American summit to a large group of senior staff, and someone had leaked it to the press. Major excerpts of the speech were already circulating online.

In crafting the speech, Stephen Miller and I had gone to extraordinary lengths to keep it under wraps. We wanted to hold the world in suspense and build anticipation right up until the moment Trump stepped onstage—not to mention, the president hadn't yet reviewed a single word of the draft. For all we knew, he would take it in an entirely different direction. Priebus was on the verge of a meltdown. Stephen calmed him: "It's no big deal. The flight is over ten hours. I'll write a new one." As we sped across the Atlantic and through the night, I kept running through the details for the trip, trying to distract myself from worrying about all the things that could go awry.

When we landed, King Salman was waiting on the tarmac to greet President Trump and Melania. Cannons boomed, and nine F-16 military jets screamed overhead, leaving behind a trail of red, white, and blue smoke. The dramatic welcome contrasted starkly with the reception that Obama received during his final visit to Saudi Arabia in 2016, when King Salman did not even greet him at the airport. After Trump and the king finished exchanging pleasantries, we made our way to the Saudi Royal Court as the president's limo was escorted by a dozen Saudi guards on Arabian stallions carrying the American and Saudi flags.

Shortly after arriving at the Saudi Royal Court for the formal welcome ceremony, President Trump and King Salman presided over a military arms transaction worth $110 billion, intended to offset the cost of US expenditure in the Middle East. That night, King Salman hosted our delegation for an intimate dinner at the Murabba Palace, the residence of King Abdul Aziz Ibn Saud, the late founder of the modern kingdom of Saudi Arabia. As Trump arrived at the palace, the doors of his limousine swung open into a sea of Saudi Arabians in their traditional white and red garments, swaying back and forth to the rhythm of a beating drum and chanting in chorus the melody of a traditional song. As the president walked toward them, the sea parted in front of him, funneling him into a spacious courtyard where our hosts commenced a ceremonial sword dance that the tribal warriors of Saudi antiquity would perform before battle.

In a gesture of respect, the king handed the president a sword and invited him to join, which he did. The resulting footage of Trump and King Salman with arms linked, bouncing up and down to the banging of drums, went viral. During a break in the action, I looked over at Rex Tillerson, who couldn't hide his enjoyment. "This isn't my first sword dance," he grinned, while commending me for how smoothly the trip was going so far. Inside the palace, we were feted with an elaborate spread of traditional Saudi Arabian food, including camel meat, which is not kosher, so I moved it around on my plate.

The next morning Trump attended a meeting of the Gulf Cooperation Council (GCC) as a guest of honor. The United States and the GCC countries of Saudi Arabia, Bahrain, Kuwait, Oman, Qatar, and the UAE entered into an historic agreement to pool resources and intelligence to block the financing of terrorism. During the meeting, economic adviser

Gary Cohn slipped the president a note: "For a change, you are the poorest guy at the table." The president couldn't contain his smile.

In the main banquet hall for lunch with the Arab leaders, MBS had seated Ivanka and me at a table with him. When we sat down, he introduced me to the other leaders at the table. These were the "Jared Kushners" of the Middle East, he told me—they didn't all have obvious titles, but if I needed to get anything done in their countries, I should call them. As we got to know each other, I was impressed by their openness to reform. At one point, the UAE's national security adviser, Tahnoun bin Zayed Al Nahyan, said to Ivanka with a smile, "Go ahead, ask MBS when he's going to let women drive." MBS overheard the suggestion and flashed a big smile. "Very soon," he said. Ivanka was surprised but also pleased. It confirmed what she had heard at the roundtable for women small business leaders earlier that day. The women had openly discussed how the driving ban wasn't the only hurdle to their success. Guardianship laws—preventing them from owning a bank account or property—inhibited their ability to start and grow businesses. The fact that the women were speaking about this challenge at an event organized by MBS was an indication that they knew he was open to reforms. One year later, MBS surprised the world and took steps to lift the ban on driving. The following year, he changed the guardianship laws.

As Trump approached the stage for his keynote address to the leaders of fifty-four Muslim majority nations, the stakes couldn't have been higher. In agreeing to make the trip, Trump had insisted on bringing a tough message that the United States was done spending trillions of dollars and thousands of American lives on endless foreign wars. Our allies must step up and contribute more to their own defense. "I don't want to go there to kiss ass," he warned. Inside the banquet hall, the leaders were seated at ornate desks beneath a thirty-foot ceiling and chandeliers the size of Buicks. The hall was so spectacularly large that it made everything else seem tiny by comparison, and it was eerily silent as the president began his thirty-five-minute speech.

"I stand before you as a representative of the American people, to deliver a message of friendship and hope," he began, and implored the Muslim world to step up their efforts to eradicate terrorism. "Drive. Them. Out. Drive them out of your places of worship. Drive them out of your communities. Drive them out of your holy land. And drive them out of this earth."

He renewed his pledge to work with the Arab nations to isolate their common enemy and the world's number one sponsor of terrorism, Iran. He called on Arabs to set aside old grievances and pursue a new spirit of partnership for the region. He then turned, in closing, to the key question for his audience: "Will we be indifferent in the presence of evil? Will we protect our citizens from its violent ideology? Will we let its venom spread through our societies? Will we let it destroy the most holy sites on earth?"

It was a tour de force. Afterward, as we were walking in the lobby of our hotel, Secretary Tillerson's top policy adviser Brian Hook overheard a few of the Arab leaders say among themselves, "Trump really gets us." After nearly two decades of fraught relations between the United States and the Middle East, we were adopting a new approach—an approach that didn't seek to remake nations in our image, but that instead sought to build coalitions based on our shared goals.

The trip was going better than I possibly could have hoped. Trump was at the top of his game, and the king's hospitality impressed him. The Saudis spared no effort in demonstrating their commitment to reform. In the thirty days leading up to the president's arrival, they had constructed a state-of-the-art facility called the Global Center for Countering Extremist Ideology and staffed it with more than two hundred data analysts to confront the Islamic radicalization online and other terrorist activity that had plagued Saudi Arabia for decades. Online extremism had also become a threat in America, contributing to the attack in San Bernardino, California, in 2015 and the brutal assault on the Pulse nightclub in Orlando, Florida, in 2016. During the election, many feared that Jihadist-motivated shootings would continue to plague the United States and countries around the world. If we could get Saudi Arabia, the leader of the Arab world, to crack down on radicalization online, other Muslim countries would follow. This historic effort improved the safety and security of American citizens and innocent people everywhere.

I did make one error: I scheduled too many engagements for Trump.

The Saudis had originally wanted the visit to be a five-day summit, but I negotiated it down to two and packed Trump's schedule with meetings to accommodate requests from the Saudis and other Arab leaders. In the forty-eight hours in Riyadh, Trump attended eight bilateral meetings, two different summits, numerous receptions, several lunches, and a state dinner. At the end of a long day, when we told him about an additional event, he turned to me and complained, "Jared, this schedule is inhumane. You know you aren't in my will. Why are you trying to kill me?"

Trump asked Ivanka to take his place at the event, a forum with social media leaders in Saudi Arabia, which has one of the highest social media usage rates in the world. This was a major initiative for MBS, and part of his strategy to foster tolerance and condemn extremism. With only a few minutes to prepare, Ivanka went out onstage before hundreds of people. Throughout the visit, she trended on social media across Saudi Arabia.

As we taxied on the tarmac the following morning, we were about to embark on the first reported direct flight from Saudi Arabia to Israel. King Salman had agreed to waive their airspace restrictions and allow our plane to fly directly to Israel, which would reduce the flight time by three hours. Just before we took off, I got a call from US ambassador to Israel David Friedman. The Israelis were nervous about Trump's expectations for his visit: they didn't have the limitless budget or lavish palaces that the Saudis had, but they had still planned a first-rate trip by their standards. I told him they shouldn't worry.

Shortly after we arrived in Jerusalem, the president and First Lady led a delegation to the Western Wall, and Trump made history as the first sitting US president to visit this holy site. As Ivanka and I walked alongside Trump, I prayed that God would protect my family, help me to live up to my potential, and give me the wisdom and strength to use the responsibilities I had been given to serve my country faithfully. The sun began to set, and Trump desperately wanted a night to relax before another busy day. But he and Melania were scheduled for a private dinner with Prime Minister Bibi Netanyahu and his wife, Sara. In what I jokingly referred to as a massive failure of Israeli intelligence, Bibi planned a multicourse meal that dragged on for hours. Bibi should have known his audience better—he would have won Trump's favor if he had simply served a hamburger and allowed him to go to the hotel to relax. Trump called me after, frustrated: "The guy kept me up for three hours and was talking my ear off. It was beautiful, but every time I thought the meal would end, another course would come out."

The following morning, the president was scheduled to meet with the president of the Palestinian Authority, Mahmoud Abbas, in the West Bank. Abbas had come to the White House in May, told the president he was ready to negotiate, and expressed confidence in Trump as the arbiter of a peace agreement between the Palestinians and Israel. We were intrigued, but still waiting to hear more. Just prior to our departure, Trump was briefed by Ambassador Friedman, who showed Trump a video compilation of Abbas making menacing threats toward the Israeli people. Friedman's message was clear: Be careful with Abbas—he tells you he's for peace in English, but look carefully at what he's saying in Arabic. Tillerson saw what was happening with the video and went ballistic, claiming it was dishonest. Friedman pushed back: "Are you saying he didn't say these things?" Tillerson had to admit that they were Abbas's words, but he was fuming because he was losing control. It was important for the president to see all sides of the issue, especially since he was hearing from several respected businessmen that Abbas was an earnest man who sincerely wanted to make peace.

During the bilateral meeting in Ramallah, Abbas recited the same talking points he'd used during his recent White House visit. It was as if the first meeting had never happened. He failed to show any progress on the issues he and Trump had previously discussed. Disappointed by the Palestinian leader's behavior, Trump grew exasperated and did not parse words: "You pay people who kill Israelis. This is an official government policy. You have to stop this. We can make a deal in two seconds. I've got my best guys on it. But I want to see some action. I want to see it fast. I don't believe you want to make a deal." Abbas became defensive and complained about Israeli security. Trump responded, "Wait: Israel is great at security, and you are saying you won't take free security from them? Are you crazy? Without Israel, ISIS could take your territory over in about twenty minutes. We spend so much on the military. Everyone in this region spends a fortune on security. If I could get free high-quality security for America and save the cost, I would take it in a second." After witnessing Abbas's intransigence, I better understood why twelve previous presidents had tried and failed to reach a peace deal.

When we arrived in Rome, Ivanka and I treated the team to dinner to thank them for their hard work on the trip. Joe Hagin, the president's deputy chief of staff for operations, directed us to one of his favorite pasta spots, Da Sabatino. After a classic Italian dinner, Dina Powell and Gary Cohn, neither of whom are Catholic, offered to give their spots to meet the pope to two Catholics on the staff, Brian Hook and Ivanka's chief of staff, Julie Radford. The Vatican adhered to strict protocols about the number of people allowed to accompany the president and the First Lady. The manifest, negotiated weeks in advance, was restricted to the president's immediate family members and seven staff members. But when the delegation passed through Vatican security the next morning, a guard stopped Radford and told her that someone else had already taken her spot—one of Tillerson's top aides had put herself in one of the front vehicles and had rushed to the entrance and signed in as Dina Powell. The Vatican officials would not let Radford in, and while she was disappointed, we later joked that at least she didn't lie, cheat, and steal to meet the pope.

Contrary to how the press reported the visit, Pope Francis was warm and gracious to Trump and Melania, and during their discussions, the pope said that he felt like we were living through World War III, only with the conflicts broken into pieces all throughout the world. While at the Vatican, Ivanka met with the Holy See to discuss human trafficking, and they agreed to work together to combat this evil form of modern slavery.

That night Ivanka and I accepted a generous offer from the Italian government for a tour of the Pantheon and its magnificent dome. We assumed we would be among the many tourists visiting the ancient temple, but when we pulled up in a Secret Service car, the police had roped off the entire area and hundreds of people were waiting for our arrival. While flattered by the kind greetings, we wished we could go back to the days when we could go almost anywhere without causing any fuss. Our lives were no longer our own.

Ivanka and I broke off from the trip after Rome, flying commercially back to the United States, while Trump continued to Brussels for the commemoration of the new NATO headquarters. Trump exhorted the European leaders to honor their NATO commitments and spend more to build up their collective defense. He revisited this concern throughout his presidency, along with his strong opposition to Europe's reliance on Russia for natural resources. He privately warned German chancellor Angela Merkel that her country's dependence on Russian gas and support for the Nord Stream 2 pipeline would enrich Putin financially and give Russia leverage over the European economies. He cautioned European leaders that pressing to include Ukraine in the NATO alliance would provoke Russia and back Putin into a corner, even as Europe was in a weak position strategically. This could lead to war.

Just before Ivanka and I took off, a press aide alerted me to a breaking story. The *Washington Post* was planning to report that I was under criminal investigation by the FBI. From the tarmac, I called Fred Ryan, the CEO of the *Washington Post*, and told him that I had not been informed that I was under investigation and didn't know what on earth I could possibly be under investigation for. I argued that they were basing their story on a lie from someone within the intelligence community who was willing to breach confidentiality standards by leaking to the newspaper. I told him it was totally irresponsible to publish such a damaging story based entirely on an unsubstantiated claim. Doing so would make me radioactive in Washington and have major implications for my life. He said there was nothing he could do. We took off and had no internet for the twelve-hour flight. I had no idea of the intensity of the storm awaiting me back home.

{ 12 }

The Art of War

Upon landing back in the United States on Thursday, May 25, Ivanka and I had an important engagement: Arabella's ballet recital. We had promised her we would attend, and we made it just in time. As we settled in to enjoy her performance, my phone rang. A reporter wanted my reaction to the *Washington Post* story, which had just posted: "Jared Kushner Now a Focus in Russia Investigation." I called my lawyer, Jamie Gorelick, who had recently delivered me a gut punch when she told me that she needed to step down from representing me due to a conflict of interest. Bob Mueller worked at the same firm, and he had been appointed special prosecutor. She told me that she was also receiving a flurry of calls from reporters. I returned to Arabella's ballet, but Gorelick called back a few minutes later with urgent questions. Throughout the performance, I kept stepping out into the hallway to take the incoming calls. A deluge of news and misinformation had begun.

The next day, a Friday, the *Post* ran another hit piece: "Russian Ambassador Told Moscow that Kushner Wanted Secret Communications Channel with Kremlin." That weekend, to escape the media camped outside our Kalorama home, Ivanka and I went to Bedminster. Even from there, two hundred miles outside Washington's beltway, we could feel the heat. CNN seemed to be going completely berserk, rotating between panels of "experts" who assumed that this unsubstantiated claim was true. Shocked and perplexed, I called Gorelick. "This is crazy," I told her. "I've got to put out a statement. They're making an inconsequential transition meeting sound nefarious. I didn't meet a single Russian during the campaign; there was absolutely no collusion with Russia." I forwarded her a statement I had drafted and asked for her legal clearance to send it out.

"I wouldn't do that," she said. When I asked why, she responded, "We're going through your tens of thousands of emails, and I found one email that you're going to want to see. I know it's painful now, but my job is for you to get to the other end in good shape. I'm thorough, I'm expensive, but there's a reason people use me. And you don't want to put out any statement until I've reviewed all the facts."

"Well, what does the email say?" I inquired.

"I'd rather show it to you in person. Come to my office when you get back."

I racked my memory, but nothing came to mind. I had received three hundred emails a day during the campaign, but I couldn't imagine what Jamie had referenced. I knew for certain I hadn't done anything inappropriate with Russia.

When I got back to Washington, a close friend flew in from Arizona for dinner to cheer me up. At my lowest moment in Washington, he encouraged me to have faith, stay strong, and keep my head up. I would make it through, he said. It was a much-needed pep talk.

The next day, I went over to Gorelick's office at the WilmerHale law firm, three blocks west of the White House on Pennsylvania Avenue. Gorelick placed a three-page document in front of me. I scanned the first page. It was a campaign email from Don Jr., asking me to stop by a meeting. This was a frequent and ordinary occurrence.

"So, what's the big deal?" I asked. She told me to go to the third page of the printout and read the end of the email thread—something I never would have done during the campaign. At that time, I was still running my businesses and joining a dozen meetings a day. I didn't scroll through long email chains when the message at the top was about logistics. At the very bottom of the thread, on page three of Jamie's printed document, I saw the initial email that was part of a chain I had not been on. It was a description of the purported topic of the meeting: to share information "that would incriminate Hillary and her dealings with Russia," as part of "Russia and its government's support for Mr. Trump."

Seeing the email triggered my memory. "The meeting was a joke," I told Gorelick.

"That doesn't matter," she said. "You can't say you never met a Russian during the campaign since this meeting took place. I don't care how innocent or uneventful it was. The media will have a field day with this. Tell me what happened in that meeting."

I remembered arriving a little late, just as a Russian attorney launched into a monologue about how Trump could win the election if he got Russia to reverse its misguided ban on US adoptions of Russian children. Immediately recognizing that the meeting was a colossal waste of time, I emailed my assistant to pull me from the meeting: "Can u please call me on my cell? Need excuse to get out of meeting." When my assistant called, I stood up, excused myself from the room, and forgot that the meeting even had occurred. Campaigns have a way of bringing out crazy people who think they know best. They're like sports fans who whine about the decisions of coaches and managers.

Gorelick and I discussed what to do, and I began to grasp the new reality. The stories I'd been dealing with on Russia weren't going away anytime soon.

After the press reported that I was the subject of an FBI investigation, I started to notice Priebus, Bannon, and McGahn excluding me from White House meetings. They seemed to avoid me in general. Years later, it was reported that during this time Bannon had sent an email to a staff member at Breitbart asking him to investigate a story linking me to a Russian oligarch I had never met. I could hardly wrap my mind around this sinister and destructive behavior. The president was beset by a false Russia collusion narrative that was hampering his presidency and causing him significant stress, and yet Bannon seemed to be deliberately stoking the narrative to hurt a colleague. It was a level of betrayal that I had never conceived possible, even in Washington.

In late June my old friend Eric Herschmann, a former New York prosecutor and a senior partner at the law firm Kasowitz, Benson & Torres, called and said, "Jared, you have a big problem." He had heard that Bannon and McGahn planned to hire a high-powered defense lawyer to help the White House legal team on the Russia defense. His name was Ty Cobb, just like the famous baseball player. "If he gets hired, I'm afraid you're dead," Herschmann warned. He explained that he thought Cobb was unwittingly part of a plan by Priebus and Bannon to push me out: hire a lawyer who would tell the president that I had become a serious liability and needed to leave. New hires typically had a honeymoon period with Trump; for the first three or four weeks of their tenure, he showed them a lot of deference. If Herschmann's information was correct, and Cobb was hired, he would be at the peak of his powers as he pushed for my removal. Bannon had masterfully choreographed a series of leaks and lies over the past few months, keeping my name constantly connected to Russia in the press. The House and Senate intelligence committees were preparing to question me.

Herschmann also flagged that Trump had an off-schedule meeting with Cobb at noon that day. I walked into the Oval Office shortly after the meeting started and sat down in one of the chairs next to the lawyers, directly in front of the Resolute Desk. Priebus and McGahn, who had not expected me to join, looked uncomfortable. Priebus stepped out of the room, and after ten minutes of discussion, McGahn politely asked me to leave, explaining that I was an official witness in the Russia investigation and therefore shouldn't be a part of their discussion. The president took his counsel's advice. When I got back to my desk, I learned that while I had been in the Oval Office, an agitated Priebus had been circling the reception area outside the main entrance to the Oval, anxious about how to get me out of the meeting.

After several agonizing minutes, I decided that I couldn't sit by while others planned my execution. The president has a private dining room in the West Wing that opens onto the main hallway through a two-foot-wide service pantry; it can also serve as a back entrance into the Oval Office. To avoid Priebus, I snuck through this back entrance. Reappearing in the middle of the meeting, I interrupted and told Trump that I really needed to speak with him. Surprised to see me and annoyed that I had disrupted an intense discussion, he reluctantly agreed, stood up from his chair behind the Resolute Desk, and followed me into the dining room.

"I understand you feel like you can't fire Bannon because you don't want him to go rogue," I began, "but don't hire Ty Cobb."

"I was literally just about to make a deal with the guy," Trump responded. "Why can't I hire him?"

"Because I'm told he has another agenda," I said.

Trump looked at me silently for a moment, told me to wait there, and went back into the Oval Office, where he called for John Dowd, his personal lawyer, who was also in the meeting. When they both entered the study, Trump asked me to repeat my concern to Dowd. It was clear that in my desperation, I had made a major mistake, and my Hail Mary attempt to save myself was about to make my situation much worse: through Dowd, Ty Cobb would now know that I opposed his hiring. Just as I was about to speak, miraculously, Trump got distracted by a Fox News segment and asked us to pause so that he could hear what the pundits were saying. I quietly suggested to Dowd that we go in the other room to talk, and we proceeded into the small study located between the Oval Office and the private dining room—referred to around the White House as the "Monica Lewinski room."

"Tell me about Ty Cobb," I began.

"He's a great lawyer."

"Do you trust him?"

"Yeah, I trust him."

"It feels like there are so many people surrounding the president who don't have his best interests in mind," I continued. "They are causing more problems than they are solving. Look, our lives would have been so much easier if we had stayed in New York, but we decided to come to Washington to help the president succeed. We didn't come here to fight with people and play political games. We care about him, and we care about our country."

Dowd, a former Marine and retired litigator who believed in Trump and saw this chance to work for the president as his final tour of duty, probed me with questions. I answered each one honestly and took him through the facts of my Russia case. After about ten minutes, he looked at me and said, "You know, from the moment I got here, McGahn, Priebus, and Bannon each came up to me individually and said that you and Ivanka were the problem and that everything would be resolved if I was able to get you out. But I think I get it now. You're the only ones who actually care about the president. I see what these self-serving bastards are trying to do. I'm onto them. I've got your back, and I'm going to get every single one of them." I was later told that McGahn, Priebus, and Bannon had drafted my resignation letter and were pushing the president to get me to sign it.

Ten days before my scheduled congressional testimony, the president and Ivanka traveled to Bedminster to make an appearance at the US Women's Open golf championship. I went with them to clear my mind and finalize my testimony as I prepared for the consequential day. The news coverage that weekend was especially salacious.

While I pored over my testimony, Trump walked into our cottage and sat down across from me. Frustrated by the negative news cycles and concerned about me, he remarked, "You're too hot right now. Did you do anything?"

I told him I had not done anything wrong. "You have to clean yourself up and fix this," he implored.

I explained that I was working on my testimony and couldn't say anything publicly until then. I could tell that he was pained that people who opposed him were coming after his family. What bothered him most deeply was that there was nothing he could do to stop it. He had my back, but he couldn't solve my problems for me. He told me to stay strong and do my best, but to be careful not to make it worse.

*   *   *

On the morning of July 24, Ivanka kissed me goodbye as I left for my hearing with the Senate Select Committee on Intelligence. Just before I shut the door behind me, she gave me one last piece of advice: "Just remember, you did nothing wrong. You have nothing to hide, so don't let them intimidate you. Keep your head held high and smile."

Earlier that morning, I had publicly released my comprehensive eleven-page statement.[11] My new lawyer Abbe Lowell had suggested that we release the statement in advance to prevent Democrats from inaccurately spinning selective parts of my testimony to fit their narrative. For months I had held back on refuting the accusations, and now I could finally respond, methodically addressing each claim and providing a thorough defense that destroyed the false narratives.

At 9:45 a.m. I walked past hundreds of cameras and reporters and mustered a smile as I entered the Hart Senate Office Building. Flanked by security escorts and my two lawyers, I entered a private room and sat down at a table with senior staff members from the intelligence committee. Based on their questions, which were surprisingly rudimentary, I could tell they knew next to nothing about how we had run our campaign. When they asked about my interactions with Russia, I told the truth: "I did not collude with Russia, nor do I know of anyone else in the campaign who did so. I had no improper contacts. I have not relied on Russian funds for my businesses. And I have been fully transparent in providing all requested information."

When I returned to the White House, drained but also relieved, I found the president preparing in his private study for a press conference on Paul Ryan's health-care bill. I told him that the testimony had gone well, and I was going out to the "sticks"—the spot in front of the ceremonial entrance to the West Wing where administration officials make formal statements to the White House press corps. Just before I walked outside, Sarah Sanders, who had recently been named White House press secretary, dusted me with some makeup powder—a first for me—and offered a tip: "When you go out there, it will be overwhelming. Take a deep breath and read really slowly because, believe me, your heart will be racing." As I faced the sea of cameras, I was intimidated and scared, but I remembered Sarah's advice, took a deep breath, and read my statement. When I walked back in, Madeleine Westerhout, the president's assistant, called to say that Trump wanted to see me. I went back to the Oval Office, and he congratulated me on the statement.

The next day I testified before members of the House Permanent Select Committee on Intelligence. When Democratic ranking member Adam Schiff kept asking me additional questions past the previously agreed time limit, I stayed and answered them. "I don't want him to be able to go out and say, 'I have a lot more questions,'" I said. "I'm going to keep going until—as long as you'd like."

Unlike court cases across America, this congressional investigation presumed that I was guilty until proven innocent. One misstep, one poorly phrased answer, or one accidental omission would mean humiliation, eviction from Washington, and possibly jail. Fortunately, the testimony was a major success. I cleared the decks of the false accusations. I hoped that Congress would accept the truth and that the public would eventually see that there was nothing to investigate.

That night Ivanka and I celebrated together over dinner at home with the kids. For the first time in months, I felt a true sense of relief. Now that the testimony was behind me and I finally had the chance to correct the record, I was eager to focus more energy on achieving things that actually mattered. Through the experience, I learned that it wasn't enough to avoid stepping on the line. I needed to stay six feet back from the line, and I needed to be more careful not to give my adversaries easy opportunities to hit me.

Other developments in the West Wing quickly swallowed up the news of my testimony. A week earlier, the president had hired Anthony Scaramucci as his new director of communications. The successful financier had come to the Oval Office to pitch himself for the job, and Trump gave it to him on the spot.

Bannon and Priebus lost it. They had a stranglehold on the communications department and used it to attack their foes and protect themselves. Shortly after making the decision, the president brought Priebus and Bannon into the Oval Office, along with Scaramucci, Hope, Scavino, Ivanka, and me. During the meeting, the phone rang: it was Mike Pompeo, the CIA director at the time, returning a call. Trump praised Pompeo for a speech he had recently given at the Aspen Security Forum, which persuasively laid out the president's America First foreign policy approach. We couldn't hear what Pompeo was saying, but at one point the president said to him, "You are a star. We have some real stars in this administration." He paused for dramatic effect, looked across the room at Bannon, and then continued, "We also have some real losers and leakers as well, but that will change."

Scaramucci's White House tenure was short-lived, but his arrival had disrupted the organization. The president decided it was time to clear the decks. Later that week, after months of work, the Senate failed to pass Paul Ryan's health-care reform bill, and the president decided it was time for Priebus to go. He unceremoniously fired his beleaguered chief of staff by tweet on July 28 as they returned to Joint Base Andrews from an event with law enforcement officers on Long Island, New York.

I sympathized with Priebus, despite the fact that he had aligned himself with Bannon. He had to contend with me, Bannon, Kellyanne Conway, Gary Cohn, Dina Powell, former *Apprentice* contestant Omarosa Manigault, and many others who had a direct line to the president. Trump was new to Washington and had an unconventional governing style. Months later, Priebus aptly described the situation: "There were so many natural predators in one zoo." He also told me that if he could do it over, he would have gone with me instead of Bannon. "You weren't the problem," he admitted. "Steve had me fully convinced that you were the problem and that nothing would work until we got rid of you. I just made a mistake. I understand you now—you play the long game."

In the same series of tweets Trump used to fire Priebus, the president announced General John Kelly, the retired four-star general who was running the Department of Homeland Security, as his next chief of staff. My interactions with Kelly had been limited during his time at Homeland Security, but I respected his reputation and his lifetime of service to our country. There was, however, something unsettling about the early signals he sent to the White House staff. After Kelly was announced as the new chief, he instructed the White House operator not to put any calls through to him, and he didn't make calls to introduce himself to members of the senior staff. In hindsight, it became clear to me that he was establishing distance and dominance—and trying to foster insecurity.

Shortly after Kelly became chief of staff, I had a drink with him while we were both at Bedminster with the president. As the general sipped his gin and tonic and I nursed a beer, I told him that I wanted to see the president succeed and walked him through my portfolio. "I just want to work on these things," I said. "And I'm here to help you with anything else—I know and understand Trump and how to make things work, but you're the boss." I then offered him two unsolicited pieces of advice. The first was to get the president to stop talking about health care—a losing issue without a cohesive plan—and to focus instead on tax reform. Our economic team was working on a proposal to cut taxes and bring relief to working families: "It's way more popular to tell people you're cutting taxes than that you're taking away their health care. Besides, Mnuchin and Gary are two of our best athletes, and they will make sure we have the best chance of achieving our first legislative victory before the end of the year."

My second piece of advice was to get rid of Steve Bannon. "He has lost his mind, wants everything to be a conflict, and he's leaking to the press all day." Kelly assured me that he had already taken steps to take care of that.

On August 18 the president fired Bannon. Stephen Miller joked to Hope and me, "I have a plan to split up Steve Bannon's extensive work-load. Hope, you leak to Jonathan Swan at *Axios*. Jared, you call Mike Bender from the *Wall Street Journal*. I'll call Jeremy Peters from the *New York Times*, and . . . we're done."

I remember a conversation at the time with a close friend. Admitting that I didn't yet have any major policy successes to show for my seven months in government, I joked, "At least I was able to get Steve Bannon fired. That partially saves the world from immediate disaster." My friend shot back, "You don't get credit for that. It's like paying your mortgage. You're supposed to do that." Those words really stuck with me. I knew he was right; just surviving wasn't enough. I recommitted myself to finding a way to make my service count.

As I learned how to navigate government, books became some of my best advisers, providing historical and factual perspective that many in government had forgotten or never even knew. An added advantage, of course, was that books didn't leak to the press. One book in particular shaped how I understood government: *The Gatekeepers* by Chris Whipple.[12] The book describes how chiefs of staff operated in different administrations and helped me realize that the power struggles in our White House were not unique. Every West Wing teems with rivals who vie for the president's ear, and there has often been tension between the "pragmatists," such as President Reagan's chief of staff James Baker, and the ideologues, like Reagan's longtime friend and counselor Ed Meese. To counter these dynamics, the chief of staff must install a strong internal process for preparing decisions for the president. I also learned that there was a big difference between leaking to the press and spinning to the press. A leak occurs when an unauthorized person divulges sensitive— and in the worst cases untrue—information to advance a personal agenda or to disparage a colleague. Spinning, on the other hand, is the sharing of nonpublic information to advance the president's agenda by helping the public better understand an issue. This is often done in full coordination with the White House communications office.

I couldn't change the game. I just needed to excel at it—adapt my approach, get smarter, get tougher, navigate the process, weather attacks, and solve problems. I needed to find the effective people within our government and in other countries who actually managed to get things done. For every Bannon, there are a thousand people who are in government for the right reasons. Most of the time, no one knows about them because they are busy doing their job, not leaking to the press.

The other book that most helped me was Sun Tzu's *Art of War*, the ancient military manual written by a Chinese general.[13] I learned three invaluable lessons from Sun Tzu. First, he writes, "The good fighters of old first put themselves beyond the possibility of defeat, and then waited for an opportunity of defeating the enemy." I was too exposed politically and needed to hunker down like a turtle in its shell and rebuild strength before I could start feeling my way out. In practical terms, this meant focusing on my files, making more friends and fewer enemies, and not trying to fight everyone else's battles. Second, "The opportunity of defeating the enemy is provided by the enemy himself." Rather than directly trying to go after those going against me, it was smarter to avoid conflict and let them create the conditions for their own defeat. This didn't ensure that they would be defeated, but it gave me the space to survive. Throughout my time at the White House, I never defeated my enemies; nearly every one of them defeated themselves. Bannon illustrated this perfectly. I never attacked Bannon. He chose to go after me, and his relentless attacks created a high-stakes situation where one of us had to go. This eventually led to his demise. And third, "Avoidance of mistakes establishes the certainty of victory, for it means conquering an enemy that is already defeated." I needed to be error-free. Working in politics was like balancing on a ball: I had to find ways to advance my goals without falling out of bounds. Every problem I tried to solve came with risks and countermoves, and the terrain could shift at any moment.

Seven months into the administration, Priebus and Bannon were finally gone. We had a US Marine general who seemed to be establishing order in the West Wing. My congressional testimony was behind me, and the whole Russia nonsense seemed to be simmering down. Now I hoped to focus more energy on a major responsibility that the president had entrusted to me: achieving the elusive goal of peace in the Middle East.

13

Great Expectations

For decades, even the most seasoned foreign policy experts had failed to broker peace in the Middle East. These experts were skeptical that I could succeed where they had repeatedly failed. What chance did a thirty-six-year-old real estate investor have? I understood that the probability of success was slim, but I was determined to search for a breakthrough. I told myself that the worst that could happen was that I would fail like everybody else.

As I talked to foreign policy luminaries to get their perspectives, I met with Richard Haass, the president of the Council on Foreign Relations, a heavyweight among the foreign policy establishment. I described the general approach I planned to pursue and then asked if he thought it had any chance at success.

"Nope," he said.

I asked him why he was so sure.

"Simple," he replied. "No one has made money betting on success in the Middle East over the last twenty-five years."

Haass was so dismissive that I began to realize how defeatist the foreign policy establishment had become. As I read books that past negotiators had written on the subject, I noticed a familiar refrain: they all explained that it wasn't that they had failed, but that the problem was too hard.

Former secretary of state Henry Kissinger was a valuable sounding board. Kissinger literally wrote the book on modern foreign policy; his 900-page treatise *Diplomacy* is required reading for students of the subject.[14] From the first time I met him on the campaign, I leaned on his wisdom, knowledge, and graciousness. Despite his advanced age of ninety-four, he was sharp and always up-to-date on the current geopolitical landscape. "Call me anytime," he offered after Trump won the election. He saw it as his patriotic duty to offer his vast experience and advice. I was deeply grateful and solicited his counsel often.

Shortly after Trump assigned me the Middle East peace file, I asked Kissinger how he would approach the job. He recommended that rather than trying to achieve a grand deal, I should focus on creating progress through short-term agreements. In 1974, as Israeli and Arab forces fought for control of the Golan Heights, he had negotiated a cease-fire.[15] The text of the deal made it explicitly clear that the agreement was "not a peace agreement."[16] Even so, Kissinger explained, it had become a new status quo over the last five decades. Permanent peace deals make for challenging domestic politics in the Middle East, he said, but if you can get rivals to agree to a short-term pact, or even a change of the status quo, it will last. Kissinger also warned me to resist efforts to run foreign policy out of the State Department. "You always have to run the foreign policy in the White House," he said. "If the White House loses foreign policy to the State Department, you will never get anything done."

After we arrived at the White House, I began spending several hours each Sunday receiving briefings from the US government's foremost intelligence, military, and foreign policy experts. The briefer who I found most insightful was CIA expert Norm Roule, who had served in the intelligence community for more than thirty years and knew the intricacies of the Middle East like few others. Roule deepened my knowledge of the history of the countries in the region, the roots of the conflicts, and the key alliances and important players. He explained that the best way to gather intelligence was to get on a plane, form relationships, and listen to people. He and others painted a picture of a Middle East in turmoil. Terrorist groups like Hamas and Hezbollah were amassing tens of thousands of rockets and aiming them at Israel. The Iranian regime was targeting American troops through its terrorist proxy organizations in the region. Syria had devolved into a humanitarian and refugee crisis.

And ISIS had a caliphate the size of Ohio and was recruiting radical disciples online.

A half century of conflict and failed negotiations had left the relationship between Israel and the Palestinians and other Arab countries in shambles. Neither side had much of an incentive to make a deal. Abbas was thought to be scared to reach a compromise, and Israel had no desire to make any concessions. Both viewed any new action as a potential threat to the fragile equilibrium they had reached over decades.

Roule's briefing reminded me of what I had just read in *The Fight for Jerusalem*, by Dore Gold, which describes why the Israelis are distrustful of the international community on the question of Jerusalem.[17] In 1948 the United Nations overwhelmingly passed Resolution 181, a proposal to partition Palestine into separate Jewish and Arab states while maintaining international control of Jerusalem. Six months later, military forces from Egypt, Transjordan, Lebanon, Syria, and Iraq invaded Palestine and occupied areas of East Jerusalem, including the Jewish Quarter in the Old City. The United Nations remained quiet during the war that followed, allowing the Temple Mount, including the Western Wall, to fall under Arab control. For the next nineteen years the Jewish people were barred from the holy site, even for peaceful prayer, while many of the other Jewish sites were desecrated. In 1967, during the Six-Day War, the Israelis were attacked and won back control over Jerusalem. Since then, they had become skeptical of any change that could once again compromise their jurisdiction.

Another book that informed my thinking was Jimmy Carter's bestseller *Palestine: Peace Not Apartheid*.[18] In the appendix, I found a draft of the 1979 peace agreement that Carter negotiated between Israel and Egypt. Up to this point, I had been so consumed with understanding the problem that I had not yet thought about what a solution would look like in technical terms. Carter's 1979 draft was not an in-depth legal document. Rather, it was a brief set of principles. Intrigued, I asked the National Security Council and State Department to pull all of the signed agreements that related to peace in the Middle East, as well as every document that had been drafted or negotiated but not signed. As I studied the stack of documents, I found that they were all scant on details. These one- to two-page documents consisted of high-level concepts that were heavily wordsmithed to avoid the most contested issues: they failed to offer specific compromises on the most pressing questions, such as where exactly to draw the lines dividing Jerusalem, what Palestinians needed to do to achieve their own state, and how to handle the hundreds of thousands of Palestinian refugees who had fled to other countries during the 1948 and 1967 wars and now wanted to return. A number of experts had alleged that the Israelis and Palestinians had been close to a deal at Camp David in 2000, but I inferred otherwise from the documents and firsthand accounts.

I was equally surprised when I studied the Arab League's 2002 Arab Peace Initiative, which the Palestinians and Arab nations cited as a basis for negotiations.[19] It was all of ten lines that didn't say much. At the time of its publication, it offered a hopeful framework for peace because it professed that the Arabs were willing to negotiate an end to the conflict and normalize relations with Israel. But it lacked viable, nuanced solutions to the major unresolved issues. On perhaps the most contentious issue—the status of East Jerusalem and its crown jewel, the Temple Mount—it had the right wording to avoid upsetting the Arab world, but it didn't include specifics. In fact, it called on Israel to accept the establishment of an independent Palestinian state with East Jerusalem as its capital without providing any detail around how it intended to define East Jerusalem, where it would draw boundaries, or how it planned to respect religious claims and keep sacred sites safe and open. After being denied access to the Temple Mount for decades, Israel would never agree to give up control of East Jerusalem. According to a Palestinian Center for Public Opinion poll, more than half of the Palestinians who lived there wanted to be Israeli citizens rather than live under a Palestinian regime that had proven incapable of caring for its own people.[20]

In searching for a fair compromise, I turned to Scott Leith, the National Security Council's top expert on Israeli and Palestinian affairs who had previously worked on Secretary Kerry's State Department team as a special adviser on Israeli-Palestinian negotiations; I perceived that he was deeply sympathetic to the Palestinians and asked him to represent their point of view in our internal debates. To my surprise, when I requested background on the origins of the Palestinian claim on East Jerusalem as their capital, Leith didn't know offhand. He asked to get back to me with an answer. Two days later, Leith admitted he could not find evidence of a formal, independent Palestinian claim on East Jerusalem as their capital that predated 1988. That was the year the Palestinians declared it as their capital in a charter of the Palestinian Liberation Organization. I couldn't believe it; Scott had spent two days delving into the research and speaking with his fellow policy specialists, and this was all he could find. This was irrational, I thought. For three decades, the international community has accepted the Palestinian claim to a capital in East Jerusalem as an immutable fact of international law. And yet the legal basis for their claim is a 1988 charter organizing a group of revolutionary fighters. This helped explain a fact of history that struck me as odd. Following the 1948 war, Jordan had annexed the West Bank and East Jerusalem, but had kept its capital in Amman.

No one had ever come close to a real solution that could be signed and implemented. Those who had gone before had made sincere efforts, but they were more focused on managing the political reaction to their negotiations than they were on producing detailed proposals that would have a practical impact. I decided to test a new approach: I wouldn't try to dodge the details. Instead, I would embrace them. I asked my team to make a comprehensive "issues list" that explained the major points of contention between the two sides. This would help me understand the granular differences between the two parties. I would then work through this issues list with leaders in the Middle East to hear their perspectives and find concrete resolutions. This was how we approached transactions in business, and it made sense to apply the same technique here.

I began formulating the working strategy of our peace plan with the advisers who composed my small but talented Middle East peace team: my deputy Avi Berkowitz, special representative for international negotiations Jason Greenblatt, State Department director of policy planning Brian Hook, US ambassador to Israel David Friedman, and deputy national security adviser Dina Powell, a Coptic Christian who spoke Arabic.

I first met Avi in 2011 at the Biltmore Hotel in Phoenix. It was Passover, and I was playing basketball with my brother and brother-in-law at the resort's court. We noticed Avi and his cousin waiting to play, and we invited them to join us. I learned a lot about Avi watching him on the court. He was a team player and a skilled passer who hustled hard, anticipated the next play, and had a great attitude. I invited him to intern at Kushner Companies. After gaining admittance to Harvard Law School, he joined our company for the gap year between college and law school. Upon graduating, he set aside other professional opportunities to volunteer on the Trump campaign, and he was the first person I hired at the White House.

Greenblatt had been Trump's real estate lawyer for twenty years, and served as the Trump campaign's liaison to the Jewish community. He had the president's confidence, deep knowledge of the subject matter, and a terrific poker face.

Brian Hook had served in the two previous Republican administrations, which made him an invaluable source of institutional knowledge, but he was also suspect among the Trump loyalists in the West Wing. I quickly saw for myself that he was a steadfast team member, and he served as our primary interlocutor with the State Department.

David Friedman rounded out the core team. A successful bankruptcy lawyer in Manhattan, Friedman had earned the confidence and friendship of Trump over fifteen years. Some criticized his selection as ambassador because he was a pro-Israel hawk with connections to the evangelical and Jewish right wings, but I saw this as an asset. He was well positioned to build relationships with the Israelis and report information back to Washington.

Early on, Friedman suggested that we should treat an Israeli-Palestinian agreement like a bankruptcy proceeding. "Israel is a secured creditor: they are the only democracy in the region with a stable government, a strong economy, and a viable market. The Palestinians are an unsecured creditor: they have corrupt leadership, a flailing economy, and no stability, and yet they think they have parity with the secured creditors. From my experience, you always end up in trouble when you let the weaker party think it can call the shots."

14

Tear Up the Talking Points

s I dealt with the barrage of false news about the Russia investigation, Secretary of State Rex Tillerson pounced on the opportunity to remove me from the Middle East file. One afternoon I received a call from US ambassador to the United Nations Nikki Haley, with whom I had developed a productive working relationship. Tillerson had told her that I was off the Middle East file and instructed her to stop routing issues and requests through my office.

Rather than confronting Tillerson, which I knew would be counterproductive, I figured it was better to assert my role. So I booked my first solo trip to the Middle East. On June 21 I traveled to Israel with Avi and Greenblatt to meet with Prime Minister Bibi Netanyahu. A war hero, diplomat, and Israel's longest-serving prime minister, Bibi was a bold defender of Israel. As the most formidable opponent of the Iran deal, Bibi had taken steps to improve diplomatic relations with a number of countries, including some Arab nations who shared Israel's concern about a nuclear Iran. During his tenure, Bibi had made Israel into an economic powerhouse, an oasis of innovation, and an ever-ready battle nation with one of the most elite and advanced militaries in the world.

I had met the Israeli prime minister many years earlier, when he spent a weekend with my family on one of his trips to the United States. At the time, he was a private citizen and was speaking around the world. My father, who met Bibi through his support of pro-Israel causes, invited him to stay at our home in New Jersey while he was in town. Bibi stayed in my bedroom, so I was relegated to the basement, where I slept with his security guards on pullout couches. Jet-lagged, Bibi couldn't sleep so he pulled a book off my shelf—*Great Expectations*, the classic novel by Charles Dickens—and got halfway through it.[21] The next morning, he asked if he could take the book to finish it on the road. Ordinarily, I would have been happy to give my book to the Israeli leader, but this book was a gift. I flipped to the front page and showed him the inscription—it was from my girlfriend at the time. Bibi smiled and graciously left it behind.

When we met in the prime minister's office on June 21, Bibi expressed his thanks for the president's recent visit. He was grateful that I had worked to defend Israel at the United Nations during the transition and was relieved that the United States was once again supporting the Jewish state after eight years of strained relations under the Obama administration. When I brought up the Israeli-Palestinian conflict, Bibi thought we should delay working on the issue. He had survived as prime minister for eleven years by appealing to his conservative base and opposing a Palestinian state. "This is not the right time," he said. He went on to explain that he didn't want Israelis to think that he was using peace negotiations to distract from an ongoing government investigation.

"This is a high priority for President Trump," I said, "and if we're going to work with you on Israel's many requests, we need Israel to work with us on this." Bibi reluctantly agreed.

As we drove from Jerusalem to Ramallah, Greenblatt reminded me, "Don't say that we're for a 'two-state solution,' because it means different things to different people." It was good advice, and I decided to avoid the term until we had defined what it meant to be a state. When we arrived, we were ushered through a maze of stairways into a small room that had regal chairs arranged for a diplomatic meeting. Palestinian president Mahmoud Abbas entered, proceeded to the front of the room, and shook our hands. He was staffed by his top negotiators: Major General Majid Faraj, a trustworthy and insightful member of Abbas's inner circle and head of the Palestinian Security Forces; and Saeb Erekat, a loquacious and always aggrieved diplomat who had been the lead negotiator for twenty-five years but had little to show for his efforts. As they served us tea, I glanced in the direction of the Palestinian leader. Abbas sat hunched over in his seat, looking every bit of his eighty-plus years. He smoked constantly, so every few minutes he would pull a cigarette from the table, put it in his mouth, and wait for an attendant to light it. I thought that Abbas seemed more like a king than the representative of an historically downtrodden refugee population.

After exchanging niceties, I started the meeting right where Abbas and Trump had left off during their May visit, and asked Abbas whether he had made progress on the details of an initial proposal.

He said they were willing to take steps that they hadn't made with anyone else—they would be incredibly flexible on the land. But they needed to know exactly what percentage of the disputed territory they would get, and they wanted us to get Israel to propose a detailed map. If we could get them a map, they would be flexible, and everything else would be easy, Abbas pledged.

I asked him if they had an initial offer on the land issue, but as I tried to drill down, Abbas wasn't willing to talk specifics. He delivered the same set of diplomatic platitudes he'd conveyed to Trump several weeks earlier. Our conversation circled back around to my request for him to share concrete details about a land proposal he could accept. Again, he refused. I started to see why people were so skeptical of our efforts: Abbas was a savvy diplomat who was unfailingly polite and expressed a desire to make progress, but he appeared unwilling to let our negotiation reach a starting point. He said repeatedly that he had a lot of new ideas and would be flexible, but he then just rehashed the same general demands the Palestinians had requested for decades.

"I'm going to go back to the president, who's not a very patient person," I said. "He's going to ask me where we are on the deal, and I'm going to tell him that the Israelis are engaged and constructive, but you guys came back and weren't willing to be flexible at all. Is that the message you want me to relay?"

Abbas insisted that he wanted to be flexible, but then it was more of the same. I wasn't sure whether he didn't know how to make a deal, or if he just didn't want to. Sensing my exasperation, Abbas made what I perceived to be a factitious offer: he seemed to imply that if I didn't like the way things were going, then he would simply give back the keys to the West Bank and let the US run things.

"Sure, I'll take the keys," I retorted. I sensed Greenblatt shifting uncomfortably in his chair, like he was trying to tell me, telepathically, *You can't say that.*

The NSC foreign policy experts had warned me not to push Abbas too hard. Over the years, he had become a valuable security partner against extremist activity in the West Bank, and they feared that he was frail, politically weak, and on the verge of resigning his position. While they considered this to be a real and dangerous concern, I saw it as an opportunity, but I knew Abbas would never follow through. If he actually turned over the keys, he would forfeit his power and relevance. And his successor would inevitably scrutinize his internal affairs, which would expose his apparent corruption and luxurious lifestyle that came at the expense of his own people. Tens of billions of dollars had been injected into the West Bank, and while there was some progress, clearly lots of the money had gone missing. He had a good life and a presidential palace in Ramallah, as well as a beautiful mansion in Amman, Jordan. While the prime minister of Israel typically flew on a commercial El Al plane, Abbas traveled the world in a $50 million private jet for meetings with heads of state. I was calling his bluff, and he knew it.

"Look, if you guys give back the keys and you resign," I persisted, "we'll work with the United Nations and our local allies and put in place a provisionary government. We'll pump in a lot of money to build out your infrastructure and grow your economy. We will create tons of jobs and establish a fair and independent judiciary. In five years, we'll draw fair boundaries, and we'll conduct an election, and your people will have a new leader, better lives, and a fresh start. I'm okay with that path forward, if you really want to do that. It might actually be the easiest way for us, and better for the Palestinian people."

Though Abbas would not admit it, he seemed content to leave things as they were. Across Palestinian territory, his picture hung next to that of the former Palestinian Liberation Organization leader Yasser Arafat. Meanwhile, each year the Palestinians received more than a billion dollars in aid from the international community.[22] If Abbas made a deal, they might lose this funding stream and the one issue that gave them attention on the world stage. The world would shift its focus away from his nearly five million people, Hamas would pounce on even the smallest concession, and Abbas's people would view him as a traitor. We had heard that Abbas often stated that he would rather go to the grave as a martyr than as a traitor. He was in the twelfth year of a four-year term. The international community didn't seem to care, so why would he risk changing the status quo?

Immediately after our meeting, confidential information about our discussion started appearing in the press, and I further understood why Abbas was so afraid to show compromise: everything leaked from his office. He had worked hard to get the entire world to stand by the Arab Peace Initiative. If he showed flexibility toward compromise, he would run the risk of frustrating his supporters. I began to lose faith that we would ever get anywhere with him. He was in a tough position and had little incentive to make a bold change. Hoping to shake loose a solution, I continued to engage with Abbas and the Israelis. I knew failure was a possibility, but I was determined to try for success.

* * *

Upon our return, I asked Jason Greenblatt to start drafting our first iteration of a peace plan. I wanted to design a plan robust enough to sustain an agreement between Israel and the Palestinians far into the future. The plan would rest on three principles. First, all religions should have access to their holy sites for respectful observance. Second, the Palestinian people should have the opportunity to achieve better and more prosperous lives. And third, Israel had a right to maintain its security.

The old way seemed like a sure path to failure, so I decided to do something untraditional: propose a highly detailed plan and try to get both sides to react to it. Until both parties could react to a substantive plan, it seemed to me that they would keep fighting over vague concessions and hypothetical solutions, rather than coming to the table and negotiating a deal that would last long after it was signed and executed. Our initial draft was a ten-page document, but it soon morphed and expanded.

I wanted to solicit input from Arab leaders, who had supported the Palestinians for decades. Because they lived in the region, they would have to live with the consequences of what we produced and could help me gauge whether my approach was viable. As I planned an August trip to visit several countries in the Middle East to update them on our approach, MBS invited me to join a previously scheduled gathering with Mohammed bin Zayed (commonly known as MBZ), the de facto leader of the United Arab Emirates, at his coastal residence near the Red Sea.

During two days of constructive meetings, the leaders described the challenges and opportunities in the region. They presided over two of the largest economies and most powerful militaries in the Middle East, and they had much to say. Just as we had heard during the president's visit to Riyadh, they reiterated that their most critical issue was the threat from Iran. They told me that after Obama signed the Iran deal in 2015, Arabs started learning Chinese because they believed China could be a better future partner than the United States. The UAE had been among the first Muslim nations to join the fight against extremism after the September 11 attacks, and many Emiratis expressed disappointment that the Obama administration did not consult their leadership before finalizing the Iran deal.

When we discussed the Israeli-Palestinian conflict, they were sympathetic to the cause of the Palestinians and expressed a sense of sadness about the situation. They were frustrated that the Palestinian leadership had failed to create a better pathway forward for the Palestinian people. They wanted the problem to be solved, whether through the framework they set forth in the Arab Peace Initiative or through a new proposal.

To my surprise, our conversation shifted to the relationship between Israel and the Arab world more broadly. We had an eye-opening discussion about the history of the region and how the conflict had reached its current state, which was far more nuanced and fair-minded than I had expected. We exchanged ideas about how we could improve the relationship between Israel and the Arab world. In the seventy years since Israel declared independence, only Egypt and Jordan had made peace and established diplomatic relations with it, in a move known in diplomatic circles as "normalization." The remainder of the Arab League, and many other Muslim countries around the globe, had refused to recognize Israel as a sovereign nation. This meant that these countries had no diplomatic relations with Israel, including no official travel, communication, business, or commerce with the Jewish state.

At one point, MBS and MBZ acknowledged that the allies of their countries were the allies of Israel, and that the enemies of their countries were the enemies of Israel. When I asked them point-blank if they would be open to normalizing, they expressed a desire to make progress on the Palestinian issue, but did not express animosity toward Israel. I sought their advice on how to approach the problem, given Abbas's intractability. They implied that if I could get Israel to agree to a credible plan that included a Palestinian state, access to al-Aqsa Mosque, and investments to improve the lives of the Palestinian people, that could change the dynamics. They indicated that if the Palestinians rejected the plan, they would be even more open-minded.

Before we departed, MBZ made one more comment: The United States typically sent three types of people to conduct diplomacy, he said. The first were people who fell asleep in meetings; the second were people who read talking points with no ability to converse; the third were people who came and tried to convince them to do things that were not in their interests. He observed that I was different. I was the first person to come asking questions and really trying to understand their perspectives. He believed we would make peace in the region.

I was honored by MBZ's observation, and I never forgot his words. After our visit, one of my Secret Service agents remarked that he had enjoyed watching a soccer game with an Emirati security guard named Mohammed. I told him that the person he had presumed to be a security guard was actually MBZ. The agent was taken aback—the crown prince had been shockingly humble and low-key, and he had taken a genuine interest in the agent's life, background, and family.

I left Saudi Arabia encouraged. I had become familiar with how diplomatic talks typically play out: the two sides sit opposite from each other, reading from note cards handed to them by their career staffers. But this was different. We tore up the talking points and engaged in a genuine discussion. The meeting clarified why it would be so critical to talk directly to the leaders of these nations: they were the ones with the authority to veer from the established talking points and make the difficult decisions on behalf of their countries.

It also helped that we quickly developed a mutual understanding. In the Arab world, politics is a family business, with members of royal families ruling for generations. As the son-in-law of the president, and a former executive of a family business, I represented something that they found familiar and reassuring. They knew that when I spoke, I did so as an extension of the president in a way that few administration officials could.

From Saudi Arabia we traveled to Qatar, Egypt, Jordan, and Israel, listening to the leaders and gauging their reactions on our approach to a peace deal. Meeting with Emir Tamim bin Hamad Al Thani of Qatar, President Abdel Fattah el-Sisi of Egypt, and King Abdullah II of Jordan reaffirmed my sense that Arab leaders were ready for new ideas on the Israel-Palestinian conflict and that the most important issue to them was expanding Muslim access to the al-Aqsa Mosque. The Arab leaders appreciated Trump's larger-than-life personality, unscripted and unconventional style, toughness on Iran, and the fact that he was more interested in working with them to solve problems than lecturing them, as previous Washington diplomats had done.

15

Great Power Competition

W here is Arabella?" the president asked as Air Force One leveled off at cruising altitude on its way to Asia on November 3, 2017.

I glanced at John Kelly, the White House chief of staff. In planning for this twelve-day swing through China, Japan, South Korea, Vietnam, and the Philippines, Kelly had made clear to Ivanka and me that he didn't want any of us on the trip. He was fine with me chasing Middle East peace, which he dismissed as quixotic, but he didn't want me to play a role in our dealings with China. This put me in an awkward spot, given the relationships I had developed through President Xi's successful visit to Mar-a-Lago.

After Arabella's Chinese poetry recitation at Mar-a-Lago, President Xi had asked if she could accompany her grandfather to China. Kelly had been particularly hostile toward the idea of Arabella joining, claiming that there wasn't enough room on the plane, even though he knew that the president wanted her to come. Ivanka and I didn't press the issue. We didn't want to pick a fight with Kelly. Plus, Ivanka and Secretary Mnuchin were in the last stretch of their push for tax reform. If she skipped the trip, Ivanka could continue to advance the president's top legislative priority.

With Kelly standing nearby, I answered the president's question. "Kelly said Arabella couldn't come, but we recorded a special video of her singing in Chinese, and I have it on an iPad, in case you want to show it to President Xi."

The president didn't mask his disappointment. "Make sure my military aide has the iPad," he said, and then he turned to Kelly. "Why did you stop Arabella from coming?"

"Sir, we didn't have enough room on the plane," Kelly replied.

The president knew this was nonsense. He looked around the spacious cabin of Air Force One and remarked sardonically, "We're on a 747, and are being followed by a support plane, which is another 747. We sent another twenty cargo planes filled with equipment, and you're telling me we don't have room for a six-year-old who's more popular in China than any of us?"

When we arrived in Beijing, the city glistened in the reflection of the pure blue sky. At the command of the Chinese Community Party, the coal-burning power plants and factories had been shut down for the three weeks leading up to our visit to allow the smog and soot to dissipate. As our motorcade entered the Forbidden City, the grounds were eerily empty. The Chinese had cleared the tourists from the vast imperial gardens and ornate palaces, which attract more than fourteen million visitors a year. The opulence, meticulous detail, and symbolism of the fifteenth-century complex was as magnificent as it was foreboding. It seemed like Xi's choice of location was intended to remind us that the Chinese had endured for thousands of years and would continue for thousands more. China was playing the long game.

Xi greeted Trump warmly and proceeded to give him a personally guided tour of the ancient city, followed by a lavish dinner and an opera performance. As Trump had anticipated, Xi was disappointed by the absence of Arabella. The Chinese leader had prepared a performance for her in an ancient pavilion that had not been used in more than a hundred years. Trump immediately called for his military aide to bring the iPad, and he played the video of Arabella singing the popular Chinese folk song "Jasmine Flower." Xi was so delighted by the video that he played it on a massive screen at the state dinner.

The next day, in a meeting on economic policy, Trump and Xi sat across from each other, facing off in a long conference room deep inside the Great Hall of the People, a massive Communist Party building on the western edge of Tiananmen Square that China uses to host foreign delegations. The two leaders were flanked by their top officials as they prepared to continue the discussion that had commenced at Mar-a-Lago seven months prior. In attendance was one key member of the US delegation who hadn't been at the meetings with Xi in Palm Beach: US trade representative Robert Lighthizer. A tough, no-nonsense lawyer from the manufacturing town of Ashtabula, Ohio, he possessed a gravelly voice to match his rust-belt roots. Lighthizer had been a force in Washington trade circles for more than four decades, serving on the Senate Committee on Finance for Bob Dole and as deputy trade representative under Ronald Reagan. Lighthizer was a thorn in the side of his fellow Republicans, advocating for protectionist trade policies that defended the jobs and livelihoods of American workers. In Trump, Lighthizer had finally found a champion of his lifelong cause.

After Lighthizer's confirmation in May, there was initially confusion about who would take the lead on negotiations with the Chinese. Each of the "trade principals"—Secretary Mnuchin, Secretary Wilbur Ross, Ambassador Lighthizer, Gary Cohn, and Peter Navarro—thought he would own the file. At one contentious meeting, Lighthizer said, "There are six trade negotiators in this room and I'm the only one with a law degree and a confirmation." We began discussing a message many of us had been hearing from our contacts in the business community: though Trump was talking tough on China, threatening unprecedented tariffs on tens of billions of dollars in Chinese imports, the Chinese didn't know specifically what he wanted in a deal. We thought it was important to put together a specific list, but Lighthizer pumped the brakes.

"They know what we want, and we're not giving them shit," he said, before providing a brief history of the economic dialogues between the United States and China. Since China entered the World Trade Organization in 2001, the US had conducted a series of technical discussions to address China's unfair trade activities. Over the course of thirteen meetings spanning the Bush and Obama administrations, the US trade deficit with China had more than quadrupled, increasing from $80 billion in 2001 to $375 billion in 2017.[23]

"These guys play us like a fiddle," Lighthizer said. "What we need to do is hit them with tariffs to show that we're not like the other idiots. And we need to stop with all these dialogue meetings, because they are a waste of time. It's their way of tapping us along."

Despite his curmudgeonly disposition, I liked Lighthizer from the outset. He was one of a handful of people who understood and agreed with Trump's pro-America agenda and also had the technical skills and knowledge to implement the changes we needed to make. Trump liked him for the same reasons and asked him to come on the China trip to give a presentation to Xi on the US-China trade relationship.

With the international press in the room, Trump opened the meeting with an effusive statement about the warm relationship that he and Xi had established so swiftly. When the opening statements concluded, the press filed out of the room, and Trump turned the meeting over to his trade negotiator. Lighthizer didn't hold back anything. He detailed a litany of China's trade abuses. They had broken nearly every rule governing modern trade: stealing American intellectual property, manipulating their currency, illegally dumping cheap products into our markets to make our companies uncompetitive, and forcing American companies to hand over their trade secrets as a precondition for entering the Chinese market.[24] Trump wanted Lighthizer to send a strong message, but Lighthizer's presentation surprised even Trump, who was typically respectful and warm to his foreign counterparts, despite his tough negotiating style. Lighthizer was neither. He later explained that the Chinese desire stability above all else. If Trump prevented them from attaining it, he would gain the upper hand in negotiations. After that, the Chinese tried to find friendlier channels to the president.

This reminded me of what I had read in Michael Pillsbury's provocative book *The Hundred-Year Marathon*.[25] During the 1992 presidential campaign, Democratic candidate Bill Clinton had talked tough on China—at one point accusing George H. W. Bush, then president, of cozying up to "the butchers of Beijing." In his confirmation hearings Warren Christopher, Clinton's choice for secretary of state, served notice that he would take a tough line with China. So, in a move that the

Chinese later dubbed "the Clinton coup," China's top diplomats developed warm relations with two of Clinton's top economic advisers, Robert Rubin and Larry Summers, who were more sympathetic to China, and worked through them to persuade Clinton to dial back his antagonism. The Chinese had used the same tactic during the Bush and Obama administrations, and they had largely succeeded.

Also on the schedule was a smaller, restricted meeting between Trump and Xi to discuss North Korea. Flanked by Tillerson, McMaster, and Kelly, Trump walked through a hallway and made his way into the restricted meeting space. As the president's military aide attempted to follow him through the hallway so that he could stand outside the meeting room, a Chinese security official closed a door to prevent him from passing through. This was an alarming diplomatic breach. It's a well-known fact that the military aide is always within earshot of the president, carrying a large leather briefcase known as the "nuclear football." It contains the codes to authorize a nuclear attack when the president is away from a secure operations center such as the White House Situation Room. As the president began his meeting with Xi, his military aide insisted that he needed to be near the president. The Chinese security officer held him back. Kelly caught a glimpse of the scuffle and rushed toward the doorway, grabbing the Chinese officer by the neck and pinning him against the wall.

"You people are rude," he screamed. "The Chinese people are rude! This is terrible! This is not how you treat your guests!"

A protocol official rushed in, realizing the security officer's mistake, and apologized profusely. But Kelly stormed away, boycotting the meeting and leaving a chair next to the president conspicuously empty. He came into the room where the rest of our staff was waiting and regaled us with the story of what had just unfolded. He bragged that he had shown the Chinese that America would no longer be bullied. In the middle of his recounting, a staff member came in and said that the head of Chinese protocol was outside and wanted to apologize personally for the mistake. Kelly paused and smirked. "Tell them I'm busy." He then turned back to us and resumed his bullshitting.

About an hour later, I saw Kelly walking next to the head of Chinese protocol with his arm around him, chummy as could be. In that moment, I finally understood John Kelly. To him, everything was a game of establishing dominance and control. He made people feel small and unimportant to establish the relationship from a place of power. Then, with his position firmly established, he would charm and disarm, leaving people relieved that they were on his good side, but fearful of what would happen if they crossed him. I thought about how he had put the entire White House senior staff on edge when he refused to take phone calls after the president announced him as chief of staff, but when he arrived on campus he had been gregarious and fully engaged. I realized that his Jekyll-and-Hyde routine would work only if the people he bullied allowed it to work. When I got back to Washington, I shared the story with Ivanka, who agreed with my perception of Kelly.

Throughout his time as chief of staff, Kelly was careful not to elevate anyone who had a close relationship with the president. The relationship that Ivanka and I had with Trump made him uneasy because he feared that we might break ranks and circumvent him. We worked hard to assuage these fears, but to no avail. He excluded us from critical policy meetings in the Oval Office relating to our own portfolios and slow-walked, or simply killed, our meeting requests or policy proposals for the president.

Kelly seemed consistently duplicitous. Normally he would shower Ivanka with compliments to her face that she knew were insincere. Then the four-star general would call her staff to his office and berate and intimidate them over trivial procedural issues that his rigid system often created. He would frequently refer to her initiatives like paid family leave and the child tax credit as "Ivanka's pet projects."

Only once did Kelly let his mask fully slip. One day he had just marched out of a contentious meeting in the Oval Office. Ivanka was walking down the main hallway in the West Wing when she passed him. Unaware of his heated state of mind, she said, "Hello, chief."

Kelly shoved her out of the way and stormed by.

She wasn't hurt, and didn't make a big deal about the altercation, but in his rage Kelly had shown his true character. An hour later he came to the second floor and paid a visit to Ivanka's office to offer a meek apology, which she accepted. Ivanka's chief of staff, Julie Radford, had been meeting with Ivanka and heard the apology. It was the first and only time that Ivanka's staff saw Kelly visit their second-floor corner of the West Wing.

When Trump hired Kelly, he asked him to be the West Wing's four-star general, and Kelly took that request way too literally. Trump was an unorthodox president who thrived on hearing from multiple viewpoints when making decisions. After Kelly came in, Trump joked that his office became so quiet and empty that he missed the action. Kelly cared more about controlling the process than producing results. Trump cared way more about results than process. Assistant to the president Chris Liddell had the best analogy for the transition from Priebus and Bannon to Kelly: "We went from a full liquid to a full solid, when we should have had something in between."

When Kelly sidelined me, it initially felt like another setback, but I gradually learned to view it differently. Because I was no longer arbitrating squabbles among staff members or putting out daily firestorms, I was able to focus my energy on my policy priorities, including NAFTA, criminal justice reform, and Middle East peace. During that period I had a conversation with Bob Lighthizer that helped me see my narrowed role as a hidden blessing. Lighthizer and I were discussing our ongoing trade negotiations with Mexico when Senator Lindsey Graham of South Carolina called me with a suggestion on immigration policy. I asked Lighthizer what he thought, and he curtly replied, "I'm not telling you."

When I asked him why not, he explained, "I have my dream job right now. I have been talking about these trade issues for forty years, and there is finally a president who understands them and has the balls to take them on. If I am great at my job, I have a one-in-ten chance of being successful given the difficulty of the task. The moment I start getting into other people's issues, these odds go to one in a million."

*　*　*

On November 1, two days before we departed for Asia, the Secret Service picked me up at 7:30 a.m. We drove to the office of my lawyer, Abbe

Lowell. I was fortunate to have hired Lowell as my attorney. He was a meticulous lawyer who examined every detail and anticipated how a partisan prosecutor might spin the case. There, in Lowell's office, I began my first interview with a team of Mueller investigators and FBI agents. I assured them that I intended to participate, and that I had nothing to hide—no smoke, no fire, no collusion. They grilled me about the structure of the transition team, my relationship with former national security adviser Michael Flynn, and the circumstances around his firing. They asked about my meeting with Ambassador Kislyak and the nature of the Trump team's interactions with the Russians during the transition. As I answered their questions, I remembered being a teenager and hearing all about independent counsel Ken Starr's investigation of the Monica Lewinsky scandal. I couldn't believe I was now at the center of a globally followed investigation. Months later, Mueller's team called me back and grilled me for another six hours. Again, I answered every question they asked.

It sometimes seemed that the whole world was rushing to convict me and other administration officials without evidence. On two separate occasions, when I was about to leave our house in Kalorama and head to the office, I heard the commotion of camera crews as the press set up in front of our house. When I looked out the window and noticed that these were not the standard *Daily Mail* photographers, who had a weird obsession with Ivanka's outfits, but instead actual news teams, I called Lowell and asked what was happening. Both times, he told me that the outlets had been tipped off that I was going to be arrested that day.

The hardest part of the ordeal was knowing the stress it was causing my family, especially my mom. With my father's experience seared in her memory, she would see news articles that claimed I was guilty and would call, worried. Throughout, Ivanka was my rock. She somehow knew exactly when I needed encouragement or just needed her by my side. She held me up while I treaded on paper-thin ice.

At my lowest point, Ivanka was at the top of her game. During the president's Asia trip, she traveled across America to sell tax reform, and her hard work paid off. She visited the congressional districts of the Republicans who were wavering in their support for the plan, and she got each one of them to vote for the bill. She delivered the two senators who no one in the West Wing thought we could get, but whose support was critical to passing the bill: Susan Collins of Maine and Bob Corker of Tennessee, both of whom were at odds with the president. Through many visits to their offices, dinner conversations at our house, and long phone calls, Ivanka became their most trusted confidant at the White House. And because of her skillful diplomacy and delicate negotiation, she got them both to a yes.

Without Ivanka, tax reform probably would not have passed, and it certainly would not have included provisions to double the child tax credit and establish a new incentive for employers to offer paid leave to working parents. These were two of the most successful aspects of the tax reform legislation, and they have given more than forty million American families an average of more than $2,000 in tax savings each year.[26]

As I navigated through all the forces that tried to take me down, I managed to maintain the confidence of the president. There were times, however, when I could feel that even his faith in me was dwindling. Soon after my interview with the Mueller investigators, I was alone with the president in his cabin aboard Air Force One. He asked me how Ivanka and I were holding up. I told him we were weathering things okay, all things considered.

"I want you to know, I wouldn't hold it against you guys if you wanted to return to New York," he said. "Washington has turned out to be a vicious place, and you guys had great lives before this, and they are still waiting for you if you want them."

I wasn't sure if this was his way of suggesting that I should leave, and I didn't want to ask. With as much confidence as I could muster, I told him that the media smears and accusations didn't bother me—I wanted to clear my name, and I was still excited by the progress I could make on several of my files. I was also concerned about what would happen if we left Trump without family in the West Wing, where Kelly and the other self-interested players would try even harder to exert more power and subvert the president's agenda. I kept that fear to myself.

16

Building Capital

N ow is the time," I said.

The president had just asked me one of the most important direct questions that I faced in the Oval Office: Should we move America's embassy in Tel Aviv to Jerusalem?

After the AIPAC speech in 2016, Trump vowed repeatedly to move the embassy—a promise that animated many evangelical and Jewish voters. He had contemplated doing this by executive order on his first day in office, but Mattis and Tillerson warned that it would result in catastrophic violence. Trump held off, and ten months into the administration, Tillerson still had no plan to move the embassy. He thought he had delayed the decision indefinitely, which seemed to be his goal, despite the wishes of the president.

By November, my team and I saw an opportunity. For more than twenty years, supporters of Israel had waited for a president to move the embassy to Jerusalem. In 1995 Congress passed the Jerusalem Embassy Act, which required the president to move the embassy. Buried in the legislation was a provision allowing the president to delay the move if he signed a waiver every six months. Since then, every president had repeatedly signed the waiver. The first time it arrived on Trump's desk in June, Tillerson urged him to sign it. As the waiver came back for a second time in late November, we planned to recommend to the president that he follow through on his campaign pledge, reverse twenty years of broken promises, recognize Jerusalem as the capital of Israel, and move the US embassy.

On November 17, Tillerson, Kelly, McMaster, Friedman, Greenblatt, and I gathered in the Oval Office. The stated purpose of the meeting was to update the president about our ongoing dialogue with the Israelis and Palestinians and to solicit his feedback. After we briefly discussed our progress, Trump asked what we were doing about the embassy. Friedman, Greenblatt, and I had expected this and were prepared for the discussion. McMaster jumped in and explained that the issue was complicated—there was a great deal at stake in the decision, and it was running through the NSC process to ensure all factors and viewpoints were considered before they brought him a recommendation. He and Tillerson were sympathetic to the decades-old logic that moving the embassy would compromise America's position as peace broker and cause the region to explode in violence. They wanted another six-month delay, at which point they would no doubt seek another, and then another, ad infinitum.

Certain that McMaster's NSC process would stall the move, Friedman took the floor. This was an early test of Trump's presidency, he argued. The whole world—from Tehran to Pyongyang—was watching to see if he was going to be a president who kept his promises, or if he was going to fall into the familiar patterns of conventional wisdom.

Trump turned to me and asked if moving the embassy would make it harder to reach a peace deal with the Palestinians.

"In the short run, it will be more difficult," I said, "but in the long run, it will be easier because we will build capital with Israel, while showing world leaders that you aren't constrained by convention. The Middle East is a rough neighborhood, where leaders respect those who do what they say and don't cower under pressure."

Concerned about the peace plan, Trump then asked if we should wait six more months and see if the plan gained traction.

"While that should be considered, this is a relatively quiet period in a normally volatile Middle East," I said. "Anything can happen, and in six months you might not have the same hand to play."

After hearing the opposing view from Tillerson and weighing the potential risks and benefits, the president made his decision: "I want to do it. Run your NSC process, and let's meet on it soon."

I was thrilled by the decision—but I didn't have time to celebrate. My assistant Cassidy Luna stopped by my desk with a note that my lawyer Abbe Lowell had called. He had received a press inquiry about the ongoing Russia investigation. I had already been interviewed by Mueller's prosecutors, submitted testimony to both houses of Congress, and had gone to extraordinary lengths to accommodate requests and to be fully transparent. But after a year of baseless investigations, I felt deflated. I couldn't imagine going through another round of inquiries.

Just then Tillerson strode into my office, huffing in exasperation and seething with anger. He had been blindsided by the meeting with the president. He had assumed it would be solely about the peace plan, and he was woefully underprepared for the embassy discussion. To compound his ire, he had spoken with MBS earlier in the day, and the crown prince had told Tillerson that he was pleased that his team was now working with the White House on a daily basis. Tillerson thought this meant that MBS and I were talking regularly. It infuriated him to think that this was happening without the State Department's knowledge.

In June, less than a week after the president returned from his first overseas trip, Saudi Arabia, the UAE, Bahrain, and Egypt cut off diplomatic relations with Qatar and halted air, sea, and land traffic. They accused Qatar of fomenting terrorism by funding the Muslim Brotherhood and working with Iran to destabilize the region.

Tillerson's sympathies were with the Qataris. Under his leadership, Exxon had invested tens of billions of dollars to build up Qatar's gas industry. He had developed close bonds with the Qatari royal family. He knew that I often spoke to the Emirati leader MBZ and had also established a friendly rapport with MBS. He speculated that the quartet of Saudi Arabia, the UAE, Bahrain, and Egypt was taking advantage of our goodwill to bully the Qataris. He claimed that I was to blame for the Saudi rift with Qatar, which was the exact opposite of the truth. In fact, I saw the hostility as counter to American interests, and when I first learned that the Saudis might take action against Qatar, I tried to convince them to delay the decision. I told them about an encouraging meeting I'd had weeks earlier with the Qatari foreign minister,

Sheikh Mohammed bin Abdulrahman, who made clear that the Qataris wanted to diffuse the mounting tensions. Sheikh Mohammed had strongly denied the allegations against Qatar and promised to immediately rectify any issues if I could bring him a specific list. My efforts to mediate were unsuccessful, so I called the Situation Room and asked to be connected to Tillerson. His chief of staff, Margaret Peterlin, intercepted the call and told me he was busy. Over the next several days, Tillerson's efforts at diplomacy drove all sides further into their corners. He had not only failed to negotiate a solution but made matters worse. From that point on, I felt like I'd lost Tillerson's trust—he seemed to stop turning to me as a confidant and ally and instead viewed me as a dangerous impediment.

"Thanks to your efforts," he snapped, "the Middle East is much worse today than when we got here. The embassy move is going to be a disaster. Between this and your relationship with MBS, you are lighting a match in a dry forest, and the whole Middle East is on fire."

He claimed that MBS would destabilize Saudi Arabia and the entire Middle East. "If you keep maneuvering around me and making these decisions," Tillerson continued, "you might as well go before the Senate for confirmation because you are going to cause a war, and I am not going to be the one to be blamed for it."

Now I was heated. I told him that he was flat-out wrong. While I did fully support MBS's vision to modernize, I hadn't spoken to him in at least three weeks, and anytime my team or I dealt substantively with the Saudis, we included NSC and State Department officials. Knowing that MBS was a 24/7 worker, I asked Cassidy to try and get him on the line. She did, and I said to the crown prince, "Secretary Tillerson told me that you implied that we are coordinating daily. I can't have this. He is the secretary of state, and while I work on the Middle East file, I report to him on foreign policy, and we try hard to stay in sync on everything."

With Tillerson standing there listening to our conversation, MBS replied that Tillerson must have misunderstood: MBS's brother Khalid, who at that time was the Saudi ambassador to the United States, had followed Tillerson's suggestion and had met in the White House that very day with the whole NSC Middle East team and staff from the State Department. MBS questioned whether Tillerson was perhaps unaware that his State Department team had been meeting with Khalid and his team regularly.

As Tillerson listened, his face turned bright red. I hung up the phone and told him to get a better grip on his team's activities before accusing me of things I had not done. Tillerson lashed back. He said the Saudis were the biggest funders of terrorism and predicted that MBS would never make the reforms he promised. "I'm selling Saudi short," he added, using a stock market term to indicate he was betting against them. Then he threw his hands in the air and screamed, "I can't operate like this! I feel like we have four secretaries of state."

Usually I avoid engaging in futile arguments, but this was enough. I could feel my voice rise. "If you actually did your job and implemented what the president asked, we would have only one."

My words had stung. The former oil titan was growing more frazzled and insecure, so I softened my tone. "Look, I know this was a misunderstanding," I said. "You are the secretary of state, and I want to work with you. If you give me any suggestions on how I can change my style or process to make you comfortable, let me know, and I will do it." He stared at me for a few seconds, offered no suggestions, and stormed out.

* * *

On November 27 Kelly directed McMaster to schedule a small meeting with the president in the Situation Room to seek a final decision on the Jerusalem embassy. Kelly invited Vice President Pence, Tillerson, Mattis, McMaster, and Ambassador Friedman. Despite the fact that I was the White House lead on Middle East policy, Kelly refused to let Jason Greenblatt or me join. This was for our own protection, he assured me. He explained that he was concerned that the decision would result in violent attacks on our embassies, and if Americans died as a result, he didn't want me to be blamed for it. Later I learned that he had given Friedman an entirely different reason: he didn't want history to show that three Orthodox Jews, who might be biased in favor of Israel, had participated in such a consequential meeting. This was another one of Kelly's power plays. By design, Friedman was outnumbered in the meeting, but he was more than prepared to respond to those who opposed the move. Based on my private discussions with Trump, I also knew where he stood on the issue.

While the meeting took place on the floor below us in the Situation Room, Greenblatt, Avi, and I waited anxiously in my office. Friedman stopped by afterward and gave us the blow-by-blow. As the president took his customary seat at the head of the table, he set the ground rules for discussion: he wanted to hear from those who disagreed with moving the embassy, and after each spoke, Friedman would provide a rebuttal. Tillerson went first, reading from a loose sheet of paper. In his Texas drawl, he argued that the Trump administration had reestablished a solid working relationship between the United States and Israel. The current US posture on the embassy had been our policy for a long time, and moving it wouldn't dramatically improve our standing with the Israelis, so why do it? From his prepared script, he walked the president through the modern history of Jerusalem, but he made an embarrassing factual blunder when he told the president that the Israelis had controlled Jerusalem since "the war in 1996."

When Tillerson finished, Trump turned to Friedman and asked him to respond. "Mr. Secretary," he said, "I'm willing to concede that in ninety-nine percent of this world, you know the issues one thousand times better than me. I'm just not willing to concede that for Israel. I didn't bring any notes with me; you, on the other hand, were just reading from talking points that someone wrote for you. And whoever wrote them should be fired, because they contain a lot of mistakes. For example, the war was in 1967, not 1996. We can have this debate, if you want it, but it's not going to be fair to you."

Tillerson looked down his nose and over his reading glasses at Friedman, slammed his notebook shut, and stated, "I've said my piece."

Next, the president turned to Mattis, who explained that he couldn't understand why so much focus had been placed on the Jerusalem issue. He had been to Israel on countless occasions, and each time he went, his meetings were in Tel Aviv. Why move the embassy to Jerusalem if Israel's defense department is in Tel Aviv?

The group turned to Friedman for a response. "Where is the Pentagon?" he asked Mattis, who replied that it was in Arlington, Virginia. Then Friedman continued: "Is our capital in Virginia? Based on your logic, it should be. When you go to Israel, you meet at their defense headquarters, which is in Tel Aviv. And when they come here, they meet at the Pentagon, which is in Virginia. But America's Congress, Supreme Court, and the White House are all in Washington, DC. Similarly, their Knesset, Supreme Court, and prime minister's residence are all in Jerusalem." Mattis conceded that point. By the end of the meeting, the president announced that he wanted to go forward with the decision to move the embassy to Jerusalem.

Tillerson spoke up again. Apparently, he really had not said his piece. "I'd like to note for the record that this is a mistake," he said. "I've got American diplomats in Muslim countries from Morocco all the way to Pakistan. And I don't know how I will keep them safe when violence breaks out."

Mattis looked at Tillerson and said, "Look, I was against this as well, but the president has made a decision, and I'll make sure we get enough Marines to your embassies to keep every single diplomat safe."

Trump wanted to be prepared if violence did erupt, and he directed me to speak to all the leaders in the Middle East and report back if there were any problems. While many leaders made clear that they were against the embassy move, they were also committed to working with us to prevent violent backlash in their countries. As we approached December 1, the day the existing waiver was set to expire, my team received an inquiry from the press asking me to confirm a report: "On Monday a major meeting took place in the WH regarding the question of Jerusalem as Israel's capital. POTUS rejected the recommendation of his national security advisers and decided to move the embassy." Someone had leaked. Concerned, Saeb Erekat and Majed Faraj, two of the top Palestinian interlocutors with Washington, paid an emergency visit to see me at the White House.

Erekat warned that the move would be a big mistake. Faraj echoed these concerns, predicting that there would be dire consequences in the region if the president went forward with the move. Critics claimed that the decision would weaken America's relationship with the entire region, including the UAE, Saudi Arabia, Jordan, and Egypt, and suggested that anarchy would break out and the United States would be disqualified from playing any role in regional mediation.

I explained that the decision was with the president and that no decision was final, but that he was a man of his word. I told them that if the president did decide to recognize Jerusalem, we would watch their actions and statements closely to judge the degree to which the United States should maintain all aspects of its relationship with the Palestinian Authority, including our generous annual foreign aid package.

On Saturday afternoon, December 2, I received an email from Kelly: "Jared, Given this is an unsecured email I'll be careful. Just got off the phone with Secretary Mattis who is in Jordan (Aqaba Dialogue) having just left Cairo. Now that he is on the ground he is even stronger in his recommendation on this issue. Secretary Tillerson and the intel community have grown stronger in their recommendations since the POTUS discussion as well. I see POTUS tomorrow and will convey. I will ask DNI and might ask Secretary Mattis to cut his trip short to report and re-engage back here in D.C."

With all presidential decisions, nothing is final until it is signed and released. During the tense forty-eight hours leading up to the embassy announcement, Trump stared down twenty years of convention, troubling intelligence predictions, and the opposition of his own secretaries of state and defense. Foreign leaders called to warn him that the Middle East would burst into flames. And the security concerns weren't his only consideration. As a businessman, he had built a real estate empire by properly identifying leverage and using it to extract concessions out of his negotiating partners. Here, it appeared that we were asking him to give Israel a big gift for free, which cut against his instincts. When a smart businessman friend called and advised Trump that he should get the Israelis to freeze their settlement activity in exchange for the embassy move, the president questioned whether it was a mistake to give something so significant to Israel without asking for something in return. He called me to get my reaction to this idea, and I assured him he was getting something in return.

"This move will build capital with the Israeli people," I said. "If we ever make progress on the peace file, the Israeli leadership will need to make some politically tough compromises, and having them trust you is invaluable. Besides, you promised to move the embassy during your campaign, and you are working hard to keep all of your promises."

17

Papier-Mâché Wall

On December 3, Trump called Ivanka and me up to the Executive Residence. He had just spent the previous two and a half hours at the White House Christmas party for the Secret Service, taking pictures with every single officer and family member who attended. They were protecting him and his family, and he wanted to thank them in return.

Trump had been 100 percent committed to his embassy decision, he told me, but earlier that day, he had received panicky calls from Tillerson and Mattis. These were two men who liked to project a tough and calm swagger in the face of danger and difficulty, so he interpreted their fears as genuine and well-founded. Both painted an apocalyptic scenario: this decision would plunge the region into chaos, violence, and extremism. The intel was coming in fast, and it was disturbing.

"Do you still feel confident this is the right move?" Trump asked.

I could tell he was still committed, but he was aware of the risks. I also knew that he was gauging the strength of my conviction, as he often did with his advisers when wrestling with important decisions.

"Yes, you're making the right decision," I said, and updated him on my conversations with the Arab leaders, who also wanted to avoid an eruption of violence in their countries.

On December 5, the day before the planned announcement, Trump called Bibi and told him the news. Bibi said he'd support the move, if that's what the president wanted to do, but he didn't sound overly enthused.

Thinking that Bibi must not have understood, Trump repeated that he was going to recognize Jerusalem as the capital of Israel and move the US embassy. This would send an unprecedented signal to the world that the United States stood behind Israel's sovereignty in Jerusalem. He added that when the time was right, the United States would expect Israel to come to the table and make a peace deal—something Bibi hadn't done before.

Again, Bibi responded with less-than-expected enthusiasm. Thrown off by the lukewarm response, Trump began to second-guess his decision. As he continued his conversation with Bibi, he wondered aloud why he was taking this risk if the Israeli prime minister didn't think it was that important.

Trump's voice hardened into a stern tone: "Bibi, I think you are the problem."

If Bibi was taken aback by this comment, he didn't show it. He coolly countered, insisting that he was part of the solution.

After the two leaders hung up, I could tell that Trump was frustrated. Anxiety about what could go wrong weighed on everyone. Many other foreign leaders called the White House to speak to the president. Trump knew what they were calling about and told his team to schedule his return calls after he announced the decision. This way, he wouldn't have to listen to the same arguments again and then turn down the callers' requests. Instead, when he called them back, he could move past the decision and on to the next set of priorities.

The next day, everything proceeded as planned. The president signed a memo to notify a handful of senior administration officials that he intended to sign a presidential proclamation officially recognizing Jerusalem as the capital of Israel and stating his intent to move the embassy immediately.

At one o'clock in the afternoon on December 6, the president stood behind the podium in the Diplomatic Reception Room. "After more than two decades of waivers," he said, "we are no closer to a lasting peace agreement between Israel and the Palestinians. It would be folly to assume that repeating the exact same formula would now produce a different or better result. Therefore, I have determined that it is time to officially recognize Jerusalem as the capital of Israel." He made it clear that he was not taking a position on the contested borders, he urged people to respond with calmness, and he asked leaders to "join us in the noble quest for lasting peace." Trump signed the proclamation, and everyone held their breath to see what would happen next.

The words of one Middle East leader I had spoken to the day before echoed in my head: "I'm not going to tell you to do it, or not to do it, but if you do it, you will find out who your friends are." Immediately after the announcement, NSC senior director Michael Bell began to convene interagency meetings twice a day to monitor developments. As it turned out, the reaction across the Muslim world was strikingly mild compared to the forecasts. The protests in the region remained peaceful, though the West Bank and Gaza were notable exceptions. Within forty-eight hours the crowds had dissipated without major violence. In fact, one of our key partners sent signals that bolstered our position. "Despite Furor over Jerusalem Move, Saudis Seen On Board with U.S. Peace Efforts," declared a Reuters headline, in a story that described the kingdom's intention to continue working with us on a peace plan. "Initial Mideast Violence from New U.S. Policy on Israel Is Limited," reported the *Wall Street Journal.*

The Palestinians were growing nervous. Perceiving that their influence on the world stage was waning, President Abbas turned to the strategy that the Palestinians had employed since Israel's founding in 1948: using the United Nations as a forum to confront Israel and the United States. He convinced Egypt, one of the nonpermanent members of the UN Security Council, to draft a resolution condemning the recognition of Jerusalem as the capital of Israel. When Tillerson learned of the UN resolution, his instinct was to offer a concession. He approached me at a White House holiday reception and explained that the situation was dire.

"We should acknowledge the Palestinian claim to East Jerusalem to give them a bone and allow them to save face," he said. "Otherwise, they will walk away from the table and not come back for a generation."

"If they don't come back, they don't come back," I said. "If you respond to their threat by offering a concession, that sets a terrible precedent. For decades our diplomats have accepted a dynamic where the Palestinians say 'Jump,' and US diplomats ask, 'How high?'"

Tillerson rolled his eyes. He wasn't convinced, but I didn't belabor the issue. Trump had told me privately that he wanted to fire the secretary by the end of the year; he thought that Tillerson was "a below-average negotiator" and was frustrated that he kept trying to promote the conventional Washington establishment foreign policy agenda that rebuked Trump's America First philosophy.

I called the ambassador to the United Nations, Nikki Haley, who I thought was an ally of the president's agenda. I knew she supported the decision to move the embassy and was unafraid to go around Tillerson.

"Nikki, we have to do something," I said. "If I was a private citizen, and all of these countries who receive foreign aid voted to condemn America at the UN, I would think we were run by a bunch of schmucks. But we're the ones in charge now, and if we allow this to happen, then we are schmucks."

Haley wholeheartedly agreed. She called Trump directly and explained that she'd like to announce that if countries voted against us, we would take away their foreign aid. Trump loved the idea of using America's leverage to defend our interests and stop allowing our supposed allies that received billions of dollars in US funding to bully the president of the United States.

Haley and I split up the list of permanent and nonpermanent members of the UN Security Council and our key allies in the General Assembly and began making calls, asking countries to abstain from the vote. The scene had a feeling of déjà vu after our fight on the anti-Israel Resolution 2334 during the presidential transition, except this time I knew the players, and I had their numbers in my phone.

The Palestinians are a force within the hallways of the United Nations, so when the General Assembly voted on the resolution, I considered the forty-four countries that either abstained or voted against it to be a positive indicator that we were forging strategic partnerships, using our leverage, and slowly shifting the paradigm in the Middle East.

Sadly, President Abbas walked away from the US-led peace process and stoked outrage. "May God demolish your house," Abbas fumed against Trump.[27] "We will not accept America as a sole mediator between us and Israel, after what they have done."

I relayed a message to Abbas through our intermediary: "We understand you need to look strong while people in your streets protest, but we will be ready to engage when you are. If you want to work with us, work with us. If not, we're not going to chase after you. We intend to move our plan forward with or without you."

Foreign policy experts had always assumed that Abbas could manipulate the sentiment on the Arab streets. But in the aftermath of the embassy announcement, it became evident that this was not the case. It was a consequential revelation for Arab leaders, who were always trying to judge the true sentiment of the region. The limited violence on the street and the mild reaction of Arab leaders proved that we could take calculated risks and question prevailing assumptions. To the growing number of Arab leaders watching, Abbas's counterproductive reaction demonstrated that the current Palestinian leadership was incapable of delivering a better life for the Palestinian people. I had previously told my team, "Our mission is to try to break through this previously impenetrable barrier," referring to the conflict between Israel and the Arab world. "Let's hit it with everything we have and find out whether the wall is made of concrete or papier-mâché." The successful embassy move confirmed that a breakthrough for peace might be more possible than conventional wisdom assumed.

18

Fighting for the Forgotten

When my father was in prison, he found an unlikely exercise partner. The young man, who I will call "Sean," was serving a drug-related sentence. He told my father that he had become a drug dealer because his dad was a drug dealer. It was all he knew—and because of it, he wound up in trouble with the law. Sean was a bright guy who scored an 1140 on his SAT, and my father believed he deserved an opportunity to redeem himself. From the prison pay phone, he called his friend Monsignor Robert Sheeran, the president of Seton Hall University. The monsignor agreed to give Sean a scholarship to Seton Hall after he was released from prison. Sean graduated from the university with a 4.0. His story was an eye-opening example of the tremendous human potential that is often wasted through our prison systems. The incarcerated shouldn't have to meet a billionaire behind bars to earn a second chance.

I hadn't exactly planned to share Sean's story with the president at a roundtable on criminal justice reform on January 11, 2018, but when US attorney general Jeff Sessions asserted that all people in prison were irredeemable, I couldn't help but think of Sean.

Prior to the meeting, I had briefed Trump on my criminal justice reform effort, and he expressed skepticism about the subject. He was a law-and-order president, and the topic was new to him. I explained that numerous conservative and evangelical leaders supported reforms; this wasn't just a "liberal Jared" cause, as he liked to joke. He asked Sarah

Sanders, a bona fide conservative, what she thought. This was a great issue, she told him. Her father, Mike Huckabee, had enacted similar reforms when he was governor of the deeply red state of Arkansas, and the reforms were effective and popular. That was not the answer Trump expected to hear.

In the Roosevelt Room, the president greeted a group of political dignitaries: in addition to Attorney General Sessions, we were joined by many conservative advocates of criminal justice reform—including Kentucky governor Matt Bevin, Kansas governor Sam Brownback, chairman of the American Conservative Union Matt Schlapp, Koch Industries executive Mark Holden, and Texas Public Policy Foundation president Brooke Rollins. At the time, I was in the process of recruiting Rollins to lead our criminal justice reform efforts in the White House. She had helped enact successful reform in Texas, but was reluctant to come to Washington since she knew the move would be hard on her four school-age children. I had to make a hard sell, joking that she would never work more hours, make less money, and be less appreciated in her life. But, like me, Rollins saw the sacrifice as an opportunity to help people get a real shot at the American dream.

At the top of the meeting, the president gave brief remarks, making clear that he would be tough on crime but was looking for a way to provide former inmates with "a ladder of opportunity to the future." After Trump dismissed the press from the room, I kicked off the discussion with a summary of the current state of the prison system in America. I explained that the United States made up less than 5 percent of the world population, but our prison systems held nearly 25 percent of the world's prisoners.[28] The federal prison population was growing at an alarming rate, having increased nearly 800 percent since 1980, with much of the trend driven by the incarceration of nonviolent drug offenders from low-income communities.[29] Nearly 75 percent of released offenders went on to commit a new crime, and 25 percent ended up back in prison within eight years.[30]

I had become convinced that we could do better. Inmates often leave prison with mental-health or substance-abuse issues that are never properly treated. To compound these challenges, they often lack money, family support, and the skills they need to live stable lives. Their criminal records make it even harder for them to gain employment and overcome the odds stacked against them. When they find jobs and stay employed, however, they are much less likely to commit future crimes. Some of America's strongest red-state governors, including Rick Perry, Sam Brownback, Nathan Deal, and Mike Pence, had reformed their prisons to provide more effective treatment and job training. In each case, the reforms reduced recidivism rates, improved safety, and saved taxpayer dollars.

We went around the room, and each participant offered supporting facts and stories in favor of reform, save for Sessions, who had opposed criminal justice reform for years. Sessions made an impassioned argument for imposing the harshest sentences.

As the meeting drew to a close, my longtime friend Reed Cordish, one of the president's senior staffers and a former real estate developer in Baltimore, turned to Trump. "When you ran for president, you promised to fight for the forgotten men and women of this country. Well, no one is more forgotten or underrepresented than the men and women in prisons."

That registered with the president. "I wasn't expecting to like this," said Trump. "But this makes sense. If we don't help these people, of course they will go back and commit future crimes. Let's do it—this is the right thing to do. But get with Jeff to make sure it's not soft on crime."

The Roosevelt Room meeting was an inflection point for my criminal justice reform effort. We had the president's approval to move forward, and the press coverage was terrific. "Trump Hosts Discussion on Prison Reform, Reducing Recidivism," CBS reported. "Trump Tackles Prison Reform: 'We Can Help Break This Vicious Cycle,'" wrote *USA Today*. Even commentators who didn't like the president admitted that this was a surprising and positive step.

* * *

I had been quietly working on the issue since shortly after inauguration, when I received a call from Pat Nolan, a former Republican leader in the California State Assembly who served a two-year term for charges of corruption. While in prison, Nolan met evangelist Chuck Colson and decided that he wanted to devote his life to helping inmates live with a new sense of purpose.

During my father's imprisonment in 2005, a friend suggested that we meet Nolan. So my mom and I flew to Washington, DC, and met him in a conference room at the airport. Nolan greeted us warmly and asked if he could begin our meeting in prayer. As he prayed, he recounted a story in the Old Testament about Joseph, who was sold into slavery by his own brothers, but whom the Lord lifted out of bondage and placed at Pharaoh's right hand to help guide Egypt through a famine and save his family from starvation. What had been intended for Joseph's evil, the Lord had used for his good. Nolan's prayer filled me with hope when I needed it the most.

A decade later I was sitting at my desk just down the hall from the Oval Office, with Nolan on the other end of the line. He asked me to help make long-overdue reforms to the federal criminal justice system that had failed to pass during the Obama administration.

Shortly after my conversation with Nolan, Senator Chuck Grassley, the powerful chairman of the Senate Committee on the Judiciary, summoned me to his office for a meeting on the issue. The eighty-four-year-old Grassley had the energy of someone half his age. At first his manner of speaking baffled me. I thought he was yelling at me until I realized that's just how he talks. Beloved in his home state of Iowa, Grassley had famously gone twenty-seven consecutive years without missing a single vote. At least one of those votes, however, Grassley wished he could take back. In 1993 he had voted for Bill Clinton's crime bill. One of its lead drafters in the Senate had been Grassley's colleague from Delaware, Senator Joe Biden. The law had led to the mass incarceration of Black men for nonviolent drug offenses. Grassley was determined to rectify that injustice.

The Iowa senator introduced me to two of his Senate colleagues: Mike Lee, a constitutional conservative from Utah, and Dick Durbin, the Democratic whip and a fixture in Illinois politics for more than three decades. Grassley explained that for several years they had been attempting to pass the Sentencing Reform and Corrections Act. They had come close in 2016, advancing the legislation out of Grassley's judiciary committee, but Senate majority leader Mitch McConnell had held it up. When I asked why McConnell had blocked the legislation from coming to a vote, they looked at each other knowingly. Jeff Sessions, then a senator from Alabama, had spread misinformation about the bill and accused its Republican supporters of being "soft on crime." Not wanting to split the Republican conference, McConnell refused to advance the bill. Grassley, Lee, and Durbin pitched me on what they called a "simple" request: convince the president to tell McConnell to bring the bill to the floor. But nothing in Washington is ever that simple.

Back at the White House, I scheduled a meeting with my team to get a full download on the status of the bill and which provisions had caused it to stall. Typically the White House Domestic Policy Council would lead this analysis, but after one meeting, in which they explained that we could never pass criminal justice reform, it was clear that its leadership was closely aligned with Sessions.

Instead I found a smart, friendly colleague in the staff secretary's office named Nick Butterfield and asked him to research Grassley's legislation and help me understand Sessions's objections. Butterfield explained that the prison reform portion of the legislation had broad support. It included provisions to reduce recidivism rates by better matching prisoners with job training, drug rehabilitation, and faith-based programs. The sentencing reform section, however, contained several controversial provisions. It shortened sentences, gave judges more discretion in sentencing, and expanded eligibility for early release. The bill divided Senate Republicans because some believed that sentencing reforms would release violent criminals back into the community. Whether or not these concerns were accurate or fair, we would have to address them.

We spent several months going through the provisions, consulting with advocates and legislators, and developing a new plan that we thought could garner enough votes. After having a bipartisan discussion with lawmakers in September of 2017, I huddled with Ja'Ron Smith, a member of the White House Domestic Policy Council who supported reforms and quickly became crucial to our efforts. A graduate of Howard University and longtime Capitol Hill staffer who had worked for Senator Tim Scott of South Carolina, Ja'Ron had strong relationships on the Hill. I asked him to be my point person on the legislative negotiations.

To help build public support for the effort, I called Sam Feist, CNN's Washington Bureau chief, and asked if the network could take a short break from its breathless Russia coverage to pay attention to criminal justice reform.

"I know you are going to think I'm crazy for suggesting this," said Feist, "but would you be open to meeting Van Jones? He hates Trump and has been a vicious critic of the administration, but he is a super guy, and no one cares more about this issue than he does. You should speak to him."

I thought he was kidding. CNN commentator Van Jones was a former high-ranking official in the Obama White House and a vocal opponent of Trump. It had been reported that I told CNN president Jeff Zucker that Van Jones should be fired, and in this instance the reporting happened to be true. But Feist insisted that I talk to him, and I was happy to try anything.

Van Jones and I had a surprisingly constructive conversation the next day. I was frank about the road ahead: I was just one person, and the president wasn't on board yet, but I was preparing to present him with the facts and try to get his buy-in. Jones seemed to appreciate my honesty and passion for the issue. He told me that he'd get killed by his liberal friends and supporters for working with us, but if I thought there was a real chance at success, then he was willing to take the arrows. "Count me in on the team," he told me before we concluded our first of many calls. I was grateful for his offer and knew I needed his help. I needed to overcome a trust deficit between Democrats and the Trump White House.

One of the lead Democratic negotiators in the House of Representatives was Hakeem Jeffries, an influential member of the Congressional Black Caucus. I had briefly known Jeffries from my time developing properties in Brooklyn, but we hadn't yet formed a strong working relationship. Jones agreed to speak with Jeffries and other Democrats and to advise me on their internal dynamics so that I could address their fears and anticipate any problems. He also connected me to Jessica Jackson, the cofounder of #cut50, his criminal justice reform organization. Jackson became instrumental to our bipartisan effort.

After Trump gave us the green light to work on criminal justice reform, I began aggressively engaging with lawmakers on Capitol Hill to draft legislation. Just as I started to make progress with conservatives, Sessions sent a formal letter to Grassley condemning his bill and simultaneously released it to the press: "The legislation would reduce sentences for a highly dangerous cohort of criminals. . . . If passed in its current form, the legislation would be a grave error." Any official expression of the administration's views on legislation typically runs through an extensive interagency review process. But Sessions had ignored the process and sent his letter directly to Grassley.

A famously colorful personality on Twitter, Grassley responded with gusto: "Incensed by Sessions letter An attempt to undermine Grassley/Durbin/Lee BIPARTISAN criminal justice reform This bill deserves thoughtful consideration b4 my cmte. AGs execute laws CONGRESS WRITES THEM!"

19

Top Secret

Kelly summoned me to his office, closed the door behind us, and delivered some bad news. "I need to downgrade your security clearance," he said. "It'll just be temporary, and I will make sure that it doesn't impact your ability to do your job."

The date was Monday, February 19, 2018. The president had recently returned from a successful trip to the World Economic Forum in Davos, Switzerland, where he declared that "America is open for business." Many of the corporate leaders at Davos were beginning to see that Trump's policies were making America's economy the envy of the world. These same leaders were initially skeptical of Trump, and had even heralded China's President Xi as the leader of the global economy. Following the president's speech, several Fortune 100 CEOs publicly praised the administration— something they wouldn't have felt comfortable doing before.

Despite the investigations and the internal battles during the first year of the administration, the White House was beginning to rack up policy victories. Trump had enacted some of the largest tax cuts in American history. He had slashed unnecessary and burdensome regulations on businesses. America's economy was adding a record number of new jobs and expanding opportunities for Black, Hispanic, and Asian Americans. And Trump had appointed a record number of new federal judges, including a Supreme Court justice. Even *Wall Street Journal* columnist Peggy Noonan, normally a sharp critic of Trump's style, wrote a column that was her version of nice: "He's crazy . . . and it's kind of working."

On a personal level, the negative news had mostly subsided. After adjusting my approach, I had found a way to operate within Kelly's system, and was making progress on my files. But the momentum evaporated on Tuesday, February 6, 2018, when the *Daily Mail* broke the first installment in a story that would dominate the White House for the next ten days. Staff secretary Rob Porter, a clean-cut Harvard alumnus and Rhodes Scholar, had allegedly abused his former wife. Porter managed all paper flow to the president. Kelly had brought Porter into his inner circle, and given him expansive authority to run the policy processes across the federal government.

During that Friday's senior staff meeting, Kelly claimed that he'd found out about the Porter allegations on Wednesday—at the same time as everyone else—and that he had immediately demanded Porter's resignation. It was a perplexing thing to say, and it left the staff dumbfounded. Kelly had issued a strong statement in defense of Porter on Tuesday, so he absolutely had known about the allegations prior to Wednesday. Everyone knew this. What purpose, then, was there in telling such an obvious and blatant lie to the staff? Whatever the reason, his deceitfulness caused a number of the senior staff to wonder what he'd known and when he'd learned it. Many staff members felt betrayed by Kelly's lies and were angry that he had failed to act on the Porter news earlier. His conflicting statements heightened the media scrutiny and frustrated the president.

As additional details emerged, it seemed evident that Kelly and White House counsel Don McGahn had known about the abuse allegations for several months, when the FBI had flagged them for the White House security office during Porter's security review. Porter had been operating under an interim Top Secret clearance—a temporary clearance granted to high-ranking officials after an initial background screening so they can perform their duties while the FBI conducts a more extensive background check. At the time, the FBI process was so backlogged that the background checks were taking more than a year to complete. After the *Daily Mail* story broke, reporters asked predictable questions. When exactly had Kelly and McGahn learned about the accusations? Why had they allowed Porter to continue serving in such a sensitive position, overriding the FBI's concerns about his clearance request? Speculation mounted about Kelly's job security, and a chorus of people began calling for his resignation. The *New York Times* posted a story: "Kelly Says He's Willing to Resign as Abuse Scandal Roils White House." Kelly vigorously denied this claim, but he sank to his lowest point as chief of staff.

What happened next was textbook 101 on how to avert blame in a scandal. On Friday, February 16, ten days after the scandal broke, the White House press office started getting a flood of calls from the media asking for comment on Kelly's new security clearance policy and whether my clearance was going to be downgraded. This caught me by surprise. I wasn't aware that Kelly had issued a new policy, and no one had told me that my clearance was being downgraded. I soon learned that earlier that day, Kelly had sent a private memorandum to McGahn, directing him to consider a number of changes to "improve" the security clearance process, including a measure to discontinue interim Top Secret and SCI-level clearances—a higher-level clearance that granted access to "sensitive compartmented information"—for individuals whose background checks had been pending since June 1, 2017. At the time, the FBI process was so backlogged that clearances were taking more than a year to complete.[31] I was among the more than one hundred White House staff members who had been granted an interim clearance while the FBI conducted its more extensive background check.

Within hours, Kelly's "private" memo was circulating publicly, and my brief period out of the spotlight was over. Even though I had learned a lot about navigating the media storms induced by surprise legal developments, I felt uneasy. As I read the stories, it struck me that Kelly was attempting to shift attention away from his own poor management of the scandal and redirect it toward me. Nearly every story pushed a narrative that Kelly had taken charge of the situation and, with a fair and impartial hand, was fixing the broken security clearance process. Though Kelly's new policy affected many White House staffers, these stories inevitably carried my photo at the top and speculated about whether I would be allowed to keep my Top Secret clearance. Bannon would have been proud of the way Kelly used me as a foil.

When Kelly finally called me into his office on February 19 and announced that he was yanking my clearance, I protested. "General, I've done nothing wrong," I said. "I'm getting clobbered in the press for something that has nothing to do with me. Why am I being penalized because you created an arbitrary policy with an arbitrary date that is only warranted because the FBI is being slow?"

"I inherited a mess here in the White House, and we can't have another Rob Porter situation," Kelly insisted.

"Has the FBI raised any concerns or red flags about my clearance, like they did on Rob's?" I asked.

"No," Kelly replied.

"Then why are you doing this to me?"

Kelly stared at me blankly, seeming to suggest that the facts of my case didn't matter.

Kelly's downgrading of my security clearance was humiliating, but I wouldn't let his power play defeat me. Ivanka reminded me about the advice we had received from one of the best politicians we'd met: Japan's longest-serving prime minister, Shinzō Abe. At a dinner in Japan, he told Ivanka: On your worst days, wear your best suit, walk with your head held high, show no weakness, and project that nothing has changed.

In the wake of the clearance downgrade, I followed Abe's advice and decided to work even harder. Since I was no longer pulled into classified meetings, I had hours of additional time. I started to realize that I could get more done by not being involved in every decision. When the president asked me about bits of intel I hadn't seen, I tried to steer him to his national security adviser, H.R. McMaster. To borrow a concept from philosopher Isaiah Berlin's popular essay "The Hedgehog and the Fox," I became less of a fox who knew many things relatively well, and more of a hedgehog who knew a few things very well. With my newfound time, I drilled down into my three policy portfolios: criminal justice reform, Middle East peace, and America's strained relationship with Mexico.

* * *

Back in April of 2017, Trump had instructed me and several others to prepare documents to terminate the $1.3 trillion North American Free Trade Agreement. I knew Trump was impatient to fix America's broken trade policies, but I wondered whether he really wanted to take this massive gamble, or if he was trying to motivate his negotiating team to work faster. Gary Cohn and I advised the president that it would be premature to terminate NAFTA; our talks with the Mexicans had been productive. Plus, we had no replacement plan ready. The economic costs of simply tearing up the deal could be catastrophic. Trump hadn't made a final decision, but he wanted us to draft an executive order right away. White House trade adviser Peter Navarro firmly believed that tearing up NAFTA would be a political win and pounced on the president's directive to prepare documents to terminate the deal. Not coincidentally, the president's request leaked to *Politico,* putting public pressure on Trump to follow through.

Mexican foreign secretary Luis Videgaray saw the article and called me. He warned that this was a fight in which Mexico would get killed, but the United States would lose a leg and an eye.

I told him I was working to find a solution.

"We're not moving Mexico and we're not moving the United States, so I guess we have to figure this out," I joked. Luis wasn't the only person caught off guard by the *Politico* story. Secretary Mnuchin, Secretary of Commerce Wilbur Ross, and Secretary of Agriculture Sonny Perdue were equally alarmed. They wanted to present the president with a set of options that would curb the offshoring of American manufacturing jobs to Mexico, but came short of terminating NAFTA. After the report leaked, Perdue and Cohn swung by my office on the way to a weekly trade meeting with White House staff. Perdue made an impassioned case against withdrawing from the trade deal. He explained that in 2016 alone, American farmers had exported nearly $40 billion in goods to Mexico and Canada. He held up an oversize map of the United States that showed all the counties across America that would be adversely affected by terminating NAFTA. Many farmers had been operating on razor-thin profit margins during the Obama administration, and any sudden market disruption could put them out of business.

The president deserved to hear from Perdue directly, so I told him to skip the trade meeting and brought him to the Oval Office, where the secretary made the same presentation. Trump found Perdue's chart so persuasive that he later had it blown up into a poster that he kept in his private dining room for the remainder of his time in office. The president wanted to protect farmers, but he wasn't willing to let Canada and Mexico string him along like a normal politician. The leak to *Politico* had backed him into a corner—anything less than a withdrawal order would appear weak and indecisive.

Sensing that Trump was looking for a solution, I suggested a short-term plan of action: "What if I get President Peña Nieto and Prime Minister Trudeau to call right now and ask you not to cancel NAFTA, and then you can put out a statement that says you will give them time to negotiate. They will feel committed to following through in good faith if you show them good faith by not terminating." Trump agreed.

I phoned my counterparts in each government and explained the dynamics, and within fifteen minutes both Peña Nieto and Trudeau called the president and urged him not to terminate, promising to speed up the negotiations if he held off. The immediate crisis abated, I had begun walking back to my office when it struck me that the regular trade meeting probably was still going on. I opened the door to the Roosevelt Room and glanced around a full room of senior staff and cabinet officials. There wasn't an empty seat.

"Is this still the NAFTA meeting?" I asked. "We just spoke to the president. The withdrawal is off for now. He is giving us a short window to make a deal."

Nearly a year after that discussion, Trump's relationship with Peña Nieto had stabilized, and Luis and I were still working behind the scenes to improve US-Mexico relations. We had coordinated policy between our two governments on more than a dozen shared interests, including addressing illegal immigration and curbing the flow of illegal guns, drugs, and cash across the border. We believed it was time to bring the two heads of state together for their first meeting at the White House.

Before we could announce the visit, which would be a politically combustible event for both leaders, we planned for them to touch gloves by phone. Luis and I arranged a call on February 20, 2018. The conversation quickly became heated when Peña Nieto raised the issue of what the two leaders were going to say about who was going to pay for the border wall—the same question that had caused such controversy during Trump's visit to Mexico in the 2016 campaign. Peña Nieto wanted Trump to announce that he had dropped his demand for Mexico to pay for the wall. Trump would not agree to this, but he offered to say that they were still working through the payment issue, and that Mexico hadn't agreed to anything. This was not enough for Peña Nieto, and after a fifty-minute phone call, his White House visit was canceled for a second time.

The lack of chemistry between the two leaders now threatened to kill our effort to renegotiate NAFTA. As a last resort, I decided that I needed to sit with Peña Nieto face-to-face and explain the dire circumstances. If he didn't start to negotiate in good faith, the US-Mexico relationship would head over a cliff.

I flew to Mexico on March 7, 2018, with State Department director for the Western Hemisphere Kim Breier and several others. Luis had tipped me off to the fact that influential members of Peña Nieto's team were getting reticent about moving forward with trade negotiations. The Mexican president was in the final year of his term, with an election coming up in July. They didn't see the benefit of taking the political risk. If they dragged discussions out for a few more months, it wouldn't be their problem anymore, and everyone could move on. As Peña Nieto contemplated whether to negotiate or run out the clock, I was prepared to deliver a simple message: There is no comfortable pathway here. If they wanted a good outcome, they needed to trust me and make a deal soon.

Given my troubles in Washington, Peña Nieto might have brushed me off, so I was surprised by his warm reception. Before I could say anything, he thanked me for the efforts to strengthen the relationship between our two countries.

The discussions between our teams culminated with a small three-hour lunch with Peña Nieto at Los Piños, the official residence of the

Mexican president. We had a friendly but intense discussion, and I made my argument for embarking on serious negotiations.

"Doing nothing is a decision," I said. "Why can't we get this done now? If we don't try, President Trump will likely tear up the deal, which will hurt both of our economies."

Mexico's secretary of economy, Ildefonso Guajardo Villarreal, then asked about our discussions with Canada.

"We don't think they want to make a deal now," I explained, "and they are holding out on too many issues. Mexico does $500 billion worth of trade with the United States annually, and only $30 billion with Canada. It makes financial sense for Mexico to strike a deal with us first. Both sides respect and understand each other, and if we both stretch a bit further, we can reach an agreement—let's finish it. Then we will offer Canada the ability to make limited modifications and join. It's not the most elegant way to do this, but it's the only one I can see, given the playing field." If the United States and Mexico announced an agreement to move forward with or without Canada, it would place significant pressure on Trudeau, who was publicly threatening to abandon trade talks.

Peña Nieto looked at me warmly, nodded his head, and motioned to a server to bring a flight of tequila shots. "It's five o'clock somewhere," he said. He made a toast, and we collectively knocked back the reposado.

* * *

When I returned to Washington, I was anxious for the FBI to finish my clearance—but I increasingly felt like I was trapped in a Franz Kafka novel, the victim of a bizarre, opaque, and irrational bureaucracy. I didn't know what, if any, concerns existed. I had no insight into Kelly's process. I had no judge, jury, or forum for due process.

When I made a rare visit to Kelly's office to see if he had an update, he said that the FBI had completed its process, and my file was now with the head of the White House Personnel Security Office, Carl Kline, a respected career professional with more than forty years in the military and the civil service. Kline was the person who had come to my office to have me sign paperwork when my clearance was downgraded. As Kline handed me the documents, he said, "Look, Jared, I'll be honest. I don't see any problems. There's media speculation about a lot of things, you've been accused of a lot of things, but there's nothing we have seen that makes me think that I won't be reading you back in very soon to your Top Secret/SCI clearance."

Several weeks later, Kelly called me into his office.

"I have good news for you," he said. "Your security file has been adjudicated positively. It was reviewed by two people at the lower levels, and then elevated to Carl. Without any influence, Carl said that you are eligible for the Top Secret clearance."

I was relieved. Kelly told me that I should receive an email within the week to reinstate my clearance. Then ten days went by without an email. Kelly eventually called me to his office and said he was concerned about how it would look if I got my SCI clearance back.

"If there's an open issue or any security concerns, they can interview me further," I said. "I know my personal life. I have nothing I'm worried about. Is there something more I should do or someone I should talk to?"

He swiftly dismissed my offer. "No, there is no need for that. Let me think about how to manage this."

Fed up, I pushed back: "You told me that I got my Top Secret clearance back through the normal course. Is that correct?"

"Yes," Kelly admitted.

"You said they would turn it back on, so if you are telling me you don't have visibility into the timing of my SCI or whatever is holding it up, then why don't we just proceed with the Top Secret clearance?"

He agreed. On May 23, the security office was scheduled to reinstate my clearance. But Kelly was still playing games. He had previously said that White House press secretary Sarah Sanders should personally give the press background on my clearance update, but that morning he told her to instead travel with the president to an event on Long Island, New York. When she said she didn't need to be there, Kelly ordered, "You're going, and you can't talk to the press about Jared's clearance."

Meanwhile, my lawyer Abbe Lowell released an off-the-record statement to the press, as Kelly had originally suggested he do after the press received the background from Sarah. The story broke in the *New York Times*: "Jared Kushner Gets Security Clearance, Ending Swirl of Questions over Delay." Later, Acting Director of National Intelligence Ric Grenell, CIA Director Gina Haspel, and National Security Advisor Robert O'Brien told me that there were no concerns or security risks with my file.

Having my clearance restored was an even bigger moment than I had expected. It became a public vindication against the false allegations that I had colluded with Russia, clearing up a narrative that never should have existed in the first place. As I emerged from the unfortunate series of events, I thought about what I could do differently to avoid being in the crosshairs of investigations moving forward. I realized that the best way to shrink the target on my back was to achieve results. From that point forward, my goal was to avoid internal battles, stick to my files, and focus on policy changes that would leave a lasting impact.

20

The Cost of Peace

I t's time to fire Rex—I'm ready to make the change," the president told Ivanka over the phone one morning in early March of 2018.

We had known it was only a matter of time. Tillerson had been on the ropes for a while. Trump had nearly fired him a week before, but several staff members persuaded him to wait because the news cycle was unusually positive. They didn't want to upend the good press coverage unnecessarily.

While Tillerson had entered the administration with sky-high expectations, his tenure was a failure by any measure. In the summer of 2017, reports surfaced from a cabinet-level meeting at the Pentagon that Tillerson had called the president a "moron." Soon after, Tillerson told the press that he'd opened a dialogue with the foreign affairs office of North Korea, but his weak messaging was out of sync with the president, who wanted to use a different tone to set the stage for negotiations with North Korea's impetuous young dictator, Kim Jong Un. In a humiliating tweet, Trump pulled Tillerson off the file: "I told Rex Tillerson, our wonderful Secretary of State, that he is wasting his time trying to negotiate with Little Rocket Man . . . Save your energy Rex, we'll do what has to be done!"

This exchange revealed something to the world that we already knew internally: the president and his secretary of state were not on the same page. Such a public rebuke from the president cast doubt among Tillerson's foreign counterparts about whether he had influence with the president, rendering him effectively useless in his role as the nation's top diplomat. Rather than fixing the relationship, Tillerson became resentful. We began to hear that he was openly undermining Trump with foreign leaders.

In early 2018, I was meeting with the president in the Oval Office when Cassidy walked in and told me that Tillerson was on the line and wanted to speak to me immediately. I excused myself and took the secretary's call. He was on a plane flying to Mexico City, and his staff was briefing him on a package of twenty-five smaller agreements we'd nearly finished negotiating, which covered a range of issues affecting the US-Mexico relationship, from immigration to drugs and weapons trafficking to energy exports.

"What is up with all these deals with Mexico?" he demanded. "Who gave you authorization to negotiate all these agreements?"

"I have been working with Kim Breier on these deliverables for months," I said, referring to the State Department's director of Western Hemisphere affairs. "Your team has been with us every step of the way. We have made more progress in one year than in the previous ten years. Which ones do you not like?"

Tillerson ended the call in a huff. I later learned from Luis that after a bilateral meeting, Tillerson pulled him aside and launched into a rant. He accused Luis of making a strategic blunder by working with me in the White House instead of someone at the State Department.

Tillerson must have known that his haranguing would get back to the White House. He was lighting himself on fire, and if that was his deliberate strategy, it worked. From what I could tell, the former oil tycoon had made it clear that he was no longer interested in faithfully representing the president's foreign policy agenda.

Trump asked Ivanka who she thought should replace Tillerson, and she strongly recommended Mike Pompeo. That was Trump's instinct too. In recent weeks, he'd been asking his inner circle what they thought about moving the CIA director over to the State Department. By all accounts, Pompeo was the perfect fit. He not only had stellar credentials but also shared Trump's foreign policy views, understood his sense of humor, and didn't try to steal the spotlight.

Shortly after the president's inauguration, Pompeo invited me to the CIA headquarters for a visit, adding, "You're a power user of our material." He was referring to my regular Situation Room briefings with CIA analysts, who were helping me get up to speed on the Middle East. I met with several high-ranking CIA staff members. I asked them if there was any noticeable difference between our administration and the prior administration. They said that in the previous administration nearly every expenditure or action, down to the purchase of a motorcycle for an agent to slip a cover, needed White House approval. By contrast, Trump had delegated more authority to Pompeo, enabling his staff to do their jobs. They made clear that they didn't always agree with Trump's directives, but they appreciated his decisiveness. Most of the top-level policy meetings under the Obama administration ended with a decision to meet again in two weeks to discuss the issue further. Our administration held fewer meetings, and the ones that did occur facilitated robust discussions and ended with decisions that provided clear direction. I was impressed. Pompeo had empowered the staff to carry out their missions, boosting morale inside the CIA—the opposite of Tillerson's reclusive approach at the State Department.

On the morning of March 13, 2018, Trump offered Pompeo the job, and he accepted on the spot. When they discussed his replacement at the CIA, Pompeo made a case for Gina Haspel, his talented and hard-nosed deputy, who had worked her way up through the ranks during her thirty-year career at the agency. Trump agreed to promote Haspel. That morning, he announced the change in a tweet: "Mike Pompeo, Director of the CIA, will become our new Secretary of State. He will do a fantastic job! Thank you to Rex Tillerson for his service! Gina Haspel will become the new Director of the CIA, and the first woman so chosen. Congratulations to all!"

Feeling blindsided and figuring that I must have known about the decision before he did, Kelly questioned me. "Do you know what happened?" he asked. "I thought Tillerson was doing a great job, and the whole cabinet loved him," he said in a daze of cluelessness.

It was clear that Kelly was rattled that the president had fired

Tillerson—a top cabinet member and Kelly's close ally—without consulting him. From that day forward, Kelly grew more insecure about his own standing.

Later, Kelly spoke with reporters off the record. In an apparent effort to be chummy, he gave them gossip so colorful and absurd that it was bound to leak: on his swing through Africa, Tillerson was dealing with a bout of "Montezuma's revenge" and had been on the toilet when Kelly called and told him he was going to be fired. Multiple members of the press broke a long-established code of journalism and gave this irresistible off-the-record nugget to another reporter, who was not in the room and thus not bound by the protocol against off-the-record disclosures. Those reporters should never have shared the story, but more importantly, Kelly should have known better. Giving this embarrassing detail to reporters accomplished nothing other than humiliating the outgoing secretary of state, Kelly's supposed friend. For someone who held himself up as the adult in the room, it was a juvenile act of betrayal. The visual of the tough-as-nails Texas oilman getting the call while suffering on the toilet was hard to forget. Tillerson's unceremonious dumping illustrated one of Sun Tzu's principles: "The opportunity of defeating the enemy is provided by the enemy himself." Tillerson had knocked himself out of the cabinet.

* * *

In Pompeo, the president finally had a secretary of state who would faithfully advance his foreign policy aims. Within his first few weeks on the job, Pompeo invited Ambassador David Friedman and me to meet with him. In his wood-paneled office, Pompeo treated us like two old friends.

"Mike, I have a bit of a problem," Friedman said with a hint of irony. "I have all these issues I'm working to get approved, and you're really slowing me down. With Tillerson, it was very easy. He hated me, and I hated him, so I did whatever I wanted. The problem is that I really like you, so I'm trying to follow all these processes, but they are slowing me down like crazy. I know you are drinking from a firehose, but can you give our files a bit of elevated attention?"

Pompeo laughed. "Okay," he said. "I promise I'll pay attention to your issues, and we'll push them forward quickly."

It was my turn now. I started by saying that he had a big job to do, and I was flexible and open to working with him in any way he thought was most productive. I then carefully walked him through our peace efforts and the latest details around the trade discussions with Mexico and Canada.

"Are you okay if I keep working on these two files?" I asked. "I'm pretty determined when I'm given a task, but I never want to overstep my lane, so if I am ever out of line, just call and tell me. If you want me to do something differently, I'll do it. You're the secretary of state. There are not two secretaries of state, and I don't want the media ever to make such a case, as they did under Rex. I'm here to support you and the president."

"Jared, this place is a mess," Pompeo responded. "Rex hollowed out the whole building, and the staff is demoralized. I have almost no political appointees, and most of the ones I do have don't like the president. I have a lot of housecleaning to do, fifty files to catch up on, and everyone around the world wants to talk. I'm working twenty hours a day, and I need another twenty. I wish I had someone like you on every file. Keep running forward. If you need me, call, and I will always get back to you fast. If I have any suggestions, I'll call you. But at least for the next thirty days, just run forward. Don't walk, don't go slow. Do whatever you need. I trust your judgment. Call me when you think I should know about something."

His response was so cool and confident that I knew the president had made the right choice and that this was the beginning of a great working relationship. I soon noticed another welcome change. Pompeo would often call to keep me in the loop and get my thoughts on an issue. When he did, he was always friendly, but to the point. The calls rarely lasted more than three minutes. Tillerson seldom had called and often did not promptly return calls—a frequent complaint among foreign diplomats. On the rare occasion when he did, the calls almost always took thirty minutes and accomplished little. I figured that from a mathematical perspective alone, Pompeo would be able to do ten times as much diplomacy.

Trump usually made changes in batches. The week after he fired Tillerson, he decided to replace his national security adviser, General H.R. McMaster. Along with many others in the West Wing, I considered the three-star Army general a friend and a devoted leader. However, McMaster found himself outside the elite four-star-generals club occupied by Marines Kelly and Mattis, both of whom had spent their recent careers telling three-star generals what to do. Not surprisingly, the four-stars were loath to defer to McMaster, despite the fact that the president charged him with running the policy process for military and foreign policy matters.

Mattis and Kelly were military heroes who had devoted their lives to America and served with sacrifice and distinction. Kelly, in particular, had paid an enormous personal cost when his son Robert, an American hero, was killed by an explosive ordnance while on patrol in Afghanistan. I never doubted their love of country. At some point, however, it seemed like Mattis and Kelly decided that they knew better than the president of the United States and made it their mission to protect the world from Trump. McMaster would complain to me that they resisted his efforts to coordinate policy on Iran and North Korea, stalled the president's request to withdraw from the Iran deal, and refused to give the president the information he needed to bring troops home from Iraq and Afghanistan.

McMaster did not always agree with the president, and he could push back forcefully. His academic style was often at odds with Trump's pragmatic approach. Unlike the four-stars, however, McMaster did his best to implement the president's directives. Because of this, Kelly and Mattis constantly knifed McMaster. When the president asked for a concrete plan to withdraw from Afghanistan, for example, Kelly and Mattis delayed and then blamed McMaster when the president expressed frustration about the holdup. They became obsessed with taking out McMaster and replacing him with the deputy national security adviser, Ricky Waddell, an experienced but lower-ranking flag officer whom they felt they could control. In one heated exchange, McMaster warned them, "You guys are trying so hard to get rid of me. Just be careful what you wish for. You might be successful and get someone like John Bolton."

McMaster's admonition proved to be a harbinger of his fate and theirs. Just as McMaster predicted, Trump replaced him with Bolton. A cantankerous foreign policy academic and TV personality who had served as George W. Bush's UN ambassador, Bolton was a neoconservative and more hawkish than Trump, but he agreed on the need to withdraw from the Iran deal. When it came to Iran, Trump saw through the bureaucratic excuses and never lost sight of the grim facts: the deal had lifted economic sanctions and handed more than $100 billion to the ayatollah and his malign regime. As a result, Iran made a fortune and boosted its military budget by nearly 40 percent. The Iranian regime built missiles capable of carrying nuclear warheads and funneled support to al-Qaeda, the Taliban, Hezbollah, Hamas, and other terrorist organizations, which were actively working to destabilize Iraq, Lebanon, Syria, Yemen, and other countries. Most consequentially, perhaps, the deal failed on the very issue it set out to address: it allowed Iran to continue to enrich nuclear material, lacked a robust inspection and enforcement mechanism, and made no mention of Iran's missile program. As a result, the world's leading sponsor of terrorism was emboldened to pursue a nuclear weapon.

On April 30, 2018, Prime Minister Bibi Netanyahu convened a press conference in Tel Aviv and revealed to the world that Mossad—Israel's intelligence agency—had broken into a secret warehouse in Tehran and obtained thousands of documents showing, conclusively, that Iran had been engaged in a clandestine program to develop and test nuclear weapons. The regime had hidden its designs from the international community and lied in claiming that it did not have a nuclear weapons program. Netanyahu's revelation provided concrete evidence that the Iranians had failed to comply with the terms of the deal—and in fact showed that they had never intended to comply. The president now had a firm basis for withdrawing from the deal, reimposing the highest level of sanctions, and asking our partners to follow his lead.

On May 8 the president announced his decision from the Diplomatic Reception Room. "At the heart of the Iran deal was a giant fiction: that a murderous regime desired only a peaceful nuclear energy program," Trump declared. "Today's action sends a critical message. The United States no longer makes empty threats. When I make promises, I keep them."

The president's announcement commenced America's "maximum pressure" campaign against Iran.

21

A Step toward Justice

Every day at 5:00 p.m., the photographers from the *Daily Mail* packed up their cameras and left our house. It was like clockwork. You could set your watch by their behavior. They were the most devoted of the paparazzi who constantly staked out our house, seeking pictures of Ivanka—and sometimes settling for me—as we came and went. Little did they know that the best action at our house in Washington's leafy Kalorama neighborhood often took place later on in the evening.

One night in the fall of 2017, shortly after the paparazzi had departed into the dusk, a black Suburban with tinted windows rolled up our street. Out jumped the third most powerful Democrat in Congress, Senate minority whip Dick Durbin. He was joined by fellow Democratic senators Sheldon Whitehouse and Amy Klobuchar and Republican senators Lindsey Graham and Mike Lee. As they gathered in our dining room for a discussion on criminal justice reform, Senator Klobuchar raised her glass for a toast.

"It's just really nice to do this," she said. "Because this used to happen in Washington all the time. And I just feel like we don't get to talk across the aisle anymore."

Ivanka and I hosted this gathering, and many others like it, at the request of White House legislative affairs, who asked us to bring together members of opposite parties in a relaxed, closed-door setting. Many Democrats were willing to engage and discuss bipartisan reforms, but a few refused to meet, including California senator Kamala Harris.

After watching the successful tax reform effort, which Secretary Mnuchin, Gary Cohn, and Ivanka led, it was clear to me that if we wanted to pass criminal justice reform, we needed to work collaboratively with members of Congress. To get any bill passed, the White House needs to engage lawmakers on the front end, ask their opinions, understand and address their concerns, and apply the right amount of pressure. Congress governs a democracy, not a company. Changing a law is not meant to be fast, and it shouldn't be easy.

By April of 2018 we had built a formidable coalition of members on both sides of the aisle. We decided to focus the legislation on prison reform, which would improve job training programs for inmates and provide better treatment for addiction, among other priorities. This was more widely supported among Republicans than sentencing reform, which would let certain nonviolent offenders out of prison earlier. If our prison reform bill passed in the House, then we could work to add sentencing reform to the Senate version.

We were ready to put our legislative strategy into motion. The first hurdle was Chairman Bob Goodlatte's House Judiciary Committee, which needed to consider our prison reform bill before it could come to the full House floor for a vote. Republican Doug Collins and Democrat Hakeem Jeffries both sat on the judiciary committee, and I had worked with them from the beginning to draft legislation and build a coalition of support within the committee. By the end of April we had persuaded nearly half of the committee's members to cosponsor the bill. Goodlatte scheduled the markup for April 25. Legislative horse-trading and negotiations typically take place until about forty-eight hours prior to a markup, but my team was confident that the bill would sail through.

On the morning of April 25, I received an urgent alert that Chairman Goodlatte had canceled that day's markup. At 7:45 a.m. I called our legislative lead, Ja'Ron Smith. "What the hell happened to our bill?" I asked. In the background I could hear the faint automated announcement— "Step back, doors closing"—of the Metrorail car; he was on his way to the White House and limited in what he could say. "Come to my office as soon as you get in," I requested. When Ja'Ron came to my office, he explained that Sessions's team at the Department of Justice had sent over several changes that Jeffries viewed as poison pills. He briefed me on the details of Sessions's edits.

"These changes are ridiculous and show bad faith," I said. "Ignore Sessions's edits, present pragmatic compromises, and see if that gets Jeffries back on board."

Ja'Ron got to work, and by the end of the day we had removed Sessions's modifications and added Jeffries's provisions to expand the application of good-time credits and ensure that prisoners were not placed in prisons more than five hundred miles from their homes, making it more feasible for loved ones to visit. Once we had made these changes, Jeffries returned to the table, despite the pressure he was facing from the left to walk away.

I still wasn't sure what to do about Sessions. I had tried earnestly to get him to a better place on the policy, but he remained intransigent. Brooke Rollins, Ja'Ron, and I met with him at the Justice Department and went through the bill, line by line, asking him to describe his objections.

"Well, my guys will tell you," he said, asking his lawyers to explain their position. Their concerns were either easy to address or didn't make sense. Finally Sessions turned to me and in his southern drawl declared, "Jared, it's very simple. If the boy does the crime, you've got to lock him up." That's just where he was. From that point forward, I realized that he would try to subvert us at every single turn, making a nearly impossible task even harder.

Rather than meet Sessions head-on, I decided to make my case directly to the most conservative members of the House. The attorney general's objections carried weight with them, and obtaining their support would send a powerful signal to the president that law-and-order Republicans backed prison reform. On the evening of May 7, I met with the House Freedom Caucus, a conservative coalition. While many of the members were open to federal reforms, some were unconvinced. One of these skeptics was Mark Meadows, the North Carolina representative and leader of the caucus. As I addressed his concerns, I could tell that he was considering my arguments and keeping an open mind. The meeting was the beginning of a great collaboration and friendship with Meadows, who later said, "I would have thought I would have died before voting for criminal justice reform. You'll never know if I voted for this because I value our friendship, or if I voted for it because I now agree with this policy." After becoming Trump's fourth and final chief of staff, Meadows would often spring to my defense when the president accused me of being a liberal, jokingly countering, "Actually, Jared is an honorary Freedom Caucus member."

On May 9 the House Judiciary Committee finally marked up the bill. Renamed the First Step Act, it passed the committee with strong bipartisan support by a vote of 25–5. Iowa congressman Steve King was the sole Republican to vote against it, but King was known to hold extreme views, so being on the opposite side of him wasn't a bad place to be. On the Democratic side, we also lost the vote of one of our original cosponsors, Representative Sheila Jackson Lee. After we incorporated her requests to provide tampons to women inmates and ensure that women were not shackled while giving birth, she made a third demand that we couldn't accept: she wanted to allow mothers to keep their babies in prison with them for three years. "Uh, I think our goal is to help get people out of prison," I tried to explain. "We're not trying to put babies in prison with this bill."

With the committee hurdle cleared, we set our sights on moving the bill to the House floor for a full vote. Nervous that the legislation was actually gaining traction, Sessions scrambled to mount an internal effort to stop the bill in its tracks. He scheduled an Oval Office meeting to try and dissuade the president. Knowing this, I brought an all-star cabinet member for reinforcement: Secretary of Energy Rick Perry, who had successfully pioneered similar reforms while governor of Texas.

As the meeting kicked off, Sessions voiced his objections. The bill was soft on crime, he told the president, and it would put dangerous criminals out onto the street. "You don't want to be accountable for the next Willie Horton situation, do you?"

Disgusted that he would try to equate our effort with the Mike Dukakis scandal, I nearly lost my cool. Instead I answered with the facts, quoting from the text of the bill and explaining why the attorney general's claims were wrong: nothing in our bill would allow violent criminals like Horton to walk early. Sessions didn't have a response, which was rare.

The president, not used to seeing me that worked up, was impressed by my passion. "You really know this stuff," he said.

"This is serious—it's about saving lives and keeping people safe," I replied. "I would never put you in a bad position, and I know that if anything goes wrong here, it's on me, so yes, I reviewed every detail."

I showed him a list of the members who supported the bill in committee, including Jim Jordan, Ron DeSantis, Louie Gohmert, and John Ratcliffe. The president looked at Sessions and glanced back at me. "I'm going with Jared on this one."

Soon after, five Democrats—Senators Durbin, Harris, and Cory Booker, and Representatives Jackson Lee and John Lewis—penned a public letter to their congressional colleagues opposing the bill, claiming that the First Step Act was a "step backwards" and would "institutionalize discrimination and likely fail to reduce recidivism." Besides the letter's unfounded claims, what disappointed me most was seeing Booker's name in the signature block. I had met Booker when I was fifteen and my father supported his unsuccessful first run for mayor of Newark, New Jersey. He became a friend, and we stuck by him for years. Now, when we had a chance to work together on an issue that we both believed was crucial to improving the lives of millions of Americans, he'd emerged as one of the effort's most vocal opponents.

The Democrats' letter was intended to convince their colleagues, particularly those among the Congressional Black Caucus like Hakeem Jeffries and Cedric Richmond, to abandon the bill. But neither Jeffries nor Richmond backed down. I called Jeffries and told him that we were working on a response. "Jared, don't worry about it," he told me. "I'm drafting a response." In a seven-page excoriation, Jeffries refuted the claims before concluding, "Ultimately, it should be our mission to improve the lives of the people we are here to represent. In this regard, the perfect should never be the enemy of the good, particularly when it comes to the least, the lost and the left behind. That is what the FIRST STEP Act is all about. Accordingly, it is my hope that the authors of the opposition letter will reconsider their position, cast aside partisan ambition and join the House's fight to fix our broken criminal justice system."[32]

It was a masterpiece of political courage. I knew it was hard for Jeffries to break with members of his own party, including American civil rights legend John Lewis, and I was grateful for his determination to do what was right in the face of opposition.

As we entered the final stretch before the big House vote, we convened an event for the president to announce his support of the First Step Act. Trump spoke about the need to break the cycle of recidivism by helping former prisoners find jobs and contribute to society. "America is a nation that believes in second chances," he said, before cracking a smile, "and third chances in some cases. And, I don't know, I guess even fourth chances." His statement marked an important moment: it was the first time that Trump, a law-and-order president, had called on Congress to pass prison reform legislation. Over the coming days, nearly every Republican in the House agreed to support the bill.

On May 22—the day before Kelly restored my Top Secret security clearance—the House passed the First Step Act by a vote of 360 to 59, with only two Republicans opposing the bill, Steve King and Bill Huizenga. The bigger test was still to come. Senator Chuck Grassley had made clear that he would not consider the First Step Act in the Senate Judiciary Committee unless it included sentencing reform. I invited him and Senators Tim Scott, Mike Lee, and Lindsey Graham to pitch the president on the idea of incorporating sentencing reforms into the First Step Act. In the Oval Office, we presented our case for sentencing reform. The House-passed bill helped prisoners who were currently incarcerated, but it was missing robust provisions to reduce the number of nonviolent prisoners serving disproportionately harsh terms. Grassley explained that the sentencing relief he proposed would be available only to nonviolent offenders, and actually strengthened sentences for domestic violence and weapons trafficking. Grassley expressed confidence that he could build on our momentum in the House and get a more comprehensive bill through the Senate. By the end of the discussion, Trump expressed interest in moving forward. Afterward Lee was jubilant about Trump's support, but I gave him fair warning.

"This is just a soft yes," I said. "The president has still only heard our side of the story. Now I need to bring in the people who disagree to make their case to him before he comes to a conclusion."

With the president increasingly supportive of criminal justice reform, I decided it was the perfect moment to bring him Alice Johnson's clemency case. I'd first learned of her case back in December, when Kim Kardashian had reached out to Ivanka. Alice was a sixty-three-year-old grandmother serving the twenty-first year of a life sentence, without parole, for drug conspiracy and money laundering. Hers was a nonviolent drug offense. In the early 1990s she had fallen on hard times and gotten wrapped up in a drug-trafficking ring, where she facilitated the flow of illegal drugs and cash. While incarcerated, Alice had transformed her life. She'd become an ordained minister, completed multiple vocational certifications, mentored fellow inmates, and maintained a spotless behavioral record.

In an Oval Office meeting in May, after working closely with Kim Kardashian to vet the file, I presented Alice's case to the president. I explained that Alice had been sentenced for a nonviolent drug crime in the 1990s, and the methodology used to calculate her sentence was unfair and wouldn't be allowed today.

White House counsel Don McGahn countered. "Her file says she also had a murder for hire," he argued. "The reason she got such a harsh sentence was because she was really the kingpin."

I couldn't believe it. Were we discussing the same person? "She's a grandma. She's in Christmas plays and gospel concerts in prison. And she never touched the drugs."

McGahn shot back, "Jared, you were in the construction business. You were in the real estate business. I'm assuming you weren't touching the hammers."

"Why?" I said.

"Because you were the CEO. Similarly, Alice didn't touch the drugs, because she was really the mastermind."

"Look, she was a low-level person who got caught up in this thing," I responded. "But even if you're right, she's served twenty-one years for a first-time crime where nobody got killed. And by all accounts, including her prison warden's, she's fully rehabilitated. Are we going to deny her a chance at life because of a mistake she made twenty-one years ago?"

By the end of the meeting, Trump said, "If you end up on the wrong side of our justice system, you don't have a prayer. Let's seriously consider the commutation."

On May 30, Kim Kardashian met with Trump in the Oval Office. She gracefully presented Alice's case to the president. She knew the details backward and forward. McGahn and one of his team members presented the counterarguments, though he was far more mild than usual because he was starstruck by Kardashian. Trump thanked her for coming. Two days later, he called me early in the morning and said, "Let's do the pardon. Let's hope Alice doesn't go out and kill anyone!"

I called McGahn to set the wheels in motion, but he kept delaying the legal documents. Shortly thereafter, the *Washington Post* ran a story saying that Trump had grown "obsessed" with pardons and that Kardashian's celebrity was influencing his views on Alice's case. I suspected that McGahn and Kelly had leaked these falsehoods as part of a last-ditch effort to foment conservative backlash and change Trump's mind. After the torturous security clearance situation, I was keenly aware of the power that McGahn and Kelly could have over me, so I was wary of taking on another fight. But I decided to keep pushing for Alice's case, regardless of the fallout, because this was about saving a life.

Eventually, Kelly and McGahn ran out of stalling tactics, and on June 6 the president commuted Alice Johnson's sentence. Later that evening, as he sat in the small presidential dining room with Ivanka, Trump watched Alice's release from prison on the evening news. Alice ran into the arms of her family, embracing them as tears streamed down her face.

With cameras surrounding the emotional reunion, she declared, "I'm free to hug my family. I'm free to live life. I'm free to start over. This is the greatest day of my life." Her emotion was raw, her joy contagious, her long suffering and love emanated from her smile.

The president called me afterward. "Jared, that is one of the most beautiful things I have ever seen. I've been around for a long time, and that was beautiful. I can tell she is a solid person. There must be more like her in prison. Let's find more worthy cases to do."

22

No Time for Triumph

Ivanka and I arrived in Israel on May 13 to witness an historic moment in US-Israeli relations: the opening of our embassy in Jerusalem. My older sister, Dara, had made a surprise trip from New Jersey to meet us there. This day was especially meaningful for her. After high school, she had studied in Israel for a year and returned to the United States with a redoubled commitment to our family's Jewish faith. Her devotion inspired the rest of our family, and I was delighted that she was with us for this moment.

This day almost didn't happen. In the weeks that followed the president's announcement on December 6 about moving the embassy, the State Department mapped out an extensive process involving land negotiations, costly construction bills, and potentially more than a billion dollars in congressional appropriations requests. Ambassador David Friedman adeptly perceived this as a stall tactic and called the president directly.

"State is going to kill the embassy move," warned Friedman. "I thought you should know. If you don't want to do it, just tell me and I won't bother you again."

Apparently the department had already gotten to the president. "They tell me it's going to cost a billion dollars and will take five to ten years," Trump replied.

These estimates flowed from an intricate plan that wasn't optimized for speed, Friedman said. It wasn't the only option. He suggested a different plan. The US government already owned a state-of-the-art building on a sprawling seventeen-acre campus in the heart of West Jerusalem. Friedman said he could convert it into an embassy for less than $200,000, and have it ready for a big celebration on May 14, Israel's Independence Day.

"Done," the president said. "But for the first time in my life, I'm going to say that's too cheap. Why don't you spend $500,000—make it nice."

As the opening day approached, I asked the president if he wanted to travel to Jerusalem to lead the historic ceremony.

"Why don't you go?" said Trump. "I know you like to be in the background, but you should speak for a change. Moving the embassy was the right thing to do and I know that it was very important to you."

I had never addressed an audience on such a grand scale. Hundreds of officials from around the world were gathering at the new embassy for the dedication, and it would be carried on live television to a global audience. Despite my nerves, I agreed to go.

When we arrived at the freshly renovated building in Jerusalem's Talpiot neighborhood, Prime Minister Netanyahu and his wife Sara greeted us warmly. Typically, I would notice every detail of any building I entered, a trait picked up from a career in real estate, but this day I was so anxious about my speech that I couldn't have told you the color of the walls. I had spent hours meticulously crafting each line of my speech. Finally, the moment came.

"The pursuit of peace is the noblest pursuit of humankind. I believe peace is within reach if we dare to believe that the future can be different from the past, that we are not condemned to relive history, and that the way things were is not how they must forever be," I said as I stood before the newly opened American embassy in Jerusalem. "When there is peace in the region, we will look back upon this day and remember that the journey to peace started with a strong America recognizing the truth."

The crowd erupted in applause as I concluded and made my way back to my seat next to Ivanka, who squeezed my hand and whispered, "You nailed it." Beaming, Dara turned to me and said, "Bubby and Zayda would be so proud. Only God could write this script." It was a special moment that I'll never forget, but like every other surreal experience during my time in government, it too was fleeting.

Minutes after the event concluded, I saw the television coverage. It was a split screen, with footage from my remarks alongside images of protesters in Gaza being hit with rubber bullets and tear gas by the Israeli Defense Forces, which killed more than fifty people. It appeared to be a harmful overreaction to predictable Palestinian opposition, though days later a Hamas leader admitted that nearly all of those killed were members of Hamas, which the State Department has designated as a "Foreign Terrorist Organization."

Leading up to the embassy opening, Abbas had given a crazed speech before the Palestinian legislative body, in which he openly questioned the circumstances around the Holocaust, claiming, according to a transcript acquired from the BBC, that the Nazis weren't against the Jewish people but against their exploitative lending and banking practices. In other words, the greedy Jews had brought the Holocaust upon themselves. Even the famously anti-Israel *New York Times* editorial board penned a scathing call for his resignation: "Let Abbas's Vile Words Be His Last as Palestinian Leader."

* * *

As the Palestinian leadership continued to prove their unwillingness to seek a constructive solution for their people, I was eager to release our peace plan as soon as possible so that the world could react to it. I hoped this would urge Abbas to consider our proposal, which would deliver prosperity and peace for the Palestinian people.

In June I traveled back to Israel to discuss our plan with Prime Minister Bibi Netanyahu. But shortly after I landed, Ivanka called me with an urgent update.

"This is as bad as I have seen. Sessions's zero-tolerance policy has created a massive crisis at the border. Kelly is refusing to admit that he made a mistake, and he is not telling my father the truth about the situation."

Two months earlier, on April 6, Attorney General Jeff Sessions had issued a press release announcing that he would enforce immigration law with a "zero-tolerance policy" against immigrants who crossed the southern border illegally. At the time, the practical implications of Sessions's prosecutorial change were a bit murky. Under a 1997 federal court ruling called the Flores Settlement Agreement, the Department of Homeland Security could not detain illegal immigrant children for longer than twenty days. Adults, on the other hand, could be detained for much longer periods while they waited for their immigration hearings to occur. If our nation's immigration statutes were truly enforced, with zero tolerance, it would cause some children to be separated from their parents. With Sessions pushing hard for his policy, Secretary of Homeland Security Kirstjen Nielsen flagged the child separation concern for Kelly. The chief of staff called an immigration meeting in the White House Situation Room in early May. Given that immigration was outside our portfolios, neither Ivanka nor I were invited, but we later learned that Kelly decided to proceed with the zero-tolerance policy. It took about six weeks for the ramifications of Kelly and Sessions's policy to filter into the press. During that period, DHS separated 2,816 children from their parents or guardians. By the third week of June, just as I departed for the Middle East, the press got hold of the story, which almost immediately erupted.

When Trump saw the breaking headlines, he quizzed his team about the veracity of the reports and asked what could be done to end child separation. Sessions and Kelly did not present him with a full range of options, and they urged him to continue with the policy, which they believed would serve as a deterrent to people crossing the border and would put pressure on Congress to fix the broken immigration laws. Ivanka became aware of this when two staff members, including immigration staffer Theo Wold, paid her an unexpected visit. Wold was concerned that the president was not getting the full picture from his leadership team. Trump could stop child separation immediately by signing an executive order directing Sessions and Nielsen to end the zero-tolerance policy and implement a more humane approach. Ivanka thanked Wold and asked him to start working on the executive order.

When Ivanka called me, she asked me what I thought she should do.

"Kelly is telling the president that there is no other option. I've been trying to raise this issue with him, but he's excluding people with differing opinions from meetings in the Oval. He's going to be absolutely irate, but I don't see any other path other than bringing this solution directly to the president."

"I wish I was there to help you," I said, "but there is not much I can do from here. You don't really have a choice. Kelly made this mess—ignore him and do what you think is right."

Ivanka was typically careful not to bypass West Wing protocols, but in this case she felt that the president wasn't being well served, and the issue was too important. She knew the president wanted to find a solution, and he wasn't being given all the options to fix the problem. She went to see her father in the Executive Residence and handed him the draft executive order that she had asked Wold to prepare.

"I know they are telling you this can't be done," she said. "They might be right, but sign it anyway and dare anyone to challenge it."

After reading the draft order, the president called McGahn.

"I'll be down in the Oval in thirty minutes, and I want an executive order ending this policy on my desk when I get there," he instructed. "Get in touch with Ivanka and review the one she showed me."

As Ivanka anticipated, Kelly was livid. Normally she tried to avoid his wrath, but this time she didn't care. Kelly had put her father, the country, and three thousand families in this terrible situation, and she was willing to face his wrath to stop the unfolding humanitarian debacle. The next day the president signed an executive order ending Sessions's zero-tolerance policy. Ivanka had defused the immediate crisis. She wasn't looking to publicize her involvement, but when the president spoke to House Republicans about why he was going to reverse the policy, he revealed her role in his decision. The next day, he publicly recognized Ivanka when he signed the executive order. This led to a series of news stories she had hoped to avoid.

This was one of the many examples of the gap between reality and the media's portrayal of Ivanka. While they were quick to criticize her for not forcefully denouncing the policy—even though every journalist knew it would have been wildly unusual for a staffer to publicly object to an administration policy—Ivanka worked quietly behind the scenes to find a constructive solution. She was in an impossible situation, but she handled the crisis with grace under pressure. This would not be the last time that Ivanka's good judgment, compassion for people, and relationship with her father resolved a big problem and helped our White House achieve a better outcome.

23

"No One Gets Smarter
by Talking"

In politics, it's much easier to kill a deal than to make one. Even if everything goes right, success is not guaranteed, and failure can happen with the slightest misstep. In the spring of 2018, our trade talks with Mexico and Canada were starting to fall apart. It took a handwritten note on a scrap of paper to rescue them.

The note rested in my suit pocket as I left the West Wing one afternoon in May and hurried across Seventeenth Street to the building occupied by US trade representative Bob Lighthizer. One of the oldest structures in the capital, it served during the Civil War as the office of Quartermaster General Montgomery C. Meigs, a Georgian who stayed loyal to the Union and coordinated the supply of food, clothes, and other items to the field. Legend has it that Abraham Lincoln used to make the same walk from the White House when he wanted to commiserate with Meigs about the performance of his generals.

My objective had nothing to do with commiseration. I was trying to save a trade deal. When I walked into Lighthizer's office, I took the paper from my pocket and handed it to him. The country's top trade diplomat studied the document, deciphering its scribbled numbers and arrows. Then he looked at me. "I have never before seen a trade agreement resolved on a three-by-five piece of paper," he said. "But if they will really do that, I think that's a fair compromise, and we should make the deal."

Back in March, I had left Mexico with a commitment from President Peña Nieto that his team would work to resolve the disputes that continued to separate our two sides. The task fell to Ildefonso Guajardo Villarreal, Mexico's secretary of economy and a former legislator. With an eye trained on his political future, Ildefonso understood the political peril of compromise. Trump's insistence on bringing jobs back to America meant any deal would force Mexico to lose jobs. Mindful of how this would look to Mexicans, Guajardo avoided a potentially unpopular outcome by delaying discussions and twisting technical issues into unsolvable deal-breakers.

From the start, we'd known that the central sticking point in our talks involved auto-industry jobs. Under NAFTA, the United States had lost 350,000 of them to Mexico, where the labor is cheaper and regulations are looser.[33] In 2018, autos alone made up nearly $64 billion of our $78 billion trade deficit in goods with Mexico.[34] To achieve a more balanced relationship and reverse the southern migration of jobs, we wanted to require vehicles made in Mexico to use more American-made parts. After months of trade talks, Lighthizer and Guajardo stood at a stalemate. Mexico simply wouldn't budge on this central issue.

Then Trump intervened. In May, he directed his trade team to prepare a 25 percent tariff on autos imported from Canada and Mexico into the United States. This threatened to devastate both of their economies and potentially push them into recession. His bold move unnerved Washington and Wall Street, but Trump was fighting for Main Street. As a former businessman, he knew a lot more than the typical politician or fund manager about imposing leverage over a rival.

After news broke of Trump's tariff threat, Luis Videgaray, the Mexican foreign secretary, flew to Washington. When he walked into my office, he got right to the point. As an emissary of President Peña Nieto, Luis came ready to address the tough issues. He pulled out a blank piece of paper and drew a chart illustrating a potential compromise. Under NAFTA, for Mexican auto imports to come into the United States tax-free, 62.5 percent of the automobile had to be made in the United States.[35] Lighthizer wanted to raise this "rules of origin" standard to 85 percent. Each percentage represented billions of dollars of potential investment and tens of thousands of jobs. Luis suggested that we meet near the middle, proposing a 75 percent threshold for both countries, while also demanding that USTR make concessions on other sectors. This was a big move. It would practically eliminate the outsourcing of American factories to Mexico. It meant that we might have a deal. And so after Luis left, I took the note across the street to Lighthizer.

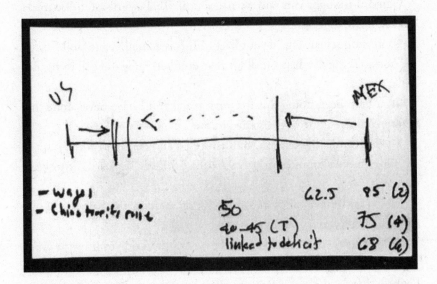

Lighthizer was on board with Luis's concept for a rules-of-origin compromise on autos, but much work remained on other contentious issues. In the middle of these discussions, on July 1, Mexico elected a new president: Andrés Manuel López Obrador, known as AMLO. His term would start on December 1. AMLO quickly named Jésus Seade, an experienced economics professor, as his preferred trade negotiator. Soon after, official trade talks resumed in Washington, and Seade joined the Mexican delegation along with Luis.

The US and Mexican negotiators packed into Lighthizer's sparse conference room, with the two teams sitting on opposite sides of a long table. Lighthizer kicked off the conversation with an optimistic tone, which was unusual for him. "I'm glad we have a deal for autos—at seventy-five percent—so now we can move forward to the other issues," he said.

Sitting across from Lighthizer, with his hands clasped in front of him, Guajardo delivered an unwelcome surprise. He rejected Lighthizer's offer and proposed moving forward at a lower percentage.

Lighthizer turned beet red with anger and shot a glare in my direction. He was stunned and asked for a break. He motioned me into his office, shut the door, and started screaming.

"Jared, I thought you said we had a deal. This is a disaster. We made our big move too soon."

"Bob, stop screaming," I shot back. "This is actually your fault."

"You told me you had a deal for that number. How the hell is this my fault?"

"You have been doing this for forty years, and I have never done this before. You shouldn't have listened to me!"

My joke broke the tension. Bob chuckled and cooled down.

"Give me ten minutes to try to fix this," I said. "I trust Luis to keep our deal."

I left Lighthizer's office and pulled Luis aside in the hallway. "Luis, what the hell is going on here?"

"I'll get us there," he said. Then he went into a side room with Guajardo. I later learned that they called President Peña Nieto, who made his orders crystal clear. After the tumultuous start, we reconvened in the conference room and moved forward with our initial agreement on cars. We had resolved a threshold issue, but several other deal-breaking concerns remained. Among other issues, we wanted stronger labor and environmental protections and a better process for resolving trade disputes.

For the next several months, Lighthizer's suite at USTR became my adjunct office, where I spent many eighteen-hour days working through the outstanding issues of the deal with the USTR staff. Though they were mostly Democratic career officials, they were energized by the fact that President Trump had empowered them to fight for American workers and businesses.

As we made progress, an important question emerged: How long should the new deal last? Like every other American trade agreement,

NAFTA had no expiration date. It existed in perpetuity, with no way to adjust or amend it if parts became outdated or unfair. We had essentially sold permanent access to our market. This encouraged corporations to move jobs overseas.

To fix this flaw, Lighthizer demanded a sunset clause, which would terminate the new agreement after five years unless the three countries agreed to renew it. This was controversial. The Mexicans, Canadians, and even Speaker Paul Ryan dismissed the idea, calling it a nonstarter. With guidance from Lighthizer, I worked with Luis on a compromise. After an intense day of negotiations, I invited the foreign minister to my house so that we could discuss the sunset clause issue privately.

We arrived after 10:00 p.m., hungry and exhausted. I found leftover Chinese food in the fridge, and we helped ourselves. When Ivanka came downstairs, she found us sitting among a pile of empty white boxes. She was mortified that I'd allowed a guest to eat our children's leftovers. "Next time, if you give me a bit more notice, I'll prepare a proper meal," she said.

Luis and I strategized past midnight. Finally I pitched him on an idea that I had previously cleared with Lighthizer: What if we included a sunset clause that automatically terminated the agreement after sixteen years, unless all three countries agreed to an extension in the interim? The parties could hold a joint review in six years to evaluate the agreement and make adjustments. If the parties agreed to an extension, the term of the agreement would reset for another sixteen years. If they didn't, a ten-year termination clock would start to tick, and pressure would build on the parties to resolve their differences as the expiration date approached.

At our next official meeting with the Mexicans, held in Lighthizer's office, I raised the matter of sunsetting. "Let me share a proposal to resolve it," I said.

Before I could get any further, Jésus Seade interjected: "No, no, I have an idea!"

Rule number one of negotiation is to always let the other side go first. "By all means," I said.

Seade pulled out his briefcase and circulated a two-page document that was strikingly similar to my idea, but with one substantive differ-ence: rather than a deadline of sixteen years, he proposed twelve. This was even more advantageous for the United States—and a case study in why it's best to let the other side make the first move.

"That's constructive, but not as good as we need," I said, trying to hide my disbelief.

We took a quick break, and I pulled Luis into a small conference room. Trying to contain my amusement, I asked what he wanted to do. We'd already unofficially agreed to a sixteen-year term, which we knew both of our presidents could approve. Seade's proposal threw a wrench in our plan. Luis had an idea: we could ask Guajardo to object and ask for eighteen years, and then we could negotiate and settle on the sixteen we'd originally planned. This would get Seade off the hook for his offer. Most importantly, it would close out the final outstanding issue of our marathon negotiations.

We all filed back into Lighthizer's office for a round of Kabuki theater. Everyone played their roles perfectly, delivering a win for all. After the Mexican delegation departed, Lighthizer and I looked at each other and laughed. That was one of the worst negotiating moments either of us had ever seen. "Just remember," Lighthizer said, "no one gets smarter by talking."

The next day, August 27, the Mexican delegation came to the Oval Office. With President Peña Nieto on the phone and the press corps in the room, Trump announced that we had reached a preliminary deal with Mexico. Shortly before the president's announcement, Seade and Ildefonso stood outside the USTR office and held their own press con-ference. Seade proudly claimed credit for the sunset clause. Here and throughout my time in government, I saw firsthand the wisdom in Presi-dent Harry Truman's adage: "It's amazing what you can accomplish if you do not care who gets the credit."

24

USMCA

P lease let the prime minister know that his negotiators are about to blow up a $600 billion trade relationship over butter."

It sounded like an outlandish skit from *Saturday Night Live*—but I was talking on the phone with Steve Schwarzman, the founder and chairman of the Blackstone Group. I had made the call from my apartment in New York, where I was getting ready to attend Trump's address to the General Assembly of the United Nations on September 25, 2018. When I learned that Schwarzman was planning to meet Prime Minister Trudeau of Canada, I asked him to relay a message. Although we had come to an agreement with Mexico, we were still waiting on a final answer from Canada—and we were nearly out of time.

Peña Nieto's term as president of Mexico would end on November 30, and we needed to sign an agreement before he left office. To complicate matters, US law required the text of any deal to be made public for sixty days before the president could sign it. This gave us a deadline of September 30—just five days left on the clock.

"We are down to the short straws," I told Schwarzman. "They are playing chicken with the wrong guy. Trump would be thrilled to go forward with Mexico and impose tariffs on Canada. He made a promise to the dairy farmers, and he isn't going to budge."

Schwarzman called back a few hours later. He said that Trudeau had gotten the message loud and clear and had instructed his team to give a final counter offer that he wanted to review himself.

It had taken a month of hard work to get to this point.

The bargaining began within twenty-four hours of Trump's announcement that we had struck a deal with Mexico. Canadian foreign minister Chrystia Freeland, Trudeau's chief of staff Katie Telford, and Trudeau's top adviser Gerald Butts flew to Washington. Upon their arrival, Telford came to my office and leveled with me. We needed to settle three issues. The first two were Trump's tariffs on steel and aluminum, and Canada's one-sided mechanism for resolving trade disputes. I knew we could solve these, so it came down to the third issue: dairy. This one would be tougher and a potential deal-breaker.

Back in the 1970s, Canada had imposed domestic price controls that allowed its dairy farmers to charge artificially high prices. At the same time, an import tax prevented American farmers from enjoying access to Canada's market. These barriers applied to a wide range of dairy products, but not to ultrafiltered milk, which is an ingredient in baby formula, cheese, and other processed foods. Because this sliver of the market remained relatively open, many Wisconsin dairy farmers had invested in expensive equipment to make ultrafiltered milk. In 2016 alone, they had rung up more than $100 million in sales.[36] To stymie these profits, Canadian policymakers came up with new restrictions on ultrafiltered milk, which threatened to put dozens of American dairy farms out of business.[37] Trump had met some of these farmers early on in his presidency, and he was determined to fight for them.

For the next three weeks, Lighthizer and I met daily with Freeland, Telford, and Butts in what became an increasingly frustrating series of negotiations. Though Telford and Butts instinctually wanted to drive the discussions toward a constructive conclusion, Freeland was in no hurry. Like Guajardo of Mexico, she was a rising star in her country's political ranks. During hours of meetings, she read from the notes she had scribbled in ink on her hand. Then she let Lighthizer spar back and forth with her trade experts on technical matters, all while refusing to commit to any substantive changes. Following this theater, she would walk to the steps of the USTR building and hold an outdoor press conference, uttering platitudes like "I get paid in Canadian dollars, not US dollars."

After three weeks of delay from Freeland, Lighthizer directed his staff to prepare two documents: a bilateral deal with Mexico and, in case our northern neighbors decided to join at the eleventh hour, a trilateral deal that also included Canada.

After I called Schwarzman and asked him to speak to Trudeau about our impasse over butter, I learned that Peña Nieto had also had a frank discussion with the Canadian prime minister. He encouraged Trudeau to consider whether his trade negotiator had brought the deal as far as she could. She had set the table, but finishing the deal would require an executive decision. When Trudeau confided that he still didn't want to do it, Peña Nieto delivered an ultimatum: he intended to sign for Mexico, with or without Canada.

Around the same time, on September 26, Trump held a press conference and a reporter asked him whether Canada would join the deal. The president seized the chance to negotiate through the media, a tactic he had mastered. "With Canada, we'll see what happens," he said. "They are charging us three hundred percent tariffs on dairy products; we can't have that. . . . So Canada has a long way to go. I must be honest with you, we are not getting along at all with their negotiators. . . . If Canada does not make a deal with us, we're gonna make a much better deal."

Less than an hour later, the Canadians gave us an offer in writing. After sixteen months of stalling, they were finally ready to talk specifics.

I knew a lot about what separated our two sides, but I was no expert in the arcane details of the dairy provisions. I sent pictures of the documents to Lighthizer and his top deputy, C. J. Mahoney, before heading into a long meeting with Prime Minister Bibi Netanyahu to discuss our peace plan.

When I called Lighthizer after the meeting, he exclaimed, "This is all rubbish! They don't want to make a deal—this doesn't work."

"Can C. J. and I get on the phone with Katie and explain why it doesn't work and give them one last chance to take our final offer?" I asked.

"No," Lighthizer shot back.

"Why not?" I questioned.

C. J. piped in. "Haven't you seen *The Godfather*?" he asked. "That's how the Godfather gets shot."

"Okay, guys, you don't have to break ranks," I said. "But I think they want to make a deal and this is a good faith offer. Let me go back to them one more time. In order to do so, I need to get every detail exactly right. Can you walk me line by line through their offer and tell me what we would accept?"

Lighthizer agreed, and the next morning, we spent nearly two hours going through the details. Then I called Telford and went through the changes we needed.

She said it was going to be tough to get Trudeau on board, but she and Butts were heading into a meeting with the prime minister shortly, and promised to call me back after and let me know his answer. Telford and Butts called an hour later: The prime minister was going to take the deal.

We had less than eighty hours before the deadline to submit the new deal for congressional review. Lighthizer and his team worked through the night to finalize the technical details. On Sunday afternoon, Lighthizer and I visited Trump in the White House residence and briefed him.

"Bob," said Trump, "why don't you go out and do the press conference tomorrow and sell the deal? I have never seen a trade deal in my life that was received positively."

Lighthizer and I were completely deflated. We had worked on this agreement for nearly two years. At times, it felt like an impossible task. But when we encountered resistance, we kept pushing forward, reaching an even better agreement than we'd expected. Now, the president wanted us to prove that the typically hostile press was going to portray the deal as positively as we described it to him.

Just as we were walking out, Trump added, "I want it to be called the U-S-M-C-A, like the US Marine Corps," he said, making a final tweak to the deal.

As the clock neared midnight, we sent the freshly inked deal to Congress, beating the deadline by just thirty minutes and ensuring that we stayed on schedule to wrap up before Peña Nieto left office.

The media reception the following morning was overwhelmingly positive. Ivanka called the president, read him the upbeat headlines, and encouraged him to embrace the victory by making the announcement himself.

A few hours later, Trump took the podium in the Rose Garden. Lighthizer and I stood behind him, along with Treasury secretary Steven Mnuchin and other members of the cabinet. Joining us onstage at Trump's request were the USTR career staffers who had worked tirelessly to draft the highly technical agreement at record speed—just one example of Trump's instinct to thank people who often did not receive enough credit. After the president spoke, he asked Lighthizer to say a few words. Though I did not expect it, Lighthizer thanked me onstage: "I've said before, and I'll say again. This agreement would not have happened if it wasn't for Jared."

Amazingly, the draft agreement never leaked to the press. In fact, days before the president's announcement, *Axios* reporter Jonathan Swan wrote, "Only a tiny circle of administration officials, including Robert Lighthizer and Jared Kushner, have full visibility of the NAFTA negotiations. They've been almost entirely leakproof." That was a high compliment and a rare accomplishment in the Trump White House.

Negotiating a trade deal is like a game of chicken, with real consequences. The other side has to believe you are going to jump off a cliff. We succeeded because Trump was absolutely prepared to terminate NAFTA—and Mexico and Canada knew it. His style made many people uncomfortable, including his allies in Congress, foreign leaders, and his own advisers, but it led to unprecedented results. After thirty years of free-trade globalism that shuttered American factories, USMCA reshaped trade to bring back jobs and achieve better wages for American workers. The $1.3 trillion deal implemented strong "rules of origin" requirements to drive manufacturing back to the United States. It opened up new dairy markets for American farmers. It included detailed and enforceable requirements to give workers a fair wage and to protect the environment—a first in the history of American trade. It took steps to counter China's malign influence in the world economy through a provision to kick any party out of the deal if it joined a trade agreement with China. It also featured an innovative sunset provision to hold Canada and Mexico accountable to the terms of the deal, and it ensured that trade disputes would be settled in American, Mexican, or Canadian courts, rather than in a globalist international forum. The USMCA changed America's legacy on trade. We set forth a new "America First" template for American officials to use in future negotiations with other countries.

*   *   *

Several weeks later, on October 18, I took a rare day off and traveled to New York for my brother Josh's wedding. We forged our close bond growing up, playing basketball and hockey together almost every day after school. As we drove to a friend's house near the wedding venue, I was reminded of what life was like outside the pressure cooker of Washington. Halfway through the car ride, however, my government phone rang. I glanced at the caller ID and saw the source of the call: "White House Situation Room." A call from this number usually meant that the president wanted to speak with me. When I picked up, the operator asked me to hold for General Kelly. He rarely reached out, so I thought this was odd.

"Where are you?" Kelly barked.

I said that I was up in New York to attend my brother's wedding.

"You need to get back down here right away," he said.

"What's happening?" I asked.

Kelly said that caravans from Central America were moving across Mexico's southern border and heading to the United States. "The president is going nuts and yelling at Secretary Nielsen. I need you to come back right away and work on this."

I said that if the situation was truly a crisis, I would charter a plane to Washington later that night after the wedding. He didn't seem satisfied, but we ended the call.

I dialed Luis and asked him for background on the situation, explaining that Kelly had called me with his hair on fire. Luis said that the caravans were still several hundred miles away from the US border and did not present an immediate crisis. And he told me that Mexico could take several measures to ramp up enforcement and confront the caravans. "Let us get these efforts in motion, and we can revisit this in twenty-four hours," he said.

His solutions seemed reasonable, and I surmised that Kelly's fire drill was designed to cater to an audience of one: the president. So I decided to go straight to the source and see if Trump approved of Luis's plan. When I called Trump's assistant Madeleine Westerhout, she informed me that he was in the Oval Office with Mnuchin, John Bolton, and Homeland Security secretary Kirstjen Nielsen as well as Kelly.

"Perfect," I responded. "Patch me through to the president and ask if he can put me on speaker."

I described Mexico's proposal to the group. Trump seemed satisfied, asked me to thank Luis, and told me to enjoy the wedding.

A few minutes later, Mnuchin called me: "You will never believe what happened. When Madeleine came into the Oval Office and said you were on the phone, Kelly jumped up and objected to your involvement."

Apparently Kelly had insisted that I should not be talking with the Mexicans about the caravan issue. The crisis fell under Nielsen's jurisdiction, he insisted, and she had it under control. Trump looked at Kelly dismissively and said, "Of course we want Jared involved in this. He's the only one who's gotten anything done with Mexico. How else are we going to stop the caravans?" Furious that the president had questioned his and Nielsen's ability to solve the problem, Kelly stormed out of the office, left the building, and didn't return to the West Wing for several days. By then, however, he seemed so checked out that no one in the West Wing really noticed he was gone.

25

The Zombie Bill

Around Washington, our criminal justice reform legislation gained an unwelcome nickname: "the zombie bill." After the bill passed in the House with overwhelming support, our opponents ratcheted up their public criticism and stalled its momentum in the Senate. The probability of a bill reaching the president's desk seemed to be diminishing, but I was determined to forge ahead.

In August, while Trump was camped out at his golf club in Bedminster, New Jersey, as the White House underwent renovations, I seized the opportunity to plan a forum on sentencing reform with several of America's most successful governors.

As we convened, the president was running behind schedule. Waiting outside his cottage, I made a call that I'd postponed for too long. Released from prison by presidential pardon two months earlier, Alice Johnson had become something of a celebrity. Initially, I had been reluctant to interject myself into her story: it was hers to tell, and the last thing she needed was a public official taking attention away from her example. But I wanted to let her know that her story was helping our efforts, and so I dialed her number.

"Thank you so much for calling!" she said immediately. "I've been hoping you would. Thank you for saving my life. Kim kept me updated along the way on every one of your conversations. I know what you were up against and thank you for fighting for me and for believing in me. I will never let you or President Trump down. I hope you know, everyone in prison loves you and is following your efforts closely. They're praying for you every day."

I was surprised and pleased to hear this—but the point of my call was to let her know how much good *she* was doing.

"Your story has touched a lot of people, but most importantly, it has touched President Trump," I said. "Your case opened his eyes and his heart. We are about to go into a meeting to get his sign-off on sentencing reform, and I think we are going to get it done."

We concluded the call just as Trump emerged from his cottage and made his way toward a fleet of twenty golf carts and what seemed like a battalion of Secret Service agents wearing tactical gear and carrying massive machine guns.

The governors presented a compelling case for sentencing reform, and I could tell that Trump was giving serious consideration to supporting the provisions that Senator Chuck Grassley had insisted on adding to the bill passed by the House. As we approached the end of August, however, the window for passing any legislation was closing fast. The midterm elections of November loomed. Every member in the House was up for election, and so was one-third of the Senate. For Grassley and a few others, passing the First Step Act remained a priority, but for most members of Congress, getting reelected took precedence over almost everything else. To complicate matters, the political forecasters were predicting that the Democrats would retake the House and Senate. Many Democrats believed that if they gained a majority, they could push for a more liberal bill. If we wanted to pass criminal justice reform the president would actually sign, we needed to get it done before the end of the year.

Meanwhile, from his perch at the Justice Department, Attorney General Sessions watched our activity and grew increasingly nervous that the president might endorse sentencing reform. With his long experience in Washington, Sessions knew that he didn't have to convince the president to oppose us. He just had to persuade him to delay a decision until 2019. This would be enough to doom our efforts. General Kelly scheduled a meeting on August 23 so Sessions could make his case to the president.

By this point, the attorney general's relationship with the president had fully deteriorated. On the day of the meeting, footage from Trump's taped interview with Fox News host Ainsley Earhardt hit the television networks. "Jeff Sessions never took control of the Justice Department, and it's sort of an incredible thing," Trump said, visibly frustrated as he sat in the Rose Garden.

The attorney general fired back with a statement: "While I am Attorney General, the actions of the Department of Justice will not be improperly influenced by political considerations."

Trump could barely stand to look at Sessions during their meeting, which I attended along with Brooke Rollins, but when Sessions warned that sentencing reform would be a jailbreak for criminals, he took the attorney general's warnings seriously and decided to wait until after the midterm elections to make a decision.

His verdict caught me off guard. Prior to the meeting, I had signaled to many advocacy groups and conservative lawmakers that Trump was going to back sentencing reform. I had gotten ahead of myself, and should have seen this coming. Brooke Rollins and I debriefed in my office. We were disappointed by the setback, but from Trump's standpoint politically, it was the right call.

Running low on options, I turned to Vice President Pence for advice. "This is a noble effort, and as a Christian I believe in second chances," he said. "I got this done in Indiana only after gaining the support of law enforcement. That way, anytime somebody would criticize me from the right, I could say 'I worked with law enforcement to do these reforms.' Look, some people are for criminal justice reform, some people are for safety, but for me it's about redemption and I believe you can be for all three."

I went back to my team. "Okay, guys, before we return to the inside-DC game, let's focus on our outside game. We need to get the police groups on board," I said. Rollins jumped into action and reached out to the major law enforcement groups. They all loved and appreciated President Trump and were willing to work with us on sentencing, as long as our reforms made communities safer.

One group that was especially helpful was the International Association of Chiefs of Police, the world's largest organization of police leaders. I arranged for Trump to speak at their annual conference in Florida on October 8, so he could thank them for their service and express his gratitude for their endorsement of the First Step Act.

On our way back from the event, as we boarded the Marine One helicopter at Joint Base Andrews for the ten-minute flight back to the White House, an aide handed Trump a draft of the remarks for the event that evening: the swearing-in ceremony for the newest member of the Supreme Court, associate justice Brett Kavanaugh, whom the Senate had just confirmed after one of the most contentious judicial hearings in American history. When accusations about Kavanaugh's alleged conduct in high school had surfaced, many in Washington called the president and begged him to pull the nomination. Trump often said that nominating a Supreme Court justice was the second biggest decision a president makes, because it's a lifetime appointment. Only the decision to go to war is more important.

The whole controversy surprised Trump. "You're a choirboy," he had quipped, shortly after nominating Kavanaugh. Trump felt good about holding strong on Kavanaugh and not caving to what he believed were false accusations. Yet he was also concerned that the experience would alter the new justice's outlook, and that he'd spend the rest of his career trying to win the approval of liberals and the media by making decisions they favored.

As he marked up the draft of the speech, Trump looked at me. "What did you think of the crying?" he asked, referring to a moment in the hearings when Kavanaugh had broken down in tears.

"I thought it worked for him," I said. "It seemed genuine, and it changed the dynamics of the hearing."

Trump paused and gazed out the window of Marine One as we flew past the Washington Monument, not more than a hundred feet away. Then he looked back toward me and said, "Jared, you go down before you cry."

* * *

Though Trump had handed Sessions a victory by agreeing to delay criminal justice reform, their relationship was rapidly nearing its end. The atorney general's vehement opposition to reform began to irritate the president and the media. Rollins and I had assembled a robust coalition of conservative support, which included many of Trump's friends and allies. They reinforced with the president that our reforms were consistent with his conservative values—and that Sessions was dead wrong.

By November we had earned the endorsement of seven major law enforcement organizations as well as more than two thousand conservative and faith-based leaders. Pastor Paula White, a longtime friend and pastor to Trump who led our outreach to evangelicals on the 2016 campaign, worked tirelessly with faith leaders to build support for our bill. The faith community's passion for the issue was key to keeping Republican members of Congress engaged.

In the Senate, we continued to lean on a group of lawmakers who brought unique skills to the table. Chairman Grassley held the judiciary committee gavel, and his principle and passion combined to make him a bulldog for reform. Mike Lee was an exceptional lawyer and carried significant sway with Senate conservative holdouts like Ted Cruz and Marco Rubio. Tim Scott, the only Black Republican senator, was an effective legislator and could speak with moral authority on the disparities in America. And Lindsey Graham, the gregarious and dogged South Carolinian, had mastered the art of getting skeptical colleagues to yes and was a fearless advocate in the press.

On the Democratic side, I had been speaking almost daily with Dick Durbin. When the House passed the First Step Act without sentencing reform, he withdrew his support from our proposal. He later rejoined our effort after he saw that we were serious about including sentencing reform in the Senate bill. We were constantly worried about losing his support. Because of the Senate's filibuster rules, we needed to secure more than sixty votes. This meant that we couldn't lose the backing of Durbin or the other key Democrats. Cory Booker, who had previously opposed the bill, came on board and proved instrumental in expanding our coalition.

At one point, we made several concessions to law enforcement groups, which required us to remove a few provisions that were important to Durbin. Ja'Ron Smith, our legislative lead, called me in a panic. He had heard that Durbin was on the verge of pulling his support. I headed straight to the Senate: "If you aren't comfortable, then I am not comfortable," I told Durbin as we met in Grassley's office. "We all started this together, and we are going to finish it together." Our teams worked through the night, and Ja'Ron masterfully led the negotiations to a consensus that held our coalition together.

The updated bill contained several breakthroughs. First, it lowered mandatory minimums for nonviolent offenders, including the life-in-prison sentences for certain nonviolent drug offenders like Alice Johnson. Second, it made sure that the penalties for possession of crack cocaine, the most prevalent form of cocaine in Black communities, were proportionate with the penalties for possession of powder cocaine—and it applied this relief retroactively, so that those currently serving unfair sentences could gain release. Third, it gave judges more discretion in sentencing, so they could impose harsh sentences on genuine threats to the community and more lenient sentences on those with minor or no criminal histories. And finally, it reformed "good-time credits" to make sure that inmates who demonstrated good behavior were not imprisoned longer than they should be.

As we incorporated these sentencing reforms into the bill, I received an unexpected call from Anthony Romero, the executive director of the American Civil Liberties Union. Knowing that the ACLU's endorsement would encourage Democrats to join our coalition, I had met with Anthony several months earlier to ask for their support.

"Congratulations, you now have our endorsement," he said. "I promised that if you included retroactive sentencing reforms, we would support it—and I always keep my promises." I had forgotten about Romero's promise, but I was grateful that he kept his word. Nearly every Democrat who had initially opposed the bill—including Jerry Nadler and Kamala Harris—immediately came on board after the ACLU endorsement.

In the midterm elections on November 6, the Democrats won back the majority in the House of Representatives, but the Republican losses were milder than expected. In the Senate, Republicans not only retained control but gained two seats—a rare achievement in a midterm election, when the president's party usually suffers setbacks.

The day after the midterms, Trump fired Sessions, removing the biggest internal impediment to sentencing reform. But even with Sessions gone, two daunting obstacles remained. First, the president still hadn't made a final decision about sentencing reform. Second, even if he did decide to support it, we still needed Senate majority leader Mitch McConnell to move the bill through the Senate, and we knew he wasn't eager to do so. Several prominent Republican senators still opposed the legislation, and McConnell—a six-term Kentucky senator and a virtuoso in the art of electoral politics—was loath to spend Senate floor time on an issue that divided the party.

On November 14, Rollins, Ja'Ron, and I organized a presidential meeting with a broad swath of our coalition: lawmakers, advocates, and law enforcement leaders. On several occasions, Trump had hinted that he was almost ready to endorse our expanded version of the First Step Act, including the sentencing reforms. I thought that if he heard from some of the most powerful conservative reform advocates, he might endorse the bill on the spot. The timing was important because the next day Trump was scheduled to meet with McConnell to discuss the legislative priorities before the end of the year.

I briefed Trump on the meeting that we were about to attend, handing him a copy of my bound, two-inch-thick book of endorsement letters from supportive groups, including law enforcement and his strongest evangelical supporters. I wanted to be ready in case he decided to come out publicly in favor of sentencing reform then and there. I even prepared a draft speech, in case he needed it.

Hoping for the best, I asked the president if he was ready to endorse. It was a big moment, and I knew that the fate of our project probably rested on what he said next.

"Let's do it," he said.

When McConnell met with the president the next day, the majority leader explained that there wasn't enough time to pass criminal justice reform. He was trying to pass other legislation, including a contentious bill to fund the government. General Kelly had excluded me from the meeting, but Trump summoned me: "Get Jared in here," he ordered. As soon as I walked in, Trump said, "Mitch, why don't you tell Jared what you just told me about his bill."

McConnell chuckled. "I've been in Washington a long time, and I must say, Jared is one of the best lobbyists I've ever seen. Mr. President, at this point, I think Jared has had every single person I know call me to lobby for this bill."

"Mr. Leader, that's not true," I quipped. "I have spoken to a lot more of your friends who haven't called you yet."

Everyone, including McConnell, laughed.

"I appreciate your passion for this issue and your persistence," said McConnell, "but it will take ten legislative days to do this. We don't have time on the calendar. We have to fund the government. Why don't we wait until next Congress?"

"If we punt until the next Congress," I said, "the Democrats will change the deal. We have carefully negotiated this bill to get everyone on board, and my coalition is already hanging by a thread."

While I knew a lot less about Senate procedure than McConnell, I was certain that I could get the Democrats to shorten the time to just one or two days. But I didn't want to fight about process in front of the president, so I made a suggestion: "Let me work and see if we can reduce the number of days this will take."

"That sounds good," McConnell said, ending the conversation. He probably believed that he had effectively delayed the vote, but just the opposite was true.

After the meeting, I updated Chairman Grassley and suggested that he call the president as we flew to Florida. Grassley did and told Trump that McConnell was dead wrong on the timing. If we applied enough pressure, he said, McConnell would take the path of least resistance and move the bill.

Aboard Air Force One, Trump drafted a tweet: "Really good Criminal Justice Reform has a true shot at major bipartisan support. @senatemajldr Mitch McConnell and @SenSchumer have a real chance to do something so badly needed in our country. Already passed, with big vote, in House." The president typed it on his phone, adding one of his signature flourishes at the end, "Would be a major victory for ALL!"

Then Trump made a comment to me that he did not share on Twitter: "McConnell only cares about staying in power. Let's do something great to help a lot of people."

Between Grassley's call, the president's tweet, and the Democrats agreeing to reduce the amount of floor time needed, McConnell relented and scheduled a vote.

Now we just had to make sure it passed.

* * *

Amid the wrangling over the criminal justice reform bill, I joined Trump on a trip to Argentina for a meeting of the G-20, a forum for the world's wealthiest countries. In the days leading up to the trip on November 29, I worked nonstop to pass the First Step Act and to prepare for what would be the signature moment on Trump's itinerary: a ceremony for the signing of the USMCA. Because Peña Nieto was about to leave office, we had to wrap up the North American trade deal in South America.

A few weeks earlier, Luis had called to tell me that Peña Nieto wanted to present me with the Aztec Eagle. Not one for awards—or the pomp and circumstance that can surround them—I thanked Luis but demurred. I thought that the signing of our unprecedented new trade deal was reward enough. Besides, I'd never heard of the Aztec Eagle. Curious, I did a quick Google search and found that it was Mexico's highest civilian award—their equivalent of America's Presidential Medal of Freedom. Luminaries like Walt Disney, Dwight Eisenhower, and Nelson Mandela had received it.

I called Luis back. "I didn't realize that this is such a big honor. I'm humbled. Thank you very much. Please tell President Peña Nieto thank you as well."

When I mentioned the Aztec Eagle to Trump and asked his permission to receive it, he joked: "After sticking it out when no one thought we would ever get this deal with Mexico done, you deserve more than a sash."

His itinerary for the two-day trip was packed, and I didn't want him to feel obligated to attend the ceremony, so I didn't invite him and requested a small, private event. But as we flew in Marine One from the White House to Joint Base Andrews, where Air Force One awaited us, Trump leaned toward me, so I could hear him over the thrumming blades of the helicopter. "Do I have to wait for you to invite me to your award ceremony?"

"I learned my lesson from overbooking you in Saudi Arabia," I said. "I didn't want to bother you."

"I want to come," he said. "This is a big honor, and you earned it. You are always there for me. I want to come and be a part of this for you." I thanked him and rearranged the time of the ceremony so that he could be there.

I spent the ten-hour flight to Argentina calling dozens of senators to secure their votes for the First Step Act. Air Force One had several telephone operators on its upper level. They could track down almost anyone in the world, and no one could turn down their announcement: "Hello, this is the Air Force One operator calling you on behalf of Senior Adviser Kushner from Air Force One, please hold while I transfer the line." Trump once joked: "These guys are so good at finding people that if I asked, they could probably get Elvis on the phone."

On the morning of November 30, less than twenty-four hours before Mexico inaugurated its new president, Trump, Trudeau, and Peña Nieto signed the United States–Mexico–Canada Agreement. The USMCA was the largest and most advanced trade agreement in the history of the world. Its thirty-four chapters, four annexes, and sixteen side letters created the highest standards in environmental and labor protections, and it was by far the most favorable trade deal for American workers ever signed.

Right after the USMCA signing ceremony, Peña Nieto presented me with the Aztec Eagle, a beautiful medallion with a golden eagle layered over a turquoise backdrop and framed by a five-pointed star. Before handing me the award and pinning a bright yellow ribbon on my lapel, Peña Nieto called me a "great ally of Mexico" and "an important actor" in the relationship between our two countries. While I felt uncomfortable being the center of attention, especially with the president sitting in the front row, I was proud of what the award symbolized: the respect and friendship I had built with Peña Nieto and Luis, and the magnitude of what we had achieved in resetting the US relationship with Mexico. Just two years before, Democrats had made the US–Mexico relationship a central issue on the campaign, accusing Trump of racism and xenophobia toward the Mexican people. Against every expectation, we had completely flipped the script, leaving both countries better off.

That evening, the leaders of the G-20 dined in the renowned Teatro Colón opera house in Buenos Aires. Before the meal, each head of state sat in an opera box with a spouse and two guests. Trump invited me and Ivanka to attend with him and Melania. We absorbed the breathtaking beauty of the magnificent theater. Gold-gilded boxes, red velvet seats, and mid-century light fixtures wrapped around the oval theater, which was crowned by an octagonal dome with a 700-bulb crystal chandelier. Built over two decades around the turn of the twentieth century, it was widely considered to have some of the best acoustics in the world.

As we took our seats, the lights dimmed, the room quieted, and the performance began. After the frenetic pace of meetings, remarks, and press conferences, the world's top leaders listened in stillness to the magnificent performance. I glanced around the room and thought about how all the leaders had to confront the burdens of their offices. The sleepless nights, constant worries, and impossible decisions were etched into their furrowed brows. While they put on a strong face to represent their countries on the world stage, the fear of future problems penetrated their gaze. They were all masters of their craft, who had outmaneuvered their opponents. But in that moment, I realized that while Trump faced enormous challenges at home, so did every other leader.

Emmanuel Macron of France was confronting yellow-vest protesters who were marching in the streets, vandalizing property, and calling for his resignation. Angela Merkel, who had indicated that she was nearing the end of her thirteen-year tenure as chancellor of Germany, could not escape the sharp criticism for her management of Syrian refugees. In the

United Kingdom, Theresa May had struggled with her country's looming exit from the European Union and was essentially a lame duck prime minister. In Russia, Putin was Putin—he always had problems but maintained his grip on power and caused chaos for others. Saudi Arabian crown prince Mohammed bin Salman was dealing with the global outcry from the death of the journalist Jamal Khashoggi, who was murdered at the Turkish consulate in Istanbul. In Japan, Shinzō Abe's popularity had plummeted after an alleged scandal within his government, and his upcoming election suddenly looked difficult. These world leaders appeared calm and in control, but they all had challenges, they all had flaws. They were all human.

At the corner of the concert hall, I caught a glimpse of Luis, a solitary figure in Mexico's box. It was November 30, Peña Nieto's final day as president of Mexico. Before the expiration of his term at midnight, Peña Nieto had flown back home, leaving Luis as his stand-in at the G-20. Luis had now served his county for fifteen years, first as chief of staff, then finance minister, and finally foreign minister. There in the presidential box, surrounded by the most powerful people in the world, he served out his final hours of a successful government career, engrossed in the performance and smiling from ear to ear, an unmistakable expression of happiness and relief.

After the formal dinner, Luis met Ivanka and me at a famous Argentinian steakhouse. Katie Telford joined us as well. At midnight, we raised our glasses of Argentinian Malbec and toasted to the end of Luis's devoted public service and the beginning of his life outside of government. I reflected on the fleeting nature of our time in government, and I remembered the advice of Canada's former prime minister Brian Mulroney: The only things that remain after our service are the changes that we bring to government and the friendships that we build along the way.

The day after the USMCA was signed, Trump was scheduled to have a globally anticipated meeting with President Xi of China. The tariff war between our two countries had intensified. Since February 7, 2018, Trump had imposed five separate rounds of tariffs on Chinese imports into the United States. Xi had retaliated in kind by surgically placing tariffs on agricultural goods from swing states. But instead of retreating,

Trump doubled down and retaliated with even more tariffs. Leading up to their meeting in Argentina, Trump was threatening to increase the tariffs on $200 billion in Chinese imports from 10 percent to 25 percent. Despite economists' predictions that such tariffs would trigger a global economic downturn, Trump rightly believed that the United States had the upper hand, and that if he continued to apply economic pressure, China would bend.

As the two leaders met, Trump sensed that Xi was ready to make a deal. Trump agreed to put a ninety-day pause on the additional tariffs and instructed the negotiators to get to work quickly. He pointed to me at the far end of the table. "Jared did an amazing job working with Bob Lighthizer on the incredible USMCA trade deal we signed yesterday. He did so well that Mexico just gave him their highest award. Now I'm asking him to get more involved and work on this China deal with Bob and Steven Mnuchin. But no pressure, Jared," Trump said as he leaned forward, looking down the long table, and caught my gaze. "If it doesn't get done, I'm blaming it on you."

26

Victory and Defeat

We were all glued to the TV. The speeches were nice, but we were yelling at the television 'Trump, sign the damn bill already!' " said Matthew Charles, as he described what it was like to sit behind bars and watch the president sign the First Step Act. He became the first inmate released because of the new criminal justice reform law. I invited him to the White House after he got out of prison, and I asked him what it had been like to keep track of the developments from afar.

"Politicians had promised us criminal justice reforms for more than a decade," he said. "We all worried that it was a mirage that would vanish at the last second. We followed every twist and turn of the legislation, and when Trump tweeted at McConnell, there was a big applause in my prison."

In December of 2018 Congress passed the new and expanded version of the First Step Act. In the House of Representatives, the vote was 358 to 36. In the Senate, it was 87 to 12. This handed the president a major bipartisan victory and one of historic magnitude—but more importantly, it would help thousands of people like Matthew Charles who deserved a second chance.

On December 21, at the signing ceremony in the Oval Office, the president was struck by the makeup of the group that had helped us achieve this remarkable victory: Republican and Democratic lawmakers, conservative and progressive advocates, law enforcement professionals, and former inmates crowded behind the Resolute Desk. This unique cross-section of America was probably one of the most unlikely groups ever to assemble in the Oval Office.

From his chair behind the Resolute Desk, the president told the story of a judge he met who had recently left the bench because he was forced to sentence a young man to twenty-eight years in prison when he believed he only deserved two. Then he made a move that few politicians would ever have the chutzpah to do: he invited his guests to speak extemporaneously. In most administrations, public comments are carefully scripted. But Trump would often take a risk and invite his guests, many of whom he'd never previously met, to give remarks that hadn't been cleared with anyone at the White House. This created raw and riveting made-for-television moments that brought his message home.

Mike Lee was quick to jump in: "It's almost hard for me to speak about this without being emotional. In the process of this, this has brought together friendships that I will cherish for the rest of my life. I'm now texting buddies with Van Jones, Dick Durbin, and with Cory Booker, and I speak to Jared Kushner about five times a day."

Trump next motioned to Van, his frequent critic.

"There's nothing more important than freedom," said Van. "And the freedom of people who are trapped in a broken system, the freedom of people who are trapped in addiction, the freedom of the people who are trapped in poverty—those are the people that your opportunity zones are targeted at, your opioid policy is targeted at, and your criminal justice policy is targeted at. And when you're trying to help people on the bottom, sir, I will work with or against any Democrat, with or against any Republican, because there's nothing more important than freedom."

Van had suffered vicious attacks from many of his political allies on the left for working with Trump and me, but through his courage and conviction he had now made a more serious impact than any of his Twitter detractors could ever have imagined.

Around this time, Matthew Charles was probably shouting at the television for Trump just to sign the damn bill, but it took Chuck Grassley to get him to do it. Growing restless, the senator leaned toward the president and whispered: "Sir, would you mind signing the bill? I need to go back to the Senate for a vote—I haven't missed one in twenty-five years!" The president signed the legislation with his usual oversize Sharpie.

As soon as the event concluded and I had said goodbye to our guests, I slipped into my office and called my dad. "God works in mysterious ways," I said. "Maybe you paid the price then, so that thousands of families could get relief now and for years to come."

"I'm nothing—my life doesn't matter," he said. "This is so much bigger than me. I'm so proud of who you are and what you've done. You just made the pain that we felt go away for thousands of families. Our family has paid a big price for your service, but to me, this alone makes everything worth it thousands of times over."

Hours after Trump signed the First Step Act, Cassidy came in and told me that the president wanted to see me back in the Oval Office. I wondered if my father-in-law wanted to reflect on what we had just achieved. But as I walked in, I saw the legislative and budget teams assembled, and immediately knew that our momentary celebration had passed: Trump had turned to the crisis at hand. The federal government was on the verge of shutting down because the congressional spending bill failed to include sufficient funds for building the wall on the southern border.

"Jared, why have you been spending all of your time on prison reform instead of working on immigration? I didn't campaign on prison reform. The wall is my number one issue."

"Sir, General Kelly has been running this issue," I explained. "He gave me strict instructions not to touch immigration." Trump sighed and told me to get involved immediately.

* * *

Earlier in the year, Congress had sent the president a sprawling $1.3 trillion government funding package with only $1.6 billion for the wall— well short of the $25 billion Trump had wanted for a project he had mentioned in nearly every one of his campaign speeches. On March 23, minutes after the bill arrived at his desk, Trump tweeted, "I am considering a VETO of the Omnibus Spending Bill, based on the fact that the 800,000 plus DACA recipients have been totally abandoned by the Democrats (not even mentioned in Bill) and the BORDER WALL, which is desperately needed for our National Defense, is not fully funded."

After a frantic call from Speaker of the House Paul Ryan, who warned about the dangers of a government shutdown, Trump decided to sign. But he wasn't happy. In televised remarks from the Diplomatic Reception Room, with the thousand-page bill stacked theatrically beside him, he made a promise: "I will never sign another bill like this again."

Now, nine months later, Congress had done it again, sending him a huge bill to fund the government but providing a measly $1.6 billion for the border wall. Trump had made up his mind to take a stand this time around. The wall became a benchmark for measuring Trump's success: if the Democrats could stop it, they would claim that Trump was all talk and no action. Tens of millions of his voters closely associated the wall with the Trump presidency, and failing to deliver on his promise would hurt his credibility. Trump often joked that the easiest way for him to get the wall funded would be to come out against it; then the Democrats would again be for it.

If Congress and the president failed to agree on a budget, funding would lapse for more than half of the government. In practical terms, this meant that nine of the fifteen major departments would shut down, along with dozens of smaller agencies. Approximately 380,000 federal employees would be furloughed, while another 420,000 would have to work without pay, including security officers at airports and customs and border officials at ports of entry and along the international frontier.

Was the president willing to risk it?

"I don't know yet, but I need to try and fight for the wall—and hopefully find a way forward," Trump said.

My reluctance to touch the immigration file extended back to the summer of 2017, when Ivanka and I had hosted a bipartisan dinner at our place in Kalorama. The group included Democratic senator Dick Durbin, Republican senator Lindsey Graham, and White House adviser Stephen Miller. Immigration had become such a toxic political issue that

Democrats and Republicans were afraid even to talk to each other about it. At the urging of Graham and with the blessing of the president, I planned a private discussion with the Democrats to identify common ground to improve our nation's immigration system and build the border wall. Watching the friendly repartee between Durbin and Miller at our dining room table, I was struck by the constructive discussion and the opportunity that lay before us. They had differences of opinion on some aspects, but both were surprised by a number of points on which they agreed. By the end of the dinner, we'd reached a general agreement to explore a deal allowing existing unaccompanied immigrant children—known as DACA recipients—to stay in the country in exchange for a fully funded border wall.

"That's the first substantive discussion I've ever had with a serious Democrat on immigration," Miller told me afterward.

The next day I called General Kelly, then still the secretary of Homeland Security, to fill him in. "The president asked me to explore a scenario where he could trade DACA for the border wall," I said. "Can you share the technical specs of what we're trying to build—how many miles of wall we need, how quickly we can build it, what else is involved to complete the wall system, and the price tag?"

"We don't really have that together yet," he replied.

This response astonished me. The border wall was the president's signature campaign promise, and six months into the administration, the secretary of Homeland Security didn't have a plan.

Kelly questioned why he was talking to me, rather than Reince Priebus, the chief of staff at the time. I explained that the president had asked me to work quietly on the issue. I didn't realize it at the time, but this exchange probably planted seeds of distrust. I was trying to protect the president and solve a problem, but in retrospect, my approach was amateurish. Had the situation unfolded in our fourth year at the White House, rather than our first, I would have asked the White House policy team to solve the dilemma for the president, and then would have helped to execute on his decision. When Kelly joined the White House staff, he probably thought I was a bad actor who operated around the chief of staff. One of his first moves as chief was to order me to stay away from the immigration portfolio. He wanted to run it himself.

One of the reasons Trump chose Kelly as chief of staff was the perception that he'd been enormously successful at cracking down on illegal immigration as secretary of Homeland Security. During the first several months of the administration, border apprehension numbers—a key indicator of illegal immigration from Mexico—dropped off precipitously, falling 75 percent from their preelection levels. The president was impressed, and everyone praised Kelly. It later became evident, however, that the steep drop in apprehensions was related not to any change that Kelly had implemented but to the deterrent effect of Trump's tough campaign rhetoric and the aggressive executive orders he signed in the first months of his presidency.

By the middle of 2018, border apprehensions were skyrocketing. Apparently human smugglers had realized that there had been almost no policy changes under Kelly or his hand-picked successor, Secretary Kirstjen Nielsen. For her part, Nielsen had a good grasp of the technical aspects of Homeland Security, but she seemed unprepared for the complexities of running a department of 240,000 employees. The positive trends from the first days of the Trump administration had reversed entirely.

By law, the president must submit a budget to Congress each year. It includes his funding requests for everything from roads and bridges to health care for veterans. In both the 2017 and 2018 submissions, Kelly and Nielsen had asked for $1.6 billion for the wall. When I asked Kelly and Nielsen why they had submitted such low requests, they argued that $1.6 billion was sufficient and implied that Trump simply didn't understand how the process worked and that there were too many bureaucratic hurdles to build the wall any faster. In a best-case scenario, it would take ten years to build the wall, they claimed. If we asked for more funding, we wouldn't be able to spend it before the next fiscal year.

A few hours later, I went back into the Oval Office, where the president was still seething over his predicament. He was particularly furious at Paul Ryan and Mitch McConnell for sending him a bill without the wall funding.

"Don't be mad at Paul or Mitch," I said. "They got you exactly what Nielsen and Kelly asked for in our budget."

Trump could have called for Kelly and reamed him out, but at this point the two were barely speaking. Around the West Wing, it felt like Kelly had checked out of the day-to-day operations for months, and it only worsened after Trump's announcement on December 8 that the chief of staff would leave at the end of the year.

"You have a terrible hand to play here," I said. "If you veto the spending bill, the Democrats will blame you for the shutdown. They won't cave on the wall and have no reason to, since they will control the House on January 3. Let's retreat today and find another way forward. We can look at ways to get the funding without a shutdown."

Trump listened, but his resolve stiffened: "You are giving me rational advice, but I'm still not going to sign the bill. Throughout my life, I have taken on all kinds of fights with bad hands, and somehow I figure it out. Jared, if I go down, I'm going down with my boots on."

At the president's direction, I jumped into a limo with Vice President Pence and rode up to Capitol Hill, prepared to negotiate with Democrats. After several hours of unproductive meetings, it was clear that Pelosi and Schumer didn't care about finding a long-term legislative fix for the unaccompanied immigrant children if the tradeoff was giving Trump a win on his biggest campaign promise.

As Secret Service drove me home, I thought about the whirlwind of the last twenty-four hours. I had hoped to celebrate the passage of the First Step Act with Ivanka and the kids that evening. But I didn't leave the Capitol until long past the children's bedtime. While I had achieved a massive success, the rest of the White House was in crisis.

Now that Kelly was on his way out, I needed to integrate back into the team and help the rest of the West Wing succeed, while making sure the incoming chief didn't feel threatened. During this time, Sarah Sanders recommended me for the chief of staff job. Trump asked if I wanted to be considered.

"You already have me," I said. "I think you would be better served if you find someone else who is more political and press savvy, and I can help them with operations and execution." I wasn't particularly excited to jump into the shutdown battle, but I knew it was important. One rule applies to both fathers-in-law and presidents. When they ask for help, there's only one answer: yes.

Around the same time, Chris Christie was aggressively lobbying to be chief of staff, telling Trump that the Russia investigation was a result of bad staffing and that he needed a chief of staff who would forcefully defend him on television. When Trump asked me what I thought, I joked that Christie might be better at Homeland Security: "If he can close the George Washington Bridge, maybe he could close the border."

Trump was concerned with Christie's track record in New Jersey and also worried about Christie's ability to keep information confidential: every time they met, the details of their meeting ended up with the same two reporters. I told Trump that Christie and I had worked well together on the criminal justice reform legislation and assured him that I felt comfortable with my role and would work well with whomever he chose. When Trump became aware that Christie had a book coming out in several months, he called Christie and asked him if there was anything that could become a problem. Christie told him that he mostly portrayed Trump well, but that he was a bit tough on me and my father.

This was a deal-breaker for Trump, who told Chris that he couldn't have a chief of staff publishing a book that attacked his family. Christie told Trump that he called the publisher to see if he could return his advance and cancel the book, but the publisher had already spent hundreds of thousands of dollars printing copies. It was too late. Months later, the book, titled *Let Me Finish*, came out. It was filled with vicious and untruthful attacks on me and my father.[38] Ironically, Christie's petty obsession with using my family to get media attention had destroyed his dream opportunity to rehabilitate his image and finish his political career.

27

The Longest Shutdown

The White House is a stunning national treasure, and it especially shines during the holidays, but anyone who spends weeks on end within its walls can start to feel like they are trapped in a museum. Sensing that Trump might appreciate an evening out, Pence invited Trump, Mick Mulvaney, and me to dinner at his residence, the Naval Observatory. Trump had appointed Mick Mulvaney as his acting chief of staff in December to replace Kelly. Before we departed, Mulvaney and I met with the president to discuss his upcoming schedule. Then Mulvaney handed Trump a document to sign. "This will end the practice Kelly started of listening to all of your phone calls," he said, explaining that Kelly had given himself the ability to listen surreptitiously to the president's calls.

"Kelly did what?" the president asked, stunned at the invasion of privacy. "End that immediately."

Over the next four weeks, I traveled up and down Pennsylvania Avenue with Pence and Mulvaney. On the Hill, we talked with Democrats and Republicans and tried to find a path forward. During one meeting with Paul Ryan and other House Republican leaders, we discussed a possible compromise to end the shutdown.

Ryan interrupted: "How do we know if the president says 'Yes,' that he's actually going to follow through?"

Taken aback, I replied: "With all due respect, I think you've misunderstood the president. If you give him all the information and brief him on the facts and the situation, he will make a firm decision. If you try to get him to agree to something without giving him all of the facts, however, he will likely change his mind when he learns them."

I attributed Ryan's disconnect with Trump to his lack of private-sector experience. He'd been in Congress for nearly two decades—since he was twenty-eight. In business, negotiators often agree to a deal in concept, and then have lawyers work out the details. New issues can surface during this second step of the process. Ryan had assumed that he could just call Trump and get him to agree to a conceptual framework without his approval on the final details. As a former businessman, Trump didn't work this way.

One of the greatest tragedies of the first two years of Trump's presidency, when we had majorities in both chambers of Congress, was that neither Ryan nor McConnell understood the president. Like many establishment Republicans, they resented his disruption of the system they had grown used to. They found themselves in a dilemma: they did not fully agree with Trump's style, but they couldn't defy him because their own voters loved him. They had become generals without an army. I often wondered why establishment Republicans didn't seem to respect the sixty-three million voters who elected Trump. Instead of working with Trump to pass legislation that delivered on his promises to voters, a Republican Congress wasted two years ducking the new leader of their party.

After a series of dead-end meetings on the Hill, I began looking for creative ways to fund the wall that didn't require approval from Congress. I collaborated with the president's new White House counsel, Pat Cipollone, a talented Washington litigator and principled conservative who had taken over when Don McGahn had departed in October, and his deputy, Pat Philbin, an understated but remarkably intelligent former Supreme Court clerk. Stephen Miller, Mick Mulvaney, and Russell Vought, who was running the budget office in Mulvaney's absence, and his deputy Derek Kan also joined the effort.

With the federal government spending about $4.5 trillion a year, we figured that we could cobble together a few billion dollars for the wall. After spending a few weeks researching the president's authorities and the federal government's budget accounts, the team came back with a list that included $600 million in a Treasury forfeiture account, $3.6 billion in an account for overseas military construction, and potentially another $6.3 billion through a general transfer and by pulling from a counternarcotics defense spending account.[39] This was incredible. They'd found the government equivalent of nickels and dimes and come up with $11 billion in existing funding in the federal bank accounts.

"Are we certain we have the authority to divert these funds?" I asked.

They explained that under an emergency powers statute, the president had the authority to reprogram military funds. He just had to demonstrate that the United States was dealing with an emergency. With border apprehensions skyrocketing, drug and human trafficking proliferating, and caravans marching toward our southern border, the president had a clear basis for this. We took the idea to Trump.

"We've got to end the shutdown," I said. "It's going to look like you're taking a loss on this, but what matters is that in June of 2020 there will be a big, beautiful wall, just like you promised. And we've now found the funding for it."

The president crossed his arms and leaned back in his chair. "Jared, if I agree to do this, then you have to personally make sure the wall gets built fast. But let's play this out a bit more with Congress and see where we get."

By the end of January, it was clear that our only path forward was the emergency declaration.

On January 24, as we prepared the declaration, my five-year-old son Joseph called: "I miss you, Dad. Can Grandpa end the shutdown so that you can come home?"

My heart sank. I hadn't made it home for dinner or bedtime in weeks. Figuring that this was one of the few problems that could potentially be solved with soft serve, I invited Joseph to come to the West Wing for frozen yogurt from the Navy Mess. Half an hour later, the Secret Service delivered him to West Executive Drive, just outside the West Wing. He had a big smile on his face, and I gave him a hug as we walked inside to the Navy Mess.

Located in the basement across the hall from the Situation Room, the Navy Mess is an intimate, windowless, wood-paneled dining room, lined with rows of mostly two-person tables bedecked with crisp linens, White House china, and fresh flowers. Since Harry Truman established the Presidential Food Service in 1951, the Mess, as it's commonly called, has been a favorite feature of the West Wing among White House staff. It includes a carry-out counter with an excellent menu of items, ranging from salmon fillet to mozzarella sticks, so that overworked staff can grab a meal without leaving campus.

Joseph and I ordered two vanilla frozen yogurts, topped with Oreos. As we ate our soft serve in my office, Pence called for me from down the hall to talk about the emergency declaration and his latest discussions with members on the Hill. I left Joseph with my trusted staff member Cassidy Luna. While I spoke with the vice president, his assistant said that the president wanted us both to come to the Oval to work on his speech for the announcement the next day. I brought Joseph with me. He hugged his grandpa, who as usual slid a piece of chocolate into his grandson's pocket. Joseph spotted the Lego replica of the White House that he had built for his grandpa. Trump proudly displayed it on the mantel in the Oval Office and showed it to world leaders.

After an hour of patiently sitting through meetings about topics he didn't understand, Joseph came up to me with tears in his eyes. "This is boring," he whispered. "I want to go home."

It was the ultimate defeat. I apologized for being so distracted and walked him out to the car. He would barely look at me. That evening, while I was still at the office preparing for the announcement, I called Joseph before he went to bed and asked if he would come back with me to the White House the next morning before school for a special pancake date. No interruptions, and unlimited whipped cream, I promised.

That next day was January 25, the thirty-fifth day of the longest government shutdown in history. For Joseph and me, it began with a 7:00 a.m. pancake breakfast in the Navy Mess. I couldn't imagine a better start to the morning.

A few hours later the president announced that he had reached a three-week deal to end the shutdown, fund the government, and ensure that the federal workers received back pay. When three weeks passed without Congress finding a solution, we handed Trump a presidential proclamation declaring a national emergency on the southern border. On February 15 he signed it, giving us access to $11 billion for the border wall.

The president had been clear: it was now my responsibility to get it built.

As I took up the project—one of the largest American infrastructure endeavors since the building of the US highway system—I organized meetings in the Situation Room with key officials from within the Office of Management and Budget as well as the Departments of Defense and Homeland Security. I had them brief me on the details of exactly what we were building. Was it concrete, steel slats, or barbed wire? It soon became clear that no one had settled on the exact type of structure we intended to erect. As a former builder, the president would have a strong perspective, so I organized a briefing. The experts recommended a thirty-foot-high barrier made of long steel slats, with anti-climb panels lining the top. Trump didn't like the look of the anti-climb panels, but he approved the design at the strong recommendation of Border Patrol.

We also needed to identify the stretches of our border that were most vulnerable to illegal crossings and to the smuggling of people, narcotics, and weapons so that we could focus our construction efforts on priority locations. Much of the land along the southern border was privately held, and the Army Corps of Engineers needed to engage in a cumbersome process of land acquisition, which at times could involve eminent domain, a less-than-ideal legal proceeding that gave the federal government the authority to force private citizens to sell parcels of their land. The Army Corps estimated that this step alone would take six to twelve months to complete. We didn't have that long.

After working through these issues, we decided to define success as building 450 miles of a new state-of-the-art border wall by the end of 2020. This was an aggressive but achievable target that would strengthen our border security in strategic locations, including San Diego, Yuma, El Paso, and the Rio Grande Valley. Some of the wall would rise in places where there were no barriers, and some would replace existing but ineffective fencing. As the construction ramped up, I held weekly meetings in the Situation Room. I always began with two questions, typical of any construction project: Why is it taking so long? And why is it costing so much?

"We are right on schedule," said Lieutenant General Todd Semonite, the impressive three-star general who ran the Army Corps of Engineers, in one of these weekly meetings. As he listed the construction numbers from the previous week, I opened my manila folder, pulled out the schedule from the week before, and double-checked the projections.

"With all due respect, General, you're not on schedule," I said. "Last week, you said that you'd be at a hundred and seventeen miles, and you're only at one hundred and fifteen."

"That's the old schedule," he said. "I'm talking about the updated schedule."

"General, unlike most of the jobs I have been assigned in government, this is one that I have a bit of experience in," I quipped. "I've never had a contractor admit to missing their schedule—they just keep revising the damn schedule."

Everyone laughed.

"I know how to do this stuff. Every time we meet, I need you to give me an update on where we were the day before, and where we were projected to be. There are a lot of moving parts, and things will go better and worse than we expect. Let's agree to have a transparent flow of information, and we will solve problems as they arise."

Out of that meeting, I created a one-page spreadsheet, with specific tracking and updates to monitor the progress, and tasked one of my top lieutenants, an affable jack-of-all-trades named John Rader, to run point on coordinating the project. General Semonite and his team stepped up to the challenge and did an outstanding job. We all accepted accountability as a team, and we started to steamroll through the project.

After Kelly and Nielsen took two years to construct 35 miles of new and replacement border wall, we worked with Acting DHS Secretary Chad Wolf to build more than 415 miles in 2019 and 2020—much faster than the experts predicted. To get it done, I applied a formula similar to the one I'd used for USMCA, the First Step Act, and the Middle East file. I defined success, developed a plan, and built a great team that was creative, agile, and focused intensely on execution. More than anything, I didn't want to let Trump down. He had promised to build the border wall to keep Americans safe and secure. Subsequent reports have found that the wall we built accomplished this goal, disrupting the flow of criminals and smugglers across the border.

Mark Morgan, the charismatic and talented leader of Customs and Border Protection, joked after one meeting that immigration hawks in conservative media like Ann Coulter and Tucker Carlson would have heart attacks if they had known that the fate of the border wall's construction depended on someone they maligned as a "liberal New Yorker."

28

Eureka

By February 2019, the prospect of peace in the Middle East seemed more distant than ever. Israeli prime minister Bibi Netanyahu was caught in a holding pattern until after the April elections, and the Palestinians were still fuming over the Jerusalem embassy and refused to talk directly. Despite the dim outlook, my small team forged ahead.

Over the past six months, we had developed the Peace to Prosperity economic plan—a blueprint for investing $50 billion into the West Bank, Gaza, and the surrounding region if the Palestinians accepted our political peace plan as the starting point for negotiations. Now it was time to gauge world leaders' reaction to the proposal. We made plans for two overseas trips. We would start in Europe with a two-day, sixty-nation counter-Iran conference that Brian Hook, who had been appointed as special envoy for Iran, had organized in Warsaw, Poland. The conference was notable in that it brought together the prime minister of Israel and the foreign ministers of the UAE and Bahrain in a public forum. Next would come the Munich Security Conference in Germany, a vaunted yearly gathering of the world's top authorities on defense and national security. Then, less than a week later, we would sprint through six Middle Eastern countries in five days on what would be my third extended trip to the region.

Vice President Pence was scheduled to deliver the keynote address at the Warsaw Conference on February 14. He invited me to fly with him to Poland on Air Force Two and join him and Polish president Andrzej Duda on a tour of Auschwitz-Birkenau, the former Nazi concentration and death camp where the Nazis had murdered a million Jews.

I had visited Auschwitz twenty years earlier as a high school senior. Back then, when I asked my headmaster to excuse my absence from classes for the trip, he refused, citing the school's attendance requirements. At a meeting in his office, I confronted him: "If you fail me, I accept that, but I believe that in ten years, I will remember more from this trip than from a week of class." He relented, and the trip indeed changed my life.

As we walked along Auschwitz's corridors of death, one of the guides shared a story about his father, who had been a prisoner there sixty years earlier. His father had smuggled in his tefillin—a small box containing Torah inscriptions secured by a leather band, and worn as an act of faith. As the Nazis forced the prisoners to line up and march in the prison yard each morning, he would duck down, take the tefillin's leather band, wrap it around his arm, and say a prayer before handing it to the next prisoner in line. Of all the things the prisoners could have smuggled into the death camp, they chose tefillin. They risked their lives to pray. As a teenager, too often I skipped the morning ritual. The story in Auschwitz inspired me to change my ways. If these prisoners were willing to die for their faith, I should embrace my freedom and make a better effort. From that day forward, I've started my mornings by wrapping the leather band around my arm and praying.

Now, twenty years later, as President Duda of Poland led our delegation through Auschwitz-Birkenau, I was struck by how little had changed since my high school visit. The sites, which together covered 470 acres, seemed frozen in time. We entered beneath the ominous wrought-iron sign—ARBEIT MACHT FREI, German for "Work sets you free." Then we moved through the carefully spaced rows between the barracks, where the Nazis packed emaciated human beings onto wooden shelves as ice, rain, sleet, and snow leaked through the ceiling. We passed the clinical room, where men, women, and children were stripped of their dignity. We stood where the prisoners had stood, in rows that separated the dead from the living, which seemed to stretch as far as the eye could see.

Visiting Auschwitz as an adult reminded me of why the Israelis constantly live in a state of fear that one day they will find themselves in a powerless position against an oppressor, and why they negotiate every little point as if it is a life-or-death issue.

As we prepared to leave Poland, taxiing on the runway in Krakow, a voice came over the plane's PA system: "Unlike the grandparents of my three sons, this Jew arrived at Auschwitz not in a cattle car but in a vice presidential motorcade. Today he doesn't depart Auschwitz as did more than a million Jews, by the night that 'transformed the small faces of children into smoke under a silent sky.' Today, he is delivered from that hell upon the wings of eagles on Air Force Two."[40]

The voice belonged to Tom Rose, a senior adviser to the vice president. His family also had survived the Holocaust. Rose put words to what I felt, and there wasn't a dry eye on that plane.

When we arrived in Munich, I peered out the window of our motorcade en route to the hotel and marveled at the immaculate buildings and public squares that showcased the most brilliant engineering in the world. My mind flashed to Auschwitz. How could people have taken their ingenuity and creativity and twisted those talents to build perfectly designed factories of death? Even good people are capable of cooperating in incomprehensible evil. It's up to each of us to choose how we will use our brief time on earth.

At the Munich Security Conference, I planned to brief European leaders on our peace plan. To this point, I'd spent the bulk of my time soliciting feedback and building support among the Arab leaders, but I also needed Europe's input before releasing the plan, which we were tentatively scheduled to do following the Israeli parliamentary elections in April. Germany had gone to great lengths to acknowledge the atrocities and try to heal the wounds of the Holocaust, so I hoped that Chancellor Angela Merkel would be receptive. I briefed her on our efforts when our delegations met. She expressed support, but wouldn't commit to backing our plan. She asked me to work with one of her lieutenants. Soon after, I learned from Ric Grenell, our ambassador to Germany, that Merkel had connected me with a paper pusher. Her handoff was a sure sign that she had no interest in rethinking her Middle East policy. French president Emmanuel Macron was no different. When we spoke, he hardly acknowledged that past approaches had failed. His talking points were straight from the traditional foreign ministry playbook, and France would not support a plan unless the Palestinians signed on.

Europe's most powerful leaders showed little interest in breaking from the failed policies of the past in search of a new path to peace.

* * *

On February 23 I departed for the Middle East, along with Avi Berkowitz, Jason Greenblatt, and Brian Hook. The trip included our first visit to the Sultanate of Oman. Strategically located along the mouth of the Persian Gulf, Oman shares a border with Saudi Arabia, the UAE, and Yemen. Across the Strait of Hormuz is Iran, a mere twenty-one miles away.

Upon arriving in the afternoon, we headed to the hotel and waited for a call from the palace with instructions on what time to arrive for dinner with the sultan. In the lobby of the hotel, we bumped into *New York Times* columnist and Middle East expert Thomas Friedman, who was there to give a speech. Over coffee, he revealed that he had followed my efforts closely and appreciated that we were approaching negotiations differently. Whereas our predecessors had tried to play the role of neutral brokers, we were unapologetically standing with Israel on the policies where we agreed, knowing that it would build trust with them. He reminded me of his first rule about the region: "In the Middle East, you get big change when the big players do the right things for the wrong reasons." He insisted that if we weren't planning to offer the Palestinians a state, our efforts would never bear fruit. Not wanting to show my hand to a journalist, I said that we were still working through the issues and trying to capitalize on our strong relationship with Israel.

Shortly after our coffee, we received a call from the palace. It was time to meet Sultan Qaboos bin Said Al Said, a towering figure in the Middle East. A fourteenth-generation descendant of Oman's founding monarch, Qaboos was the longest-tenured leader in the region and the only founding member of the Gulf Cooperation Council still living. Since overthrowing his father in a British-backed coup d'état in 1970, he had implemented significant reforms at a methodical pace. Over his nearly half-century reign, he had abolished slavery, recognized women's right to vote, built modern infrastructure, and transformed his country from a land plagued by poverty and isolation into a prospering and diverse economy respected by its neighbors. Like most Arab nations, Oman did not have formal diplomatic relations with Israel, but the sultan had recently hosted Netanyahu for a visit. This was big news, and even surprised the US intelligence community, which interpreted the overture as a sign that our efforts were changing the sentiments in the region.

As we entered the palace, an official escorted us into a magnificent reception room, where we met three high-ranking ministers dressed in traditional Omani attire with muzzar-style turbans and heavily jeweled daggers on their belts. We exchanged niceties as we waited expectantly for the sultan. Half an hour went by, then an hour. We tried not to show our hunger and exhaustion as we made small talk, but maybe we should have, because they did not give us any indication of when the sultan would arrive. Finally, at 10:00 p.m., two hours after our expected start time, an official announced that the sultan was ready. We were escorted into a windowless mahogany-paneled conference room lined by chairs. Not a trace of food was in sight. Not even a dining table. Hook whispered to Avi, "I guess we aren't having dinner."

The sultan, a small-framed man with a neatly cropped beard and a regal turban, greeted us warmly. His proud and unhurried bearing seemed to convey a battle-tested aura earned from five decades in the world's roughest geopolitical neighborhood.

As we introduced ourselves, I asked questions about the history and personalities in the Middle East. Whereas most of the leaders in the region tended to be animated and even emotional, I was impressed by the sultan's calm demeanor, especially as he told a story about one of his neighbors who tried to kill him, but then coolly stated that they had resolved their issues and he didn't hold a grudge.

His matter-of-fact statement astonished me, but perhaps it shouldn't have. He had survived by picking his battles wisely and taking steps forward at his own speed. He knew his strengths and vulnerabilities, and he was focused on the long game.

When we reached the topic of the Palestinians, the sultan shared a view that I had heard from nearly every leader in the region. Yet he captured the essence of the issue with impressive precision and clarity: the most crucial element of Israeli-Arab peace was access to the al-Aqsa Mosque on the Temple Mount. Going further than other leaders, he expressed his disappointment that for years the Arab media had spread a false narrative that Israel wanted to destroy the mosque. This lie was commonly believed in most Muslim nations, and it needed to be addressed. The sultan clearly sympathized with Abbas, explaining how, for years, Arab leaders had deliberately stoked the conflict between Israel and the Palestinians to deflect attention from their own domestic shortcomings and rally popular support.

He was candid about the fact that, in the past, Arab countries had publicly put pressure on the Palestinians to stand up to Israel and not be traitors. Then, to my surprise, he admitted that these public statements often contrasted with what Arab leaders would say privately, when they were much more willing to admit the benefits that Israel brought to the region. He predicted that the hypocrisy would end only when leaders said publicly what they said privately. As our discussion continued, the sultan placed some blame on Abbas for his inability to find solutions and for his role in perpetuating the conflict. "We are supposed to learn from history," he said, "but you can't live in history."

I was shocked by what I was hearing. Coming from the longest-serving ruler in the Arab world, these statements gave me hope that the sultan would support our peace plan or, even better, establish diplomatic relations with Israel. He told me about how much he had enjoyed his dinner with Bibi and how he saw tremendous opportunities for Oman and its neighbors to collaborate with Israel.

When we finished walking the sultan through the plan, I asked if he thought we had a chance at peace between Israel and the Palestinians. If we don't start, we we will never accomplish or change anything, he said.

Abbas has his limitations, but his heart is in the right place. Then regret crossed the Sultan's face as he spoke: *I feel badly for the Palestinian people; they carry with them the burden of the Muslim world.*

For more than two hours, the sultan shared stories and gave insights from his unique perspective. I was so wrapped up in the conversation, I had forgotten my hunger. My team, however, was growing tired. Avi and Hook were fighting to keep their eyes open and readjusting in their chairs to stay awake. The meeting finally wrapped up after midnight. As we stood to leave, the sultan asked: "Shall we eat?" There was only one acceptable answer. I could see the deflated looks on the faces of Avi, Hook, and Greenblatt as I answered in the affirmative. They just wanted a little sleep before our 8:00 a.m. departure to Bahrain.

The sultan's staff opened the doors into a magnificent dining room, lined with grand columns and archways. At the center stood a hand-painted marble table, adorned with gold trim. Three tuxedoed waiters stood behind each chair. The sultan of Oman was legendary for hosting dinners more formal than those at Buckingham Palace, but nothing could have prepared us for what followed.

Glistening silver plate covers dotted the table, accompanied by voluminous menus with descriptions of thirty different courses, separated by categories such as "soup" or "fish," written in elegant English and Arabic calligraphy. I expected to choose a single selection from each category, as is common in America. Before I could decide, however, a waiter delivered cold avocado soup, followed by cold lamb jelly soup and tomato basil soup. As each new dish arrived, the sultan described where he had discovered the original recipe. "Will you have some?" he asked, over and over. I couldn't refuse my generous host and sipped from each. After serving seven different soups, the waiters began to bring the seafood courses: grilled prawns, shrimp scampi, fresh lobster, grilled kingfish, fried cod. After fourteen courses, I peeked at the menu and saw that we weren't even halfway done.

I tried to avoid nonkosher food and took small bites so that I could make it through the meal as the sultan continued to explain the dynamics in the region. On multiple occasions, I was so engrossed in the discussion, I forgot to try a new dish. Eventually I noticed Hook and Avi glaring at me, and caught on: the waiters would not serve the next course until we stopped talking. The guys wanted me to shut up so we could keep the dinner moving.

As the sultan regaled us with stories of conquest and intrigue, he displayed a remarkable grasp of history. When a date slipped his mind, he looked to one of his ministers. "Was that in 1942?" "No, it was 1943," came the prompt response. This routine happened several times. "Was that in 1973?" "No, it was in 1974." For months, this was a running joke for my team. Hook would ask, "Was that in 1942?" Avi would shoot back, "No, it was in 1943."

Four hours and thirty delicious courses later, we finished the meal. It was after 4:00 a.m. I couldn't have been more delighted by the productive and riveting discussion. In six hours, I had built a new relationship and gained tremendous insight into the world's most complex diplomatic issues. I felt I had a new partner. As the sultan walked us to the door to bid us farewell, he casually asked: "Would you like to see my car collection?" Knowing that he owned one of the best car collections in the world, with more than three hundred antique vehicles, I was about to agree. Then I looked at Avi, who shook his head. "We better not," I said. "I will look forward to seeing it on our next visit."

As soon as the doors of our SUV shut, I turned to Avi and Hook with a smile. "I wanted to see the car collection." They lost it. "That was eight freaking hours of opulence torture!" Avi said. "We haven't slept in thirty hours, and we take off for Bahrain in less than four hours." I sympathized, but we had traveled halfway across the world to meet with the sultan of a country, and he had clearly appreciated our company. If he hadn't, he wouldn't have kept us so long. I was happy to forgo a few hours of sleep to build greater trust and give us a better chance to make peace. Plus, I had enjoyed every minute of the experience. When else would we have the chance to talk through the night with the sultan of Oman? As it turned out, this would be our only meeting. The sultan died of cancer in January of 2020, at the age of seventy-nine.

On the flight to Bahrain, I couldn't stop thinking about the previous evening's conversation with the sultan. One line played over and over in my head: *I feel badly for the Palestinian people. They carry with them the burden of the Muslim world.* It made me wonder who had appointed Mahmoud Abbas, with his incompetent band of negotiators, to represent the entire Arab world on the issue of the al-Aqsa Mosque.

This led to a eureka moment: maybe the reason the Israeli-Palestinian conflict hadn't been solved was because it is two separate conflicts, not one. There is the territorial dispute between Israel and the Palestinians about where to compromise and draw the borders in Jerusalem and the West Bank. Then there is the broader conflict between Israel and all Arabs about access to the al-Aqsa Mosque. For decades, conflating these two issues had made the conflict unsolvable. If we focused on each issue individually, perhaps progress would be possible.

Two years after the Allies defeated the Nazis in World War II, the United Nations called for separate Jewish and Arab states, while retaining international control of Jerusalem. The Jewish people in Israel supported this plan, including its Jerusalem proposal, but the Arab world rejected it. When British rule ended in 1948, the Jews declared their independence, announcing it on May 14. The next day, the nations of Egypt, Transjordan, Iraq, Syria, and Lebanon attacked.

Surrounded by enemies and outnumbered by the Arab forces, the newborn State of Israel miraculously won what came to be called the 1948 Arab-Israeli War. At the outset of the Arab invasion, thousands of Palestinians fled the area, believing that they would be able to return and partake in the spoils of an impending Arab victory. But when their side was defeated, they could not return to their homes and became refugees. A similar scene played out during the Six Day War in 1967. Instead of calling for the over fifty Muslim and Arab nations to welcome these refugees and grant them citizenship, Egyptian general Gamel Abdel Nasser and his allies refused to admit defeat and pledged that one day the refugees could return to Palestine. These refugees became geopolitical pawns, used to promote the continued anti-Semitic quest by the then leaders of the Arab world to justify their opposition to Israel's existence. This failure to resolve the refugee situation has continued for seventy years, leading to regional instability and turmoil. While all other post-WWII refugees have been resettled, today only displaced Palestinians still live in refugee camps across the Muslim world.

Following Nasser's humiliating defeat, the Egyptian general directed significant ire toward Israel and the Jewish people. As a result of this and similar rhetoric from other Arab leaders, some eight hundred thousand Jews, who had lived peacefully with their Muslim neighbors for centuries, were driven from their homes in Baghdad, Cairo, Fez, Damascus, and Tehran. They all resettled, and many found refuge in Israel. Unlike the Palestinian refugees, Jewish refugees were not given special designation by the United Nations.

When I met with leaders who objected to Israel's current position in Jerusalem, I would remind them that three times—in 1948, 1967, and 1973—the Arabs had attacked Israel and lost. After the 1967 and 1973 wars, the United Nations passed resolutions that called on Israel to return any land gained through the wars to the Palestinians. Anti-Israel internationalists ignored the fact that Israel had agreed to the 1947 UN resolution that created two sovereign states, with international control of Jerusalem. The real violators of international order were the invading parties. In most historical cases, there is a consequence to losing an offensive war. And they had lost three.

After two years of exploring every angle of this seemingly unsolvable conflict, I felt like I had finally reached a conceptual breakthrough: perhaps the way to achieve peace and reduce regional tension was to narrow our focus to the issue of access to the al-Aqsa Mosque. I was optimistic that this approach aligned with the sentiment of the Arab people—not just that of their leaders. Months earlier, I had commissioned State Department focus groups in the West Bank, Egypt, Jordan, and the UAE. When Arab respondents were asked to describe the source of the Arab-Israeli conflict, the vast majority cited access to the mosque. The issue of territorial sovereignty, which was the fixation of "experts," hardly came up.

If Israel would guarantee Muslim custodianship of the holy site, and expand access to Muslim worshippers, then we could address the issue of greatest concern to Arabs. And if these nations made peace with Israel, flights to Israel would open up, making it possible for hundreds of millions of Muslims to make pilgrimages to the mosque. In order to do this, our peace plan would need to demonstrate a serious commitment to solving the Israel-Palestinian conflict. We were ready to offer a plan that would require compromise, but still maintained Israel's security while improving the lives of the Palestinians.

A detailed proposal would put Abbas in a tough negotiating position. If he accepted the offer and ended the conflict, he would risk losing billions per year in international aid. But if he rejected our proposal for a pragmatic two-state solution, which included a massive investment plan for the Palestinian territories, he would reveal his true indifference to the wellbeing of his own people. This would strengthen the argument I was making to the leaders of the Muslim countries—that it was time to focus on their national interests and move forward with normalization.

In the twilight of his tenure as secretary of state, John Kerry gave parting words of advice to a Washington audience. "There will be no separate peace between Israel and the Arab world," he said at the Saban Forum. "I want to make that very clear to all of you. I've heard several prominent politicians in Israel sometimes saying, 'Well, the Arab world is in a different place now, we just have to reach out to them and we can work some things with the Arab world and we'll deal with the Palestinians.' No, no, no, and no. I can tell you that reaffirmed even in the last week as I have talked to leaders in the Arab community. There will be no advance and separate peace with the Arab world without the Palestinian process and Palestinian peace. Everybody needs to understand that. That is a hard reality."

This was the conventional wisdom for decades, and I initially accepted it as fact. But as I listened and learned, I felt like the reverse might be true. If we could make peace between Israel and the Arab world, then more likely than not, a path to making peace between the Palestinians and Israel would eventually open as well.

As our flight approached Bahrain, I leaned toward Jason and asked him to make two changes to the peace plan. First, he should reframe the issue of access to the al-Aqsa Mosque, removing it as a subject of negotiation with the Palestinians and turning it into the centerpiece of broader normalization agreements between Israel and the Muslim world. Second, we needed to finalize the boundaries in Jerusalem and the West Bank in a rational way that was based on the modern reality, not a UN resolution from 1967. Both concepts were rooted in finding a pragmatic solution that could end the conflict and move beyond the failed paradigm of the past.

If the Palestinian leadership rejected this approach, which they almost certainly would, the Arab leaders would recognize that Palestinian intransigence was undermining their own interests in a time of increased common threats and shared opportunities.

Our dinner with the sultan of Oman, and my subsequent eureka realization, crystallized our strategy and paved the way for the Abraham Accords. As we pursued a new paradigm, we began to see an enormous opportunity that had been hiding in plain sight.

29

A New Cliché

May I ask you a potentially inappropriate question?" I asked King Hamad bin Isa Al Khalifa of Bahrain, ruler of the thriving island nation off the coast of Saudi Arabia.

The king smiled and nodded.

"Yesterday in Egypt," I continued, "at the Arab League meeting with the European Union, you called for a Palestinian state with East Jerusalem as its capital and the borders established in 1967. Those are the same Arab Peace Initiative talking points that everyone has been using since 2002. You know that's not happening. I know that's not happening. Everyone in that room knows that's not happening. So why do you and others keep saying it?"

The king paused, gathered his thoughts, and spoke: It had become a cliché, he said, before conceding that perhaps it was time for a new cliché.

The king added that he had no ill feelings toward Israel and predicted that the region would move forward only when the three Abrahamic faiths reunited—and this was why he wanted to see progress for the Palestinians.

This was the best conclusion I could have hoped for after constructive meetings with the king and his son, Crown Prince Salman bin Hamad Al Khalifa, a forward-thinking leader who had taken great strides to modernize and diversify Bahrain's economy. Both the king and the crown prince appreciated that our Peace to Prosperity economic plan included a detailed blueprint to bring jobs to the West Bank. They offered to assist in any effort to help the Palestinian people.

From Bahrain, we flew to Saudi Arabia and went straight to the Saudi Royal Court for a meeting with King Salman bin Abdulaziz. The eighty-three-year-old monarch expressed his appreciation for Trump's work to stabilize the region, and fondly recalled the Riyadh summit he had hosted nearly two years earlier. He stressed that access to the al-Aqsa Mosque was the most important issue for all Muslims and directed his team to continue working with us to see what we could do to resolve the conflict. Knowing that Israel was not King Salman's favorite topic, I asked Brian Hook, our special envoy for Iran, to give him an update on our actions against the regime in Tehran. I included Brian in most of my meetings with Arab leaders not only because he was an astute policy adviser, but also because his presence reminded Arab nations that we were aligned on a key priority for them, and we expected them to engage constructively on our other priorities. In the past, they had cherry-picked the issues on which they were willing to engage, while saying that American priorities would be too destabilizing for them internally. I wanted this to stop. If they hoped to reap the benefits of the policies they supported, they couldn't run away from the issues they preferred to avoid.

After our meeting with the king, we entered another beautiful building in the Royal Court compound: the offices of Crown Prince Mohammed bin Salman. It was our first in-person meeting since Saudi agents had murdered *Washington Post* columnist Jamal Khashoggi in October of 2018.

I made sure that the communications team released an official readout of the group meeting to the press, so it was clear that I wasn't hiding the meeting and that the United States was standing by its ally in a low moment. I walked MBS through our Peace to Prosperity economic plan, which included a $50 billion economic investment into the West Bank, Gaza, and the surrounding region.

For decades, Arab countries had invested billions of dollars in these areas, with very little return. "If the investments were made directly into the Palestinian economy, rather than through their leadership, and conditioned upon making reforms, the people would benefit more," I said. "Right now, the current system is a massive uncapped liability."

MBS thought the plan made a lot of sense and asked how I was able to pull it together, observing that it looked like the work of a hundred McKinsey consultants.

I explained that I'd assembled three of the smartest people from the White House who had backgrounds in finance, and that we'd spent several months researching the problem and running the numbers.

I told him that his Vision 2030 plan for the Saudi people had inspired our blueprint.[41]

He predicted that our natural critics would claim that I was trying to buy the Palestinian people, but noted that this criticism would happen either way. It was a solid plan, he said, and the people would judge for themselves whether it would help them achieve a better life. The same thing happened in his country with Vision 2030.

When we discussed the murder of Khashoggi, the crown prince took responsibility for the fact that it happened on his watch, though he said he was not personally involved. He said that he was conducting a thorough investigation and planned to address the murder publicly as soon as it was complete.

We also discussed ways to resolve the Gulf rift with Qatar, which was entering its second year and was responsible for instability and economic damage in the region. "Let's put aside the reasons for why this started," I said. "Qatar now has to rely on Iran for groceries. The longer this fight continues, the more animosity there will be in the region, lowering your chances of achieving your ambitious dreams for your country."

"I'm open to finding an agreement," MBS said. "But it has to be a deal that really solves the problem. Past agreements between our countries only made the problem worse."

I offered to speak with Sheikh Tamim, the emir of Qatar, and probe the Qataris to see whether a resolution that addressed Saudi concerns was possible. MBS agreed.

Several months later, MBS addressed the murder of Khashoggi during a CBS *60 Minutes* interview, telling the world what he told me: "This was a heinous crime. But I take full responsibility as a leader in Saudi Arabia, especially since it was committed by individuals working for the Saudi government. . . . And I must take all actions to avoid such a thing in the future."

While this situation was terrible, I couldn't ignore the fact that the reforms that MBS was implementing were having a positive impact on millions of people in the kingdom—especially women. Under his leadership, Saudi Arabia began allowing women freedom to travel, participate in the economy, and own property. It had loosened cultural restrictions on dancing and concerts. And it had dramatically scaled back its religious police, which for years had harshly enforced a stringent form of Islamic law. All of these reforms were major priorities for the United States, as they led to further progress in combating extremism and advancing economic opportunity and stability throughout the war-torn region. The kingdom was poised to build on this historic progress, and I believed it would.

* * *

In Turkey, I met with President Recep Tayyip Erdoğan in his massive presidential palace, which he recently had completed at the cost of $600 million, and walked him through our peace plan. It didn't go well, but I didn't think it would.

A superbly talented politician and populist Islamist, Erdoğan expressed solidarity with the Palestinians suffering in Gaza and showed zero willingness to support my proposed compromises. When I suggested that Hamas had caused this suffering through its terrorism and political mismanagement, Erdoğan paused, looked at me incredulously, and changed topics. He was much more interested in discussing Turkey's economic relationship with the United States. He wanted to double the annual trade volume to $50 billion. I told him that I would try to encourage more trade, but that Turkey's earlier purchase of antiaircraft missiles from Russia would trigger statutory US sanctions on Turkey. Erdoğan thanked me for being willing to try. Then he looked at his finance minister, Berat Albayrak, who also happened to be his son-in-law, and said that advisers could sometimes let down their presidents—but not sons-in-law.

That wasn't his only comment about family. Before our meeting concluded, Erdoğan encouraged me and Ivanka to have more children, and expressed his sincere love for his own. He joked that at his political rallies, he always encouraged his supporters to grow their families.

I never expected Erdoğan to support our peace plan. After Iran and the Palestinians, he had been the harshest and most vocal critic of our decision to move the US embassy to Jerusalem. But it was worth a shot, and perhaps my visit would cause him to tone down his rhetoric or even remain neutral.

Our February trip through the Middle East confirmed my sense that we should release our economic plan first, followed by the political peace plan soon thereafter. After two years of building trust in the region, I was encouraged that Arab leaders seemed ready to move forward with a new paradigm for the Middle East. If Israel's prime minister endorsed the plan following the elections in April, the Arabs and Israelis would be closer than ever on several key issues, including the path to a Palestinian state and access to the al-Aqsa Mosque. This would shake up the status quo and put in motion our newly refined strategy to encourage the Palestinians to come to the table, while was pursued a parallel track of normalization between Israel and the Arab nations.

30

Exoneration

I t really was a hoax all along—and on March 24, 2019, even CNN had to admit the truth. It was a Sunday afternoon, and I was about to leave the house when Avi called.

"Turn on the television!" he said.

As I reached for the clicker, I barely had time to wonder what new crisis loomed. The first thing I saw on the screen was a breaking news chyron: "DOJ: Mueller Did Not Find Trump Conspired with Russia."

The Department of Justice had reviewed the report of special counsel Robert Mueller and concluded that neither Trump nor his presidential campaign had colluded with Russia to influence the election in 2016. Investigators had spent two years and tens of millions of dollars searching for evidence of a link—and as I'd expected, they'd turned up nothing.

So it wasn't a crisis at all. It was a relief. We'd waited years for this moment. I knew that we had done nothing wrong, but Trump was always concerned that Mueller would exaggerate some random fact of the case, handing his detractors something to seize on to claim proof of collusion. My mom also worried constantly. Sometimes it seemed like she tracked the press reports speculating about my legal status more closely than my lawyer did. I called her immediately and told her to turn on the news.

"I'm now one hundred percent in the clear," I said. "See, I told you that you shouldn't have been so worried."

As we spoke, my eyes wandered back to the television screen, where CNN correspondent Manu Raju was discussing the announcement.

"Nancy Pelosi and Chuck Schumer just issued a joint statement about the Mueller report. They said that the fact that the Special Counsel Mueller's report does not exonerate the president on a charge as serious as obstruction of justice demonstrates how urgent it is that the full report and underlying documentation be made public without any further delay given Mr. Barr's public record of bias against the Special Counsel inquiry."

This whole thing is a sick game, I thought. We came to change Washington and serve our country. The media and the Democrats challenged the legitimacy of the election with a vengeance. Trump was right all along. This whole investigation had been nothing but a witch hunt.

After the announcement of Trump's exoneration, Senator Richard Burr, a North Carolina Republican and chairman of the Senate intelligence committee, sent me a letter asking me to answer another round of questions about collusion with Russia. My lawyer Abbe Lowell pushed back, noting that I had already answered the committee's questions in July of 2017. Plus, the Mueller report closed the case. Yet Burr refused to abandon an investigation into which he had invested so much time. So he threatened me with a subpoena.

Having nothing to hide, I agreed to go. But I wasn't happy that it pulled me away from my other priorities. When I arrived at the Hart Senate Office Building, Burr slid up to me outside the hearing room. "Thank you for coming today," he said. "These investigations have been incredibly fruitful. We've found stuff that will keep our intel community busy for the next ten years."

I turned to stand nose to nose with Burr. "Senator, are you serious?" I asked, without hiding my frustration. "This investigation is an embarrassment. What you are seeing has a simple explanation. Unlike everyone else in the Washington system, Donald Trump was an unknown entity to foreign governments. You picked up a high volume of unusual intel signals because the president's victory caught the world off guard. These countries know how to influence all of the long-established politicians, so you don't see intel traffic on them, but they had no idea how to try and influence Trump, so they were scrambling to figure it out. Your exhaustive investigation has found nothing. Mueller has found nothing. This thing is over. Go and look at how we are getting our butts kicked on intelligence by China and provide oversight on something that is actually a real problem. Stop wasting my time and the taxpayers' money with this bullshit."

I was pissed. Normally I'm composed, and I surprised myself with the tone and force of these words. But Burr had struck a nerve. It was one thing to deal with critics in the media and the other party. Yet I'd faced two years of baseless investigations from some Republicans as well. They also seemed reluctant to accept the truth.

Burr was surprised to encounter something other than the customary deference accorded to senators. Rather than responding to what I had said, he replied with a typical Washington nicety: "I'm just very appreciative of the respect you showed to the committee by coming today. Let's catch up soon." Then he walked away. The senators and their staff grilled me for four hours.

I'd come to see Burr as the sort of establishment politician who valued job security more than anything else. A year later, Burr would find himself entangled in a scandal over stocks he had allegedly dumped after receiving an intelligence briefing about the severity of the coronavirus threat. While the Department of Justice ultimately dropped its months-long investigation of the senator, he stepped down as chair of the intelligence committee and opted not to run for reelection in North Carolina.

* * *

As I prepared for Burr's committee interview, Ambassador David Friedman called from Israel with a request: "Bibi is asking if the president can recognize the Golan Heights."

He was referring to the mountainous plateau that spans nearly seven hundred square miles along Israel's northern borders with Lebanon and Syria. From its position three thousand feet above sea level, the Golan Heights offers a strategic platform for the Israelis, giving them the defensive high ground on the Syrian border and a greater buffer between

Israel's population centers and its northern neighbors. During the Six-Day War in 1967, the Arabs used this high ground to bombard Israel. In an act of defense, the Israelis drove their enemies from the area. After the war, they held on to it, despite demands from other countries that they return it to Syria, and the area had remained quiet for more than forty-five years.

Much like the acknowledgment of Jerusalem as the capital of Israel, I saw recognition of Israel's sovereignty over the Golan Heights as a powerful opportunity for America to stand for the truth. After eight years of civil war, Syria was barely a country. It couldn't control what was happening within its borders, so expanding them was not an option. Acknowledging the reality that the Golan Heights belonged to Israel was the right thing to do. It also would help us build credibility with the Israelis as we prepared to ask them to support a two-state solution with the Palestinians.

Even before Friedman called to convey Bibi's request, National Security Adviser John Bolton and I had raised the issue with the president. Five minutes into that meeting, however, Trump stopped us: "I have done too much for Bibi already. Let's see what he does with the peace deal first."

So when Friedman called in late March—a few weeks before the Israeli elections—I told him that I had already run into a presidential brick wall. But we agreed that he should call Trump and explain why it was so important to the Israelis and unlikely to annoy Arab leaders.

When Friedman called and made the pitch, Trump asked a valid question: Why hadn't any of his predecessors done it? The answer was that past presidents had dodged the issue. They had wanted to avoid condemnation from the international community and also expected that Israel's de facto control would continue indefinitely.

"Then why is it controversial?" Trump asked. He wanted to make sure that he understood both sides of the issue. Friedman briefed him on the history and politics. Trump asked more questions and then arrived at his decision: "Let's do it."

"Should I call Bolton?" Friedman asked.

"I have a better idea," Trump said. He turned to his ever-present adviser Dan Scavino, who was one of the few people Trump trusted with his Twitter passcode. I joked that Scavino carried the real nuclear football, as he could start a war at any time by firing off an errant tweet. "Stay on the line," the president told Friedman. Then he called out to his assistant: "Get Jared in here."

Minutes later, I walked into the Oval Office. Trump had already drafted a tweet on a piece of paper, which he slid across the Resolute Desk so I could read it: "After 52 years, it is time for the United States to fully recognize Israel's Sovereignty over the Golan Heights, which is of critical strategic and security importance to the State of Israel and Regional Stability."

It was a classic Trump tweet. It didn't say he had made a formal decision to recognize the Golan Heights, but simply signaled his intention to do it—a vague enough statement to allow him to dip his toe in the water and see how people reacted before he took definitive action. This was another important role of Scavino: in addition to occasionally recommending against some draft tweets that could cause unintended backlash, he told the president how his tweets were received—and never sugarcoated his observations.

"What do you think?" Trump asked.

"That's perfect," I responded. "This will go over well and be an historic action."

"David, are you sure about this one?" Trump asked one last time, wanting to make sure he didn't detect any hesitation in the ambassador's voice.

"One thousand percent, sir," Friedman said. "This will get a great response."

"Dan, put it out," he said.

Scavino pushed the button.

"Now turn on the TV and see how long it takes before the fake news covers it," Trump said.

It took mere minutes for the cable networks to break from their normal coverage.

Trump was willing to take calculated risks and often enjoyed his role as assignment editor for the news organizations. Caught off guard, reporters scrambled to research the issue, find their experts, and report on an historic policy. As I had anticipated, the announcement also surprised the troublemakers of the Middle East, and the blowback was minimal. This was one of many instances in which Trump's decisiveness pushed forward a commonsense policy that would have never made it through a snail-paced and risk-averse bureaucracy. Soon after his tweet, Trump instructed us to move forward with a presidential proclamation formally recognizing Israeli sovereignty over the Golan Heights.

31

An Unexpected Visit

On Sunday, May 26, 2019, Ambassador Yousef Al Otaiba of the United Arab Emirates called and asked to speak with me in person. Elegant and accomplished, at the age of forty-five, Yousef had become one of the leading players in Washington's social and diplomatic circles. On any given day he could be spotted at an important meeting or reception, or hosting his own dinner party with a guest list that could double as a who's who of Washington. With his close ties to his boss, Crown Prince Mohammed bin Zayed (MBZ), the de facto ruler of the United Arab Emirates, Yousef was one of the most powerful men in his home country and one of the most influential in the entire Middle East.

I invited Yousef to come to our Kalorama home, and as we sat in the living room, he told me that MBZ had asked him to personally deliver to me an important message: the Emirati leader was ready to move forward and fully normalize relations with Israel.

I tried to hide my excitement behind a poker face, but my mind was spinning. The UAE and Israel were the two most advanced countries in the Middle East from a military, economic, and technology perspective, but they had no formal ties. Taking this step had the potential to unleash positive forces that the region had not seen in decades and change history in ways that were hard to fathom.

He noted that we had been gently pushing Arab nations to take this step. Our efforts had changed the region for the better, he said. Based on the trust we had built, MBZ believed that normalization was possible and he wanted to be first.

Yousef explained that normalized relations with Israel would carry tremendous risk for the UAE, both internally and externally. The Emirati government had chosen to hail 2019 as the Year of Tolerance. To celebrate it, they had invited Pope Francis for the first papal visit to the Arabian Peninsula in history. It was a great and generous act, but it triggered an outcry among religious clerics, regional activists, and the Arab media, most notably on Al Jazeera.

Yousef predicted that if the UAE took this step, others would follow.

The next afternoon Pompeo dropped by my office, lugging his oversize secure briefcase, which was always packed with the extensive briefings he loved to consume. He wore a big smile. Yousef had visited him at the State Department that morning and shared the concept. "It's definitely a long shot, but crazier things have happened," he said.

Pompeo's skepticism was merited: lots of things could go wrong. First, we had to keep it quiet. Any leak about normalization would force both parties to issue public denials and make continued talks politically untenable. Second, Israel had just finished its elections and was still forming a government. Until this was complete, Bibi wouldn't have time for diplomacy.

* * *

Days later, I left for my next trip to the Middle East. My first destination was Morocco, a country I had not yet visited during my time in government. The visit came with some trepidation. The previous summer, the United States, Canada, and Mexico had competed in a joint bid against Morocco to host the 2026 FIFA World Cup. Trump had tasked me with leading the effort, which involved working closely with the US Soccer Federation and corralling dozens of unusual technical commitments from the departments and agencies. As FIFA prepared to vote, Trump asked me to call MBS and request Saudi Arabia's support for our bid. The crown prince agreed. The Saudis' critical vote marked a turning point in our effort and helped persuade many other Arab countries to back our bid over Morocco's. Winning the bid to host the globally watched soccer tournament was a major diplomatic and economic success for the president. There was one downside, however: Morocco was the runner-up, and we were worried that our victory had come at the cost of a strained relationship with the country's ruler, King Mohammed VI.

King Mohammed VI came from a noble lineage of Alaouite leaders—direct descendants from the family of the Prophet Muhammad, the founder of the Islamic faith. Accordingly, he had enormous credibility in the Muslim world. He served as a chairman of the Al Quds Committee within the Organization for Islamic Cooperation, a respected body that helped preserve the Muslim holy sites in Jerusalem, as well as other interests.

As I prepared for the trip, our internal foreign policy experts described the king's popularity and his savviness as a businessman and predicted that he would want to discuss the question of sovereignty over the Western Sahara, the expansive desert territory on Morocco's southwestern border. Moroccan rulers had presided over the area for centuries, and King Mohammed viewed it as essential to his country's national security. When the modern government of Morocco gained independence from France and Spain in 1956, it immediately staked a territorial claim on the hundred thousand square miles of mineral-rich desert, which remained in the possession of Spain. A local group of desert-dwelling people, the Sahrawis, also asserted jurisdiction over the area through their nonstate organization, the Polisario Front, leading the United Nations to include the Western Sahara on a list of "Non-Self-Governing Territories."

The more I researched the history, the more I believed Morocco had a legitimate claim. From a security perspective, it already controlled two-thirds of the territory, and it clearly served US interests for Morocco to control the rest. Most of America's military footprint in Africa was in response to the violent ambitions of ISIS, Al-Shabab, and Boko Haram, which were expanding into areas that were left ungoverned by corrupt, struggling, or failed states. The last thing we needed was for the Western Sahara to become a haven for chaos and conflagration. Unlike most of its neighbors, Morocco had a stable government, a sound economy, and, despite our World Cup rivalry, a warm relationship with the United States. Morocco's presence in the Western Sahara would keep the area from becoming a vacuum that left room for terrorism and instability.

When I asked our experts what stood in the way of recognition, they gave me one name: Jim Inhofe. The eighty-four-year-old chairman of the powerful Senate Armed Services Committee opposed Moroccan sovereignty over the Western Sahara. Inhofe's committee held sway over the Pentagon's $700 billion budget, giving him immense power over American foreign policy. For whatever reason, he'd been traveling to the Western Sahara for twenty years and become a powerful patron of the Polisario Front's quest for independence. I had tremendous respect for the senator and figured there must be a smart reason for his position, so I made a note to reach out to him upon my return after I spoke to King Mohammed.

Upon arriving in Casablanca, I was surprised and delighted to see my beloved friend and rabbi, David Pinto. Through the highs and lows of my life, Rabbi Pinto had always inspired me to find solace and strength through my relationship with God. Rabbi Pinto was a proud French Jew of Moroccan descent, who often said that he prayed for the Moroccan king each day because of the heroic deeds during World War II of the king's grandfather, King Mohammed V. When the Nazis asked King Mohammed V to identify and hand over the Jews in his country, he is said to have responded in defiance: "There are no Jewish citizens, there are no Muslims—they are all Moroccans."

King Mohammed VI had discovered that Rabbi Pinto's great-great-grandfather, Rabbi Haim Pinto, was buried in Casablanca's historic Jewish cemetery, a revered pilgrimage site. To my delight and astonishment, the king had arranged for Rabbi Pinto to meet me in Morocco so that the two of us could pay our respects at his grandfather's tomb.

That afternoon, the American attaché in Casablanca informed me that the king had invited me and a guest to dinner at his private residence—a rare honor, as he almost always met with guests at the Royal Palace. It was Ramadan, so Jason Greenblatt and I arrived after sundown and were escorted to a regal outdoor dining area by the pool, where we discovered a massive buffet of kosher food. As we sat with the king for an Iftar dinner to break the Ramadan fast, I thanked him for his thoughtfulness in setting up the cemetery visit. My trips to the Middle East never included sightseeing, but if there was one place I would have wanted to visit, it was that cemetery—and somehow the king had known before we ever met.

As dinner commenced, I sat on the king's right. To his left, across from me, was his son, a sixteen-year-old with the bearing of someone twice his age. The warmth between father and son was obvious. While some children of dignitaries lack the maturity to carry themselves in official settings, Crown Prince Moulay Hassan was fully attentive and engaged. The king and I discussed the Israel-Palestinian conflict, and he emphasized the importance of ensuring harmony and access to the sacred sites in Jerusalem. After carefully listening to his perspective on the Western Sahara, I was more convinced that recognition of Morocco's sovereignty was the logical policy and promised that I would take the issue back to Washington and explore how to change it.

Foreign Minister Nasser Bourita also explained why the king had not visited the White House as planned in 2018: he said that John Bolton would not agree to include language in the trip readout saying that the United States would work with Morocco to find a peaceful solution on the Western Sahara. Trump had been looking forward to the visit, and to my knowledge, Bolton never informed him why they had canceled. Bolton had a long history of opposing Morocco. In the 1990s, when Secretary of State Jim Baker brokered the settlement plan that perpetuated the crisis, Bolton was his negotiator.

32

Mexican Standoff
amid Peace Talks

W e've got a really big problem," said Treasury secretary Steven Mnuchin, on a call from across the ocean.

It was Wednesday, May 29, 2019. I was at the King David Hotel in Jerusalem, dealing with my own problem: Prime Minister Bibi Netanyahu faced a midnight deadline to form a coalition government, following national elections the previous month. If he succeeded, he'd be able to endorse the peace plan publicly and potentially accept the UAE's offer to pursue normalization. Failure would trigger a new round of elections and more delays. I was supposed to meet with him the next day, but as the deadline approached, it still wasn't clear whether Bibi would succeed in forming a government.

Mnuchin yanked my attention back to Washington.

"You know the president has been threatening to put tariffs on Mexico due to the caravans at the southern border," said the secretary. "He just dictated a statement that says he is imposing tariffs immediately. He wants it to go out tonight."

I dialed the president's assistant and asked if Trump was alone. I had learned early on that it was much more effective to speak with him privately when I disagreed with him. Otherwise, someone in the West Wing would leak to the press that the two of us had clashed, which lowered the probability of persuading him. I felt strongly that staff should never put the president in a position where the public would know he changed his mind based on their advice.

Fortunately, Trump was alone. I caught him between meetings and in a relaxed mood. I briefed him on our trip, conveying the warm regards from the king of Morocco. Then I turned to the real purpose of my call: the tariffs on Mexico.

As soon as I mentioned them, Trump interrupted: "Jared, I'm tired of waiting. Everyone keeps telling me that they are working on a plan to stop the caravans and that Mexico is going to help. I think everyone is full of shit."

"I know you're frustrated," I said. "You should be. But just know you are playing with a powder keg here. Our team is really close to completing a plan with the new Mexican government that will work. AMLO gave me his word that he will help. I feel confident the Mexican president will come through," I said, referring to President Andrés Manuel López Obrador, who had assumed office on December 1, 2018.

Trump was unconvinced. "I have been hearing that they are going to help for months," he said. "I don't think they are going to do it."

"Give me a few more weeks," I pleaded. "AMLO may have left-leaning policies, but like you, he is proud and savvy. He has shown you a lot of respect to date, but he is tough. If you push him into a corner, he might come out swinging. If you put out that statement and announce the tariffs, you're putting all your chips on the table and going all in."

"I know you worked hard on USMCA," Trump said. "I don't care about the politics. I have lost patience with the border, and there is nothing anyone can do to stop me. Do me a favor: focus on Israel and let me handle this one."

For several months, I had known that a continued surge in illegal immigration might cause Trump to do something drastic. Right after Mexico's presidential elections in July of 2018, Secretary Pompeo and I had hosted AMLO's soon-to-be foreign secretary, Marcelo Ebrard, a skilled politician who was the former mayor of Mexico City, for dinner at the State Department. At the meeting, Pompeo slid a document across the table. It outlined the monthly illegal border crossings data.

Pompeo bluntly conveyed that the US-Mexico relationship would be very simple. If the border numbers went up, there would be problems. If they went down, the United States would be an incredible partner and would help Mexico with its priorities.

Ebrard took the message back to Mexico, but failed to get results. By March of 2019, the influx of illegal immigrants had grown worse. I made a twelve-hour trip to Mexico City to deliver my message directly to AMLO: if Mexico didn't act immediately to reduce illegal border crossings, all bets were off—including the recently negotiated USMCA, which was still pending congressional approval.

AMLO promised that he would give Ebrard whatever he needed to confront the crisis. I viewed this pledge as a success. But at the time of Mnuchin's emergency call to me in Jerusalem, we still had not yet presented a solution to the president. And Trump was fed up.

When I told Mnuchin that I had failed to steer the president away from announcing tariffs on Mexico, I tried to joke about our predicament: "Steven, I bet you never thought you'd be involved in a real Mexican standoff."

We knew the matter was serious. Fortunately, Mnuchin and Larry Kudlow, the director of the National Economic Council, persuaded Trump to slow down the implementation of his announcement, making the tariffs effective the following week, instead of the next day. The secretary pointed out that it would take at least that long to change our customs systems so that we could collect the tariffs. Trump consented to the one-week delay, a small but significant win that bought us a few days to try to broker a deal. Soon after, Trump tweeted: "On June 10th, the United States will impose a 5% Tariff on all goods coming into our Country from Mexico, until such time as illegal migrants coming through Mexico, and into our Country, STOP. The Tariff will gradually increase until the Illegal Immigration problem is remedied."

As I sat in that room at the King David Hotel, it felt like I was standing on the precipice of a dual disaster. Just as we were preparing to release the peace plan, it looked like Netanyahu would not form a government.

Now Trump was about to start an economic war with Mexico that could upend two years of work on the USMCA.

No more than ten minutes after Trump posted his tweet, Ebrard called from Mexico. "What is the president doing?" he asked.

I didn't want to tell him that I had opposed the tweet. My job was to represent the president's views rather than my own. "I've been warning you for weeks that the president is at the end of his rope," I said.

"Can I please come to Washington as soon as possible?"

"I'm in Israel, and I'll be back soon, but Secretary Pompeo and others can meet with you. If you come, you will have one shot—at best—to close the deal. Bring every resource you have."

Ebrard accepted the invitation, and I called Trump right away.

"Your bluff worked," I said. "The foreign minister showed me his cards. Mexico is folding."

"It wasn't a bluff," said Trump. "They'd better come up with something good, or I will go forward with these tariffs. I want the border solved."

* * *

By the morning of May 30, it was clear that Netanyahu's government had fallen apart. Israel was headed for another election in September. I expected Bibi to cancel our meeting. But his staff confirmed that it was still on, with one minor change: our breakfast meeting had been pushed back to lunch. When we sat with the beleaguered prime minister, the deep rings under his eyes told the story of the night before. Other than that, however, he remained composed, like the political master he was. Instead of jumping into a technical discussion about the peace plan, I tried to lighten the mood by asking questions about his political career.

Bibi told me that early in his political career, he had learned that the most important thing was momentum. Whenever he was down, he would find any bit of good news and would make it the biggest thing. In politics, wins beget more wins.

He grabbed a napkin and drew a triangle, separating it into three levels to illustrate "the pyramid of politics." Pointing to the sketch, he expounded. All the people in the middle level were the politicians who want the leader's job at the top. They didn't give the leader power. They tried to take the leader's power. Then he pointed to the bottom part of the triangle. The way for a leader to stay in power was to keep the relationship with the supporters strong. Deliver for them, and they never forget it. Even without the support of the press or the politicians in the middle, if a leader remained connected and loyal to these supporters, the group in the middle would be less likely to defeat the leader. It was a memorable statement at a moment when Bibi's own political survival was in doubt.

He soon turned to the reason for our meeting, telling me that he was not ready for us to release the peace plan. He was concerned about the upcoming election. But he assured me that he wanted to make peace, and he believed that under President Trump it was possible.

After the lunch—always prepared by a special chef to showcase Israel's national cuisine—I asked if I could speak to Bibi one-on-one. He led me into his private study, a small room with a desk covered in books. He displayed pictures of his family, including one of him and his brother Yonatan, who had been killed in the famous rescue of 102 Jewish hostages from a hijacked plane in Entebbe, Uganda, in 1976.

Sitting down, Bibi picked up his pipe, which was already loaded with tobacco. As he lit it and puffed, a sweet, musty smell filled the air.

"What I am about to tell you is completely real and needs to be kept between us," I said. "The UAE is ready to normalize. I believe this is a real offer."

He said he didn't believe they were serious.

"Trust me," I countered, and I described Yousef's visit to my house. "I haven't led you astray yet. If you're willing to be flexible and not make it political, they are ready."

Bibi explained that everything was political at the moment. He was in an election again, and he needed to focus on that. But he wanted to keep talking. If this was real, then he was in. He told me to let him know what he needed to do, and he would find a way to get there.

33

Peace to Prosperity

rom Israel, we traveled to Montreux, on the Lake Geneva shore-
line in Switzerland, to attend one of the world's most secretive
gatherings: the Bilderberg Meeting, an annual gathering of top
leaders in government, industry, and academia. The meeting was established in 1954 to strengthen ties between the United States and Europe.

I was skeptical of these sorts of functions because they rarely produced
tangible outcomes. I'd declined invitations to speak the two previous
years, but I thought the meeting in 2019 could be a useful venue to
explain our approach to Middle East peace and build support among an
influential class of people.

I had another objective in mind as well. Between the productive meetings and stimulating panel discussions, I went to see someone who wasn't
on the Bilderberg guest list: Kirill Dmitriev, chief executive officer of
Russia's Direct Investment Fund. Up to this point, I had avoided interactions with anybody connected to the Russian government. After Mueller
exonerated the president, however, I felt that it was finally time to reach
out. Historically, the Russians had played a role in Middle East peace efforts, and I wanted to open a line of communication and make sure they
didn't oppose our proposal.

Secretary Pompeo had suggested that US ambassador to Russia Jon
Huntsman Jr. could help me identify the best interlocutor. "Russia's a
proud country," he said in a call. "So if they're not consulted, they'll
be against it." Huntsman offered to talk to Putin's chief of staff. "You're

going to get one of two people," he said. "If you get Mikhail Bogdanov, their Middle East guy in the foreign ministry, that basically means no interest. You'll have a pro forma meeting, and they'll do nothing for you. Their foreign ministry is old-school, and many are stuck in the Cold War mentality. If you get Kirill Dmitriev, however, that means that Putin's interested and actually understands that this can be a way for Russia to work with the United States."

A few days later, Huntsman called with the news that the Russians wanted me to connect with Kirill, a Stanford University and Harvard Business School alumnus with strong ties to the American business community. The ambassador worked to arrange a meeting in a neutral setting, and we settled on Montreux during the Bilderberg event. I'd learned from Mueller's investigation to avoid meeting one-on-one with Russians. The media would obsess about it and engage in thoughtless speculation. So I asked Matthew Pottinger, the National Security Council's top China expert—and soon-to-be deputy national security adviser—to join me for the meeting. A former *Wall Street Journal* reporter and retired Marine Corps officer, Pottinger had served under three different national security advisers and enjoyed the unusual distinction of being liked and respected by them all. He had been invited to the conference to discuss China.

"I hope you understand that the last time somebody came with me to a meeting with a Russian, they ended up with a lot of legal bills," I joked.

During our meeting with Kirill, I briefed him on the key elements of our peace plan, and he thought it was a framework Russia could support. Before we concluded, I asked him to guard against leaks.

He said that, with Russia, if we showed them trust, they'd give it back. When they were disrespected, however, they didn't take it kindly.

That confirmed what I had suspected, and I was glad we had met.

* * *

On June 22 we released the Peace to Prosperity plan, the most comprehensive economic framework ever created for the Palestinians and the broader Middle East.[42] Its 140 pages outlined a detailed strategy to turn around more than seventy years of economic malaise and political abuse in the West Bank and Gaza. Billions of dollars in foreign investment had flowed into the territories through the United Nations Relief and Works Agency (UNRWA), but these funds had done almost nothing to improve the lives of Palestinians. When their corrupt leaders weren't stealing the money, they were wasting it on dead-end and low-impact projects. In Gaza, the international investments were used to pay for programs that indoctrinated the youth to hate Israel and the United States. The funds also built secret storage facilities to hide Hamas military equipment, which the Israeli forces would try to destroy during skirmishes. These fundamental flaws deterred business leaders from investing in the West Bank or Gaza and denied the Palestinian people a better future. After we moved the American embassy to Jerusalem, the State Department informed me that the US approval rating was just six percent in Gaza. When I asked how high it was before the move, they admitted it was only nine percent. At the same time, USAID's approval rating was about 70 percent, which further reinforced my point that America's current aid to Gaza made no sense.

Our plan proposed a $50 billion investment in the Palestinian territories and the surrounding region, which would be released in tranches over the course of a ten-year period with strict accountability measures in place. It set forth a business framework for improving access to the Palestinian territories, which included building new roads and railways, demilitarizing and modernizing border crossings, and connecting the West Bank and Gaza. The plan also called for improving critical infrastructure, such as water treatment facilities, power plants, and telecommunications networks. Palestinian schools weren't equipping workers with the skills they needed to fill open positions in the local labor market, so our plan called for job training, curriculum changes, and a brand-new world-class university. We also included robust reforms to establish the rule of law and prevent corruption.

While proposals to resolve the political dispute between Israel and the Palestinians often seemed esoteric, a plan to improve the lives of the Palestinian people was much more concrete. The plan detailed 179 specific projects. It included charts with cost estimates and implementation timetables meticulously calculated by Thomas Storch, an analytic savant who had been my classmate at Harvard before going on to a successful career on Wall Street. He was my right-hand man on the economic plan, and worked closely on it with John Rader.

None of these investments would matter unless they were part of a political peace agreement—the second part of our plan, which we hoped to release once Israel formed a government. A flourishing Palestinian economy depended on regional peace, and without it, we couldn't ask the Israelis to loosen their security protocols at the border and allow the free flow of goods and people from the West Bank and Gaza. Israel's vibrant economy represented tremendous economic opportunity for the Palestinians if the leaders could resolve their old political disagreements. It was like having a Silicon Valley that was disconnected from the rest of California.

In the first week of its release, the Peace to Prosperity plan was downloaded more than a million times, sparking discussions and debate throughout the region and around the world. Before we even released the plan, the Palestinian leadership rejected it, which I had expected, and President Mahmoud Abbas announced that the Palestinians were boycotting the workshop in Bahrain.

On June 25 I landed in Bahrain for the long-awaited Peace to Prosperity workshop. The purpose of the workshop was to build momentum for our economic plan by engaging with world leaders and business titans who had the ability to invest in the projects our plan recommended. I wanted to illustrate that our blueprint could quickly become a reality as soon as a political peace agreement was reached. My team had spent months planning the summit. They had coordinated every detail with the Bahrain government and the renowned marketing legend Richard Attias, who had converted a ballroom at the Manama Four Seasons Hotel into a sleek 360-degree oval stage.

The workshop drew an impressive cast of attendees that validated the seriousness and viability of our plan. Among the hundreds of participants were Bahrain's crown prince Sheikh Salman, Steve Schwarzman of the Blackstone Group, Emirati real estate tycoon Mohamed Alabbar, former British prime minister Tony Blair, IMF director Christine Lagarde, World Bank president David Malpass, Randall Stephenson of AT&T,

Masayoshi Son of SoftBank, and FIFA president Gianni Infantino. In total, more than twenty-five countries were represented, including Saudi Arabia, the UAE, Jordan, Egypt, Qatar, Morocco, and Russia. Treasury secretary Steven Mnuchin led the US delegation.

The Bahrainis agreed to waive their traditional visa restrictions and allow Israeli businessmen and a few members of the Israeli media to participate—a significant development, given that Israel and Bahrain did not have formal relations. With each small gesture like this, we were giving Arab leaders another chance to test the waters on normalization.

The media delighted in pointing out that few Palestinian businessmen were in attendance, but Mohamed Alabbar, the CEO of Emaar, the largest development company in the region, perfectly captured the evolving perspective of the Arab leaders toward the Palestinian resistance: "Every one of us, we are really Palestinian at heart. Because the Palestinian issue is our issue. So, unfortunately, they are not here. It would have been great to have them. But I feel like I represent them," he said during a panel discussion.

Ultimately, Abbas's stubbornness may have backfired. To those in attendance, his refusal to participate and his ban on other Palestinians attending seemed to be self-defeating decisions. The conference dominated the airwaves in the Middle East for three days. When reporters asked Abbas what he thought about the workshop, he called the plan "a big lie that Kushner and others invented to make fools of the people." His words insulted the leaders who had just attended. Far worse, I later received reports that the Palestinian Authority had imprisoned, intimidated, and tortured the few Palestinian businessmen who did defy the threats and attend the conference.[43] It was a deeply troubling display of Abbas's brutal retaliation against his own people.

Two years earlier, it would have been unthinkable for these Arab ministers to attend a public conference with Israelis that the Palestinians had openly attacked. The metaphorical wall between Arabs and Israelis was beginning to dissolve before our eyes. Through the workshop, Bahrain had taken a courageous step toward normalization, and the praise far outweighed the backlash. The Gulf leaders began to consider what a bigger step could look like—not just for Bahrain, but for their own countries.

34

The Demilitarized Zone

It's a shame we've come all of the way around the world and aren't meeting with Kim Jong Un," Trump said to Ivanka on a morning phone call while we were in Japan for the G-20. "My team says it's hard to communicate with him. They tell me that you can communicate only through formal letters, which have to be translated and then flown over to North Korea. But I've heard that Kim follows my Twitter account, so maybe I'll just tweet that I'd like to meet him when I'm in South Korea tomorrow. Who knows?"

"Dad, that would certainly be your way of doing things," said Ivanka.

That morning, Ivanka was putting the finishing touches on her remarks for a G-20 session on women's economic empowerment. Prime Minister Shinzō Abe of Japan had asked her to host a session on the topic, which he had taken up as a priority in Japan. Ivanka thought it would be a low-key session on the sidelines of the main conference, but nearly every world leader at the G-20 decided to attend, including the heads of state from Canada, France, Germany, and Saudi Arabia. Ivanka's father was especially proud to join.

Minutes after their call, Trump tweeted: "After some very important meetings, including my meeting with President Xi of China, I will be leaving Japan for South Korea (with President Moon). While there, if Chairman Kim of North Korea sees this, I would meet him at the Border/DMZ just to shake his hand and say Hello(?)!"

Ivanka took the stage at her event. With clarity, warmth, and strength, she urged the assembled leaders to "elevate one of the most undervalued resources in the world: the talent, the ambition, and the genius of women." She discussed her program, the Women's Global Development and Prosperity Initiative (W-GDP), which aimed to empower fifty million women in the developing world by providing vocational education, access to finance for women entrepreneurs, and reforms to the legal and cultural barriers that prevent women from participating in the economy.

Soon after, we learned yet another lesson in how the media can trivialize and distort real events. The social media department of the French government innocently posted a twenty-second video of an informal conversation between Ivanka, President Emmanuel Macron, Prime Minister Theresa May, Prime Minister Justin Trudeau, and IMF director Christine Lagarde. Ripped from its context, the video made it appear as though Ivanka was inserting herself into the conversation and that Lagarde was snubbing her. This was entirely inaccurate. Lagarde and the other leaders were attendees at an event that Ivanka had led at the request of the G-20's host, Prime Minister Abe. Further, Ivanka and Lagarde had a warm relationship. Just a few months earlier, Lagarde had even asked Ivanka to introduce her at an awards ceremony.

Despite these facts, snarky and dishonest detractors used the video, which was viewed more than twenty million times, to create the false narrative that Ivanka was unwelcome at the G-20. This was a painful and disheartening moment for Ivanka. A short video clip obscured two years of hard work, her incredible speech, and the fact that she succeeded in making women's economic empowerment a central pillar of nearly every world leader forum during Trump's time in office. Beneath her stoic smile, Ivanka has a big heart, and the media's petty attacks could sting. I wished I could do more as a husband to help her feel proud of her important work, even in the face of unfair criticism. Back when Ivanka was running a mission-driven business, she was universally praised throughout the media, even by publications like *Vogue* and *Vanity Fair*. Now, even though her government work was positively impacting millions of women globally, the media looked for every opportunity to criticize her efforts. It was often tempting to fight back, but I admired Ivanka for always opting to take the high road and stay true to herself and her service. We both had to learn to let go of the things we couldn't control and to keep perspective on what mattered most: our faith, our family, and what we were trying to achieve for the country.

From Ivanka's event, Trump moved immediately into his meeting with Xi, which was widely expected to be a showdown on trade. China's president opened with a story about the famous ping-pong diplomacy that had thawed US-China relations in the 1970s and ultimately led President Richard Nixon to open diplomatic relations with China. Xi told Trump that he spent the majority of his time thinking about his country's relationship with the United States, which he wanted to be based on mutual respect and mutual benefit. He observed that some in the United States were calling for a new cold war, but that he felt like the relationship could improve. Trump agreed, pointing to their friendly dynamic as a reason for optimism. This was always Trump's negotiating posture with Xi: he would lead an honest and tough discussion on the issues, but would do so with charm, drawing upon his natural chemistry with the Chinese president.

When Xi raised the topic of tariffs, Trump made clear that he thought they were a great thing, and might leave them in place even if the United States and China reached a trade deal. Then Trump made one of his classic, offbeat remarks intended to put Xi off-balance. He mentioned what a great job Abe had done hosting the G-20, adding that Japanese fighters were among the best in the world, dating back to the era of the samurai. He then casually observed that the United States had saved China from Japan during World War II.

The moment the translator finished conveying Trump's impromptu comment, Xi's cordial manner gave way to anger. Xi emphatically disagreed with Trump's description. After fourteen years of fighting, China had liberated itself, and they had lost twenty million people in the process, he stated.

Realizing that he had touched a nerve, Trump redirected the meeting back to trade. Xi told Trump that he understood the president's concerns about the trade deficit, which had risen to $400 billion annually, and that he was willing to take steps to create more balance. Xi said that he knew that Trump carried the farmers in his heart, and that he was their guardian.

This comment referred to the unprecedented subsidies Trump had given farmers to offset the impact of China's tariffs. Not missing a beat, Trump told Xi that American farmers had great pride, and that they didn't want the aid, but that he would continue to give it to them for as long as it took to work out a deal with China.

Toward the end of the bilateral, Xi raised the issue of North Korea. He commended Trump's previous two meetings with Kim Jong Un and offered a piece of advice: the United States should be prepared to make concessions. This would show flexibility compared to past administrations and bring North Korea closer to making a deal to denuclearize. Xi said that China was prepared to nudge North Korea to make a deal with the United States, but that Trump should be careful not to back Kim into a corner. Xi warned the president to make strong security assurances to Kim, so that Kim could feel confident that North Korean denuclearization wouldn't lead to a "Libya situation," referring to the Bush administration's approach to Libya's denuclearization, which ultimately led to the assassination of dictator Muammar Gaddafi.

The following morning, in Seoul, Trump asked Ivanka and me to join him for the formal bilateral meeting with President Moon Jae-in and the South Korean delegation. Trump intended to discuss sharing the cost of housing twenty-eight thousand American troops in South Korea. In his view, the South Koreans benefited the most from having a strong American presence defending them against North Korea, and he wanted the South Koreans to pay $5 billion annually to defray the expense. They had agreed to increase their yearly contributions to $1 billion, but Trump wanted more.

He told Moon that he viewed this deal as month-to-month. He wanted them to get to $5 billion, and he would give them five years to do it. South Korea was a great country, but the United States didn't want to continue spending billions of dollars on military costs, for the privilege of losing money to their country on trade. He wouldn't hesitate to pull American troops out of South Korea if they couldn't reach a resolution.

Trump privately remarked that even though he didn't get all that he was asking for, he did bring in an extra $500 million for American taxpayers by making a few phone calls. The meeting wrapped up, and the two leaders began walking to a press conference at the bottom of the stairs in the Blue House, the executive residence of President Moon.

Meanwhile, National Security Adviser John Bolton, who had been in the meeting with the two leaders, spotted Acting Chief of Staff Mick Mulvaney and made a beeline for him.

"This DMZ thing is off," said Bolton, pointing a finger at Mulvaney's chest. "This is fucking off."

"What happened?" asked Mulvaney. "Why is it off?"

"Because the North Koreans are insisting it be a one-on-one meeting between the two leaders, and that breaks every protocol rule in the book."

Mulvaney excused himself from the press conference and called deputy chief of staff for operations Dan Walsh, who was in the middle of impromptu negotiations with his counterpart in North Korea.

"I'll tell them that we're canceling if they insist on a one-on-one," said Walsh. "They'll back down because if this thing falls through, I'll probably lose my job, but the other guy will be executed."

Minutes later, the president announced that he was going to the DMZ and that he would meet with Kim. This infuriated Bolton, who never had approved of the president's overtures.

"I'm not going to this fucking thing," Bolton said.

He made good on his promise: as Trump prepared to fly to the DMZ, Bolton took a separate airplane to Mongolia for unrelated meetings. The White House national security adviser abandoned his boss during a high-stakes moment with one of America's major adversaries.

As we boarded Marine One for the twenty-minute flight to the DMZ, none of us knew what to expect. When we arrived, the president greeted the American troops stationed there, and the military commanders escorted him to a platform overlooking North Korea. Across the border, the North Koreans had positioned heavy artillery weapons. Guards with large machine guns stood post, as if they expected an attack at any mo-

My parents raised me and my three siblings, Dara, Nicole, and Josh, in Livingston, New Jersey, a middle-class suburb forty-five minutes west of Manhattan. Here we are on a family camping trip in 1998. *(Courtesy of the Kushner family)*

I spent much of my childhood visiting jobsites with my father, Charles Kushner. He taught me the value of hard work. During my senior year of high school, we woke up at 4:30 each morning to train together for the New York City Marathon. *(Courtesy of the Kushner family)*

Ivanka and I met in 2007 after her father encouraged her to see if I was interested in buying a Trump property. Our business lunch quickly led to a first date. We were married in Bedminster, New Jersey, in October 2009.

(Brian Marcus/Fred Marcus Photography)

With Ivanka in 2014. We share many unlikely interests, including our love for New Jersey diners.

(Courtesy of the Kushner family)

Backstage with Steve Bannon (*left*) and deputy campaign manager David Bossie (*center*), watching Trump's second debate against Hillary Clinton on October 9, 2016. Bannon and I were allies on the campaign, but our relationship quickly deteriorated in the White House.

(Courtesy of Douglas Coulter)

Watching the election returns from Trump Tower on Election Day, November 8, 2016. Trump's victory shocked the world. *(Courtesy of Dan Scavino)*

Inauguration Day was a frenzy of activity, but Ivanka and I paused in the Lincoln Bedroom to light Shabbat candelabras and pray before we departed for the inaugural balls. We were told that it was the first time Shabbat candles had been lit in the White House residence. *(Courtesy of the Kushner family)*

When President Xi Jinping and Madame Peng of China visited Mar-a-Lago in April 6, 2017, Arabella recited Tang poetry in perfect Mandarin. President Xi was deeply impressed by our five-year-old. *(Courtesy of the White House Photo Office)*

With Trump at Yad Vashem, Israel's national memorial to Holocaust victims—a particularly meaningful moment for me, the grandson of Holocaust survivors. During his visit to Jerusalem in May 2017, Trump became the first sitting US president to visit the Western Wall. *(Courtesy of the White House Photo Office)*

I traveled to the Middle East in the summer of 2017 after learning that Secretary of State Rex Tillerson was secretly trying to remove me from the file. Here I'm meeting with Mohammed bin Zayed, the de facto leader of the United Arab Emirates. His message was surprising. *(Courtesy of the UAE government)*

The Russia investigation, later debunked as a baseless partisan attack, derailed the first fifteen months of the presidency. I briefed the White House press corps after testifying before the Senate intelligence committee on July 24, 2017. Though I put on a strong face, the investigation placed enormous stress on me and my family. *(Courtesy of the White House Photo Office)*

Through my highs and lows at the White House, Ivanka was my rock. She managed to always make time for our kids while advancing critical reforms to lift up forgotten men and women across our country. *(Courtesy of the Kushner family)*

Greeting President Xi of China on November 9, 2017, during President Trump's state visit to Beijing. The two leaders had a warm personal dynamic, but Trump was a tough negotiator, and he made historic gains to rebalance the US relationship with China. *(Courtesy of the White House Photo Office)*

In November 2018, Mexican president Enrique Peña Nieto awarded me the Aztec Eagle—Mexico's highest honor for foreigners. Contrary to media reports, the US relationship with Mexico reached unprecedented heights during Trump's presidency. *(Courtesy of the White House Photo Office)*

Checking in with Prime Minister Justin Trudeau of Canada and President Enrique Peña Nieto of Mexico shortly before the USMCA signing at the G20 in Argentina on November 30, 2018. Last-minute disagreements nearly killed the deal that day. *(Courtesy of the White House Photo Office)*

My family's experience during my father's imprisonment in 2005 inspired me to fight for prison and criminal justice reform in Washington. I advocated forcefully for the First Step Act, which reformed unfair prison policies that disproportionally hurt Black Americans. *(Courtesy of the White House Photo Office)*

Matthew Charles (*center*), the first prisoner released under the First Step Act, visited my office in January 2019 with his girlfriend, Naomi Tharpe. Brooke Rollins (*right*) and Ja'Ron Smith (*far right*) were instrumental in getting the legislation through a deeply divided Congress. *(Courtesy of the Kushner family)*

With the president aboard Marine One after flying over the New York skyline. As we looked out the window, Trump proudly pointed out buildings his father had constructed across Brooklyn. *(Courtesy of White House Photo Office)*

With Mohammed bin Salman, Crown Prince of Saudi Arabia, in February 2019. After Trump visited Saudi Arabia two years earlier, the Saudis strengthened their partnership with the US and brought many positive changes into the region, including unprecedented measures to counter extremism and reform society. *(Courtesy of the Saudi Arabian government)*

I always admired President Trump's intense focus during the most significant moments of his presidency, but he never took himself too seriously. Here I am with him on February 5, 2019, hours before his State of the Union address.

(Courtesy of the White House Photo Office)

Shaking hands with Kim Jong Un of North Korea on June 30, 2019, during our visit to the DMZ. Trump first began negotiating with the dictator after the North Koreans reached out to me through a previous business contact. Our unconventional diplomacy calmed a very tense global challenge. *(Courtesy of the White House Photo Office)*

Our three children—Arabella (*left*), Joseph (*center*), and Theodore (*right*)—adore their grandpa. Despite the demands of his job, Trump always made time to give them a big hug—and plenty of presidential M&M's. This picture is from Halloween, 2019. *(Courtesy of the White House Photo Office)*

On March 11, 2020, Vice President Mike Pence asked for my help with the COVID response. I called my friend Nat Turner (*left*) and Adam Boehler (*far right*), successful healthcare entrepreneurs who helped me procure lifesaving supplies and equipment from around the world. Avi Berkowitz (*center*) was a critical source of counsel throughout our government service. *(Courtesy of the White House Photo Office)*

Walking with the president and vice president to the State Dining Room in June 2020, for a roundtable with law enforcement and criminal justice advocates. In the wake of the George Floyd riots, I engaged with these groups to fund and train police throughout America. *(Courtesy of the White House Photo Office)*

On this August 13, 2020, phone call, Crown Prince Mohammed bin Zayed of the UAE and Prime Minister Bibi Netanyahu of Israel agreed to normalize relations between their countries, the first such peace agreement since 1994. We called it the Abraham Accords. *(Courtesy of the White House Photo Office)*

Facilitating the first official meeting between Israel and the UAE on August 31, 2020. Israel was represented at the historic meeting by national security adviser Meir Ben-Shabbat, and the UAE was represented by national security adviser Sheikh Tahnoun bin Zayed. *(Courtesy of the UAE government)*

Signing the Abraham Accords at the White House on September 15, 2020. I negotiated the pact between Israel and two Arab nations: the UAE and Bahrain. For the first time in twenty-six years—and only the third time in history—Arabs and Israelis had made peace. Morocco and Sudan would soon follow.
(Courtesy of the White House Photo Office)

In my preferred place behind the scenes at a Trump campaign rally on October 29, 2020. After recovering from COVID, the president made a superhuman push to close his reelection campaign, hosting seventeen rallies in eight states over the final four days. *(Courtesy of the White House Photo Office)*

Flying on Marine One over the Thanksgiving holiday in 2020. Despite our demanding schedules, we made many fond memories together. *(Courtesy of the White House Photo Office)*

On December 21, 2020, Prime Minister Bibi Netanyahu established the new Kushner Garden of Peace in Jerusalem's Grove of Nations. Bibi was a tough negotiator, and we disagreed sharply at times, but peace would not have happened without him. *(Courtesy of the Israeli government)*

Qatar and Saudi Arabia ended their diplomatic rift on January 5, 2021, further advancing regional peace. It was the sixth peace agreement I brokered.

I am pictured here in the office of Sheikh Mohammed of Qatar (*right*) and with Brian Hook (*center*), at the end of seven hours of intense negotiations. Sheikh Mohammed played a central role in helping us reach the deal.
(*Courtesy of the Kushner family*)

As our time at the White House drew to a close, our children were looking forward to seeing more of their mom and dad. We were all excited about the road ahead.
(*Courtesy of the White House Photo Office*)

ment. But the small United Nations complex within the DMZ didn't seem like such a hostile place. The buildings were modern and clean, and nothing like the 1980s Communist war zone I had expected. As the president led our group into the Inter-Korean House of Freedom, a four-story glass-enclosed building on the South Korean side of the DMZ, Walsh walked Trump through the plan.

"You're going to walk out there and greet Chairman Kim right at the demarcation line, which is a short six-inch curb that separates the two countries," he said. "If you step over that curb, you will be the first American president to enter North Korea."

"What do I do if he invites me over?" the president asked.

"If he does," Walsh said, "you can take a step or two into North Korea but Secret Service has almost no control of what happens once you cross that line."

Trump turned to Ivanka.

"Should I go if he invites me?" he asked.

"Why don't you play it by ear and see how it feels?"

Seconds later, Trump began his solitary walk toward North Korea. When he reached the line of demarcation, he stopped and waited for Kim, who was walking energetically in his direction. The two leaders shook hands, and Kim invited the president to step into North Korea. With cameras snapping so fast they sounded like machine guns, the two men turned and walked about twenty paces into North Korea. This sent the Secret Service into a frenzy. The plan was for the president to take a step or two before turning around, but here was Chairman Kim leading Trump further into North Korea.

After a few tense seconds, they turned around, walked back toward the South Korean side, paused in front of the media scrum, and said a few words before greeting President Moon, who had insisted on playing a visible role in the visit. Then they walked into the House of Freedom, where Trump and Kim met for nearly an hour. When I was introduced to Kim, he thanked me for my role in connecting him with Mike Pompeo.

I had kept it quiet, but I had played a central role in establishing the initial line of communication between Trump and Kim. Shortly after Trump took office in 2017, Kim began a series of provocative missile tests that increased tensions with the US. The president refused to let Kim push him around. During his first annual address to the United Nations General Assembly, Trump declared, "Rocket Man is on a suicide mission for himself and for his regime. The United States is ready, willing, and able, but hopefully this will not be necessary." The president had decided to insert the "Rocket Man" line into his speech just a few minutes before he went onstage to deliver it. Contrary to public perception, he'd been very careful with his word choice. He'd thought about calling Kim "*little* Rocket Man," but felt that could be too incendiary. When he delivered the line in his speech at the United Nations, there was a four-second delay for the interpreters, and then everyone in the General Assembly Hall turned and looked at our delegation's box with expressions of disbelief. When Trump met Kim at the Singapore summit, he was disappointed to learn that the North Korean had never heard the famous song by Elton John.

Around that time, Gabriel Schulze, a past business acquaintance of Ivanka's, reached out. In earlier years, before the sanctions were tightened, he had built deep relationships with key North Koreans. "One of my old North Korean business contacts who I trust," said Schulze "is telling me a very senior official wants to open a channel to the Trump family on behalf of Kim Jung Un. I've checked it out with my other contacts over there and this is serious. They want to explore a deal with Trump, and they believe you're the best person to talk to." Schulze pointed out that the North Korean government was a family business, having been led by the Kim family for three generations, so they naturally assumed the best place to start was with a family member on the other side. "How do you want to handle this?" Schulze asked.

At the time, I was walking on eggshells around Secretary Tillerson, who was supposed to take the lead on the North Korea relationship, but it was apparent to everyone that he was getting nowhere and was out of sync with the president. The Russia investigation also had made me radioactive. So instead of engaging directly, I suggested to Schulze that the North Koreans work with Pompeo, who was then the CIA director. "Tell them that Pompeo has the president's confidence," I said. "Meeting with Pompeo is as good as meeting with me. I will stay involved in developing the relationship, but behind the scenes."

Presented with such an opportunity, previous administrations would have passed it on to overcautious bureaucrats in the State Department. But Pompeo followed up with Schulze and established very productive contact with Kim's government. This led to several meetings in Pyongyang, where Pompeo set the table for the 2018 Singapore summit between the United States and North Korea.

While Trump and Kim met in the Freedom House, Ivanka and I walked outside and went into one of the small blue wooden shacks straddling the border. Built to facilitate dialogue between the two parties, the simple structures were half on the North Korean side and half on the South Korean side. While everyone else was preoccupied, we stepped into North Korea. We didn't linger, though.

On the flight home to Washington, Trump called the parents of Otto Warmbier, an American college student who had visited North Korea on a guided tour in 2016 but was arrested for removing a poster from the wall of his hotel. He had suffered a brutal and catastrophic brain injury while in a North Korean prison cell and passed away shortly after they released him to America in 2017.

"Each time I meet with Kim Jong Un, I think of Otto, and I think of you too," he said to Cynthia and Fred Warmbier. "It's a tough situation. I feel like I have an obligation to hundreds of millions of people to try to get them to deescalate. But when I see the images from today's meeting on television, I think about both of you at home watching it, and I know it's so tough for you. I need to try to make a deal, but anyone with even a little bit of heart knows how hard this must be on you. Seeing us walk and talk and smile—it might look hunky-dory, but it's not. I need to act like that for diplomacy, but it's hard. I don't know how you handle losing your son. You are amazing people."

The president invited them to the White House, so he could personally update them on what he'd learned. He also made them a promise: "If I get a deal done, Otto will be honored like never before."

35

The Enemy from Within

After the president threatened to impose sweeping tariffs on Mexican imports back in May of 2019, the Mexican government strengthened its immigration enforcement. Illegal border crossings dropped from a peak of 144,000 in May to 52,000 in September. The numbers were continuing to fall, but Trump wanted to reduce them further.

I called Mexican foreign minister Marcelo Ebrard. "My number one rule about working with Trump is that you have to proactively keep him informed about your efforts," I told him. "Otherwise, he will feel like nothing is happening and potentially take matters into his own hands. I operate under the assumption that if he calls me for an update on something, it's too late. Why don't you come to Washington for a working session, and then I can bring you into the Oval to brief the president on the steps Mexico is taking to curb illegal immigration. Come soon, or we could be back to square one with tariffs."

I could tell Ebrard was nervous as we walked into the Oval Office on September 10. Trump greeted him warmly and motioned for him to sit in one of the chairs facing the Resolute Desk. "I don't know if you saw the news," he said. "But this morning I fired John Bolton."

Ebrard had not expected Trump to mention Bolton, and he replied to Trump's statement cautiously: "Yes, I saw the news."

"John was crazy," said Trump. "He was constantly trying to go to war with everybody. He wanted to go to war with China, Russia, Venezuela,

North Korea, and Iran all at the same time. In the beginning, I didn't mind his aggressiveness. Having him on staff made me look like the rational one for a change. It also kept our adversaries off-balance."

Trump paused and looked directly at Ebrard. "This morning, John came to my office and said, 'Mr. President, everything is ready. We have to invade Mexico; they aren't doing enough at the southern border,' and I said, 'John, that's too much. That's the last straw. I would never do that to my friend AMLO or the great people of Mexico. You're fired.'"

Ebrard was dumbfounded. But when he saw me laugh at the joke, he cracked a smile and relaxed.

Trump was pleased with Ebrard's work on curbing illegal immigration, which involved Mexico's deployment of national guard troops to police its side of the border. Our two countries also had designed a "remain in Mexico" agreement, which kept asylum seekers in Mexico while the US immigration courts reviewed their cases, rather than releasing them into the American interior, where they often vanished into an underworld of illegal work and residency.[44] The previous irrational system encouraged hundreds of thousands of migrants from Latin America and elsewhere to travel to the southern border and made it nearly impossible to figure out which were making phony asylum claims. With the remain-in-Mexico agreement in place, illegal migrants stopped coming. It wasn't worth paying tens of thousands of dollars to human smugglers for a treacherous journey if it would likely result in getting sent back home. As part of the agreement, the Mexicans also cracked down on these smugglers, known as "coyotes," who abused the women and children under their charge, in what was often a modern form of slavery.

Ebrard described these efforts, and Trump gave the minister his complete attention.

"I appreciate all your efforts," said Trump, after Ebrard's briefing. "Your actions have saved many lives, but Mexico can do more."

Ebrard promised Trump that Mexico would continue to improve border enforcement and combat human trafficking. We were finally making significant progress to curb the dangerous flow of smugglers, traffickers, weapons, and drugs—and it wouldn't have been possible if Trump hadn't pursued the paradoxical strategy of playing hardball with tariffs while building a positive relationship with AMLO and Ebrard.

Through these efforts, we had dramatically strengthened America's relationship with Mexico, improved the lives of people on both sides of the border, and increased American jobs. This contributed to a growing number of Hispanic Americans supporting Trump's policies. Unfortunately, Trump didn't have much of an opportunity to enjoy this success. A storm was quickly brewing in Washington.

* * *

Late one afternoon in August, I got a call from an old friend who was a major donor to Nancy Pelosi, the Democratic Speaker of the House.

"I know I told you last year that Pelosi saw impeachment as a political loser and had no plans to pursue it," he said, "but now she's under so much pressure from the far left that I think she's going to do it." He mentioned Representatives Alexandria Ocasio-Cortez and Maxine Waters as being especially aggressive. "They are threatening moderates with primary challenges if they don't get on board. You should get ready."

Just nineteen minutes after Trump took the oath of office in 2017, the *Washington Post* published an article with the headline "The Campaign to Impeach President Trump Has Begun." Five months later, Democratic representatives Brad Sherman and Al Green filed an article of impeachment against Trump for his firing of FBI director James Comey. And on January 3, 2018—the day that Democrats gained control of the House of Representatives—freshman congresswoman Rashida Tlaib pledged to "impeach the motherfucker." The Democratic base would not be satisfied with anything less than impeachment—they never accepted the results of the 2016 election and Trump's very presence in the White House was an affront they could not accept.

On August 8, 2019, Jerry Nadler, the chairman of the US House Judiciary Committee, announced that his committee had commenced an impeachment inquiry into the president, but it wasn't clear what for. By September, a growing chorus was demanding impeachment. I didn't realize how widespread the effort was until Representatives Hakeem Jeffries and Eliot Engel, who had been reasonable in the past, suddenly announced that they favored impeachment. They were trying to stave off far-left primary challengers.

On September 24, shortly after Trump delivered his annual address to the United Nations General Assembly, Pelosi announced a formal impeachment inquiry. It was a low blow, striking at Trump as he represented the United States on the world stage. She easily could have waited twenty-four or forty-eight hours to launch her attack.

The Democrats' stated cause for the inquiry centered on comments the president had made two months earlier, on a phone call with the newly elected president of Ukraine, Volodymyr Zelensky. Trump had asked Zelensky to investigate whether Hunter Biden's appointment to the board of Burisma, a Ukrainian natural gas company, was an act of corruption. Hunter had no experience in the energy sector, a long history of questionable business dealings, and a checkered past that included being kicked out of the US Navy for cocaine use.[45] At the time of the appointment, his father was vice president of the United States, and Hunter netted a consulting fee of $83,000 per month.[46] The Ukrainian government's top prosecutor had tried to investigate the appointment, but he was ousted after Vice President Joe Biden allegedly pressed for his removal.[47] Trump viewed this as a potential violation of public trust and wanted to learn more about the circumstances around it.

At the same time, Trump was fighting with Congress over roughly $4 billion in foreign aid, including $250 million for Ukraine. The president habitually disapproved of wasteful foreign aid programs. He thought the money would be better spent in the United States, rather than in foreign countries that were often rife with corruption. Trump had notified Congress that he intended to "impound" these funds and return them to the public coffers unless the legislative branch overrode his decision. This was not the first time Trump had used the impoundment mechanism to rein in foreign aid. In 2018 he had cited it to avoid spending foreign aid dollars, and he even asked lawmakers for permission to return the funds to the Treasury, but Congress rejected his request.

The Democrats had a whistleblower claiming the administration was withholding funds from Ukraine, and they accused the president of a quid pro quo: denying the funds unless the Ukrainians restarted their investigation of Hunter Biden. Pelosi's sidekick Adam Schiff, chairman of the House intelligence committee, somehow obtained a rough transcript of Trump's July 25 call with Zelensky. Schiff presented his view as fact: "The notes of the call reflect a conversation far more damning than I or many others had imagined. . . . The President of the United States has betrayed his oath of office and sacrificed our national security in doing so."

After a failed two-year search for a reason to impeach the president, they settled on the best bad option they could find.

The day after Pelosi's announcement, we were still in New York for meetings with foreign leaders. As I sat in the president's secure holding room between meetings, Acting Chief of Staff Mick Mulvaney handed me a folder marked "Secret." It contained a transcript of Trump's call on July 25.[48]

"What do you think about this?" he asked.

I read the transcript.

"It doesn't seem like a big deal to me," I said. "This is Trump being Trump."

Top economic adviser Larry Kudlow was sitting next to me. He also read the transcript and felt the same way. We debated whether to release it. The press had worked itself into a frenzy of speculation based on Schiff's distorted framing of the president's call. Releasing the transcript would punch a hole in Schiff's alarmist narrative. Mulvaney and I favored this approach. But it was a tough decision.

As we weighed the potential benefits and downsides of publishing the transcript, White House counsel Pat Cipollone called Mulvaney and argued that doing so would set a bad precedent. Foreign leaders would be less likely to speak candidly on future calls if they thought their private words could become public. And it was always better to err on the side of caution on a legal matter such as this. Cipollone was giving sound advice—the sort of technical legal guidance that one would rigorously follow in a courtroom. But we weren't dealing with a court of law. We were dealing with the court of public opinion, where the rules were different.

When the president reviewed the transcript, he instantly sided with me and Mulvaney, and he announced his decision by Twitter: "I am currently at the United Nations representing our Country, but have authorized the release tomorrow of the complete, fully declassified and unredacted transcript of my phone conversation with president Zelensky of Ukraine. You will see it was a very friendly and totally appropriate call. No pressure and, unlike Joe Biden and his son, NO quid pro quo! This is nothing more than a continuation of the Greatest and most Destructive Witch Hunt of all time!"

The same day the president released the transcript, he was scheduled to meet in New York City with President Zelensky. During the meeting, Zelensky was straightforward. Trump appreciated that the Ukrainian president was trying to fix a broken situation and that he didn't want to be in the middle of US politics. As the two leaders sat before a throng of press, a reporter shouted a question at Zelensky about whether he was pressured to start an investigation.

"Nobody pushed me," Zelensky said, confirming Trump's message.

Between the substance of the transcript and Zelensky's comment, the Democrats had made a tactical error by going all in on such a thin case. But as they embarked on formal impeachment proceedings in Congress and pummeled the president in the press, the White House had no communications strategy for refuting their attacks.

Unlike his predecessor, Mulvaney established a collegial culture within the West Wing. His door was always open, and he often played classic rock music while he worked. Most people felt comfortable collaborating with him. He respected and understood the president, and he asked me to get involved on issues when I could be helpful. Many aspects of the White House had improved as a result, but he had developed a sort of rivalry with Cipollone, which was becoming concerning.

Mulvaney began holding an impeachment planning meeting each morning in his office. As White House counsel, Cipollone should have been giving his advice in these meetings, but he rarely uttered a word. When I asked him why he was so quiet, he said that he suspected Mulvaney was leaking on him. He didn't want to divulge sensitive information that might find its way into the press.

A familiar routine began: Mulvaney would come to my office and complain about Cipollone. Later, Cipollone would walk in and complain about Mulvaney. I understood the costs of engaging in interoffice squabbles, so I mostly listened and didn't take sides. After surviving the rivalries of the early years, I was effectively free of enemies inside the West Wing—and I wasn't looking to make any.

I was fond of both Mulvaney and Cipollone and tried to bring them together, but I couldn't bridge the gap. Their personal differences were hurting the president's defense, so I began tracking the impeachment response more closely. While outside my purview, the issue was potentially fatal for the president, and I felt like I had a responsibility to watch my father-in-law's back. That was one of the main reasons I moved to Washington in the first place.

Impeachment was coming at Trump like a freight train. To gain perspective, I read several books about the Clinton impeachment and realized that his team had also dealt with competing power centers and personality clashes. Chief of Staff Leon Panetta and Harold Ickes, the deputy chief of staff, set up a response operation separate from the White House Counsel's Office and installed a young lawyer named Jane Sherburne to lead it. This arrangement led to brutal infighting. Panetta, whose work as chief of staff I had grown to respect, ultimately steered the effort back on course by integrating the legal and communications teams into a single unit focused on defending the president.

This was clearly the model to follow. I was confident that the White House legal team, along with outside counsels Jay Sekulow, Alan Dershowitz, Eric Herschmann, and Pam Bondi, would handle the nuanced legal arguments before the Senate. But after studying the Clinton impeachment, I knew we needed to assemble an equivalent communications team that would work in lockstep with the lawyers.

History showed that impeachment cases turned on public sentiment. Members of Congress were not impartial judges—they were political by nature and swayed by the viewpoints of their constituents. The facts were on our side, but we needed to win the battle in the court of public opinion.

36

Fight to Win

Amid the West Wing infighting, I received a call from congressmen Mark Meadows of North Carolina and Jim Jordan of Ohio. "You guys are blowing this thing," said Jordan. That got my attention. Both men were savvy politicians who cared about Trump. When they offered to come by the White House, I accepted.

Less than an hour later, they showed up with two of their fellow Freedom Caucus members, Lee Zeldin and Matt Gaetz. We assembled in Cipollone's office on the second floor of the West Wing.

"The way I see it, this case is simple," said Zeldin. "Number one, the White House released the transcript, proving the president has nothing to hide. Number two, the aid was released. Number three, the investigation in Ukraine never occurred. And number four, when asked, the president of Ukraine said there was no pressure applied. No matter what any whistleblower or Democrat says, these four facts will not change. If we all stick to them and communicate effectively, we will win big. There was no quid pro quo, and they have no case."

Meadows and Jordan asked why the White House communications team was missing in action. Not wanting to criticize my colleagues in front of members of Congress, I flipped the question.

"What do you think we need to be successful?"

"Right now," said Meadows, "no one from the White House ever calls us to coordinate your message. When I try to find out where the White House stands on a topic, Stephanie doesn't answer her phone or get back to me. Get us someone who is available twenty-four/seven to work with me and my members on this. We will help you amplify and win this battle."

Meadows was referring to White House communications director and press secretary Stephanie Grisham. Trump had promoted her after Sarah Huckabee Sanders moved back to Arkansas in June of 2019. Before then, Grisham had served as the First Lady's press secretary. Grisham was an unconventional choice for the top White House communications job. The role demanded a level of skill and commitment that far surpassed her responsibilities in the East Wing, where the pace was slower and tended to revolve around ceremonial events.

After it became clear that our communications and legal teams were not going to work together, I began searching for a senior communications person who could focus solely on impeachment—someone who would wake up every morning ready for this fight. It quickly became clear that the best person for the role was Tony Sayegh, a longtime communications pro at the Treasury Department who had helped us pass tax reform. Unlike the first year of the administration, when I made enemies unintentionally, this time I made a conscious decision to intervene, knowing that Grisham would likely turn against me. I wish I had seen another option, but protecting the president was more important than Grisham's opinion of me.

Trump was aware of the internal dysfunction, and he gave me the go-ahead to hire Sayegh. When I told Grisham about the decision, I tried to be as gracious as possible. I said that bringing in Sayegh and letting him focus on impeachment would free her up to manage the rest of the communications operation. I was pleasantly surprised when she said that it was a great idea. She said that she liked Sayegh and considered him a dear friend. She asked that Sayegh report directly to her and not have an office in the West Wing, which I immediately agreed to. I thought that perhaps things between us would not be so bad after all.

An hour later, the president called. Grisham had dashed to the Oval Office and claimed that Sayegh would imperil Trump's impeachment defense. That's when I knew that she had turned on me. At the president's request, I followed up with Grisham to discuss her concerns, at which point she accused me of trying to run her department.

"I don't want to run comms," I replied. "But you can't seem to get along with the legal team, and we need to have a senior comms person dedicated solely to impeachment to prevent the president from getting removed from office."

I felt confident the White House had the high ground on impeachment. Now we just needed to shell the Democrats rhetorically—and Sayegh would be our lead artilleryman.

* * *

In the White House it's impossible to deal with one problem at a time. Just as our team was coming together to handle the greatest domestic challenge of Trump's presidency, a conflict arose overseas. At the direction of the president, the Department of Defense was drawing down the number of American troops on the Syrian border with Turkey, where they were policing a controversial area. He warned President Recep Tayyip Erdoğan that he would "totally destroy and obliterate the economy of Turkey" if the Turks did anything "off-limits," but Istanbul's strongman ignored Trump's threat. As US troops departed the area, Erdoğan launched a massive military offensive in northeastern Syria against Kurdish fighters who had been critical partners in the US fight against ISIS. Trump asked Pence to negotiate a cease-fire with President Erdoğan.

Before Pence departed for Turkey, he stopped by my office. I had met with Erdoğan a few times, and Pence asked for advice in dealing with him.

"You don't need my advice," I said, "but I will tell you that from my experience, when you sit down with him, he's going to air all of his grievances. It will go on for a while, and I would just listen. My sense is that he values his relationship with Trump, so your success will depend on the degree to which you can convey that Trump is dead serious about a cease-fire. Erdoğan is very stubborn. I'm not sure how you solve this one, but I will root for you from here."

On October 17, while Pence was in Turkey, Trump traveled to Dallas for the grand opening of a massive Louis Vuitton manufacturing facility. Soon after the 2016 election, I introduced Trump to my friends Alexandre and his father Bernard Arnault, head of LVMH—the Louis Vuitton parent company and largest luxury goods corporation in the world. Afterward, Arnault announced that he was looking to build another Louis Vuitton factory in the United States. Two and a half years later, he followed through on his promise with a hundred-thousand-square-foot facility that would employ a thousand American workers to make the finest leather products in the world. Arnault joined us for the trip to Texas on Air Force One—a thrilling experience, even for the third-wealthiest person on the planet. As Trump and Arnault talked on the flight, Pence called.

"We made a deal," the vice president said. "I was firm in telling Erdoğan that you love him, and that you were his friend before this, you're his friend now, and you'll be his friend always—but that he needed to stop this war he started, immediately, or there will be massive economic sanctions. He raised many objections, but after ninety minutes and seven versions of the same message, he said they'd stop it, and we went into the other room and finalized a deal."

"That's great," said Trump. "This area has been a powder keg for a while, and I hated having so many troops there. We were there because both sides wanted the territory, and we could never broker a compromise. Sometimes you just have to let the two sides fight it out a bit, and when they realize neither has a great situation, then it's much easier to make a deal. If I didn't do this, America would be stuck there for the next hundred years—or even worse, end up in another war over a piece of sand that no one in America has ever heard about." Trump knew how to deal with big personalities to prevent combustible situations.

While he was on the phone with Pence, Fox News flashed to the White House press briefing room. Mulvaney was at the podium, taking questions about the ongoing impeachment inquiry. As a former elected official, Mulvaney was a sharp and effective communicator. Afterward, I dialed Mulvaney and commended him on a solid performance.

Shortly after I hung up, my phone rang. It was Cipollone.

"Did you hear what Mulvaney just said?" he asked.

I said that from what I had seen, Mulvaney had done a good job.

"That was an absolute train wreck," Cipollone moaned. "He just said that the president engaged in a quid pro quo. That blows up the entire case. We need Mulvaney to correct that statement immediately."

I had missed the most important thirty seconds of the briefing. ABC reporter Jon Karl had asked Mulvaney point-blank whether the president had committed a quid pro quo.

"To be clear what you just described is a quid pro quo," Jon Karl said as he continued. "It is: funding will not flow unless the investigation into the Democratic server happened as well."

"We do that all the time with foreign policy," Mulvaney responded. "We were holding up money at the same time for . . . the northern triangle countries, so that they would change their policies on immigration."

This was indeed a disaster. Mulvaney's point was fundamentally valid. Presidents regularly leverage foreign aid to extract concessions from their foreign partners. Trump was especially good at it. In a high-stakes moment where our messaging needed to be tight, however, it was sloppy. It energized the Democrats, who claimed that the president's own chief of staff had just provided a smoking gun.

Mulvaney tried to walk back the statement, but it was too late. Just as we were gaining momentum, our own White House team handed the advantage to our opponents. Now we were going to have to regroup and fight it out to the brutal end.

37

Hospital Negotiations

There's an old story about Robert Lighthizer, the US trade representative. In the 1980s, when he was working in the Reagan administration, he participated in trade talks with the Japanese. One day, after receiving an unsatisfactory proposal from his counterparts, he took the page that contained the proposal, folded it into a paper airplane, and tossed it back to the Japanese. The incident earned him a nickname: "Missile Man."

Some three decades later, he was Trump's top trade negotiator—and he hadn't lost his sense of humor. In 2019, as our bargaining with China entered a new phase, he emailed a limerick:

We are talking to President Xi
Whether progress is made we shall see
Should cheating continue
Beyond this brief window
Tariffs there surely will be.

To the best of my knowledge, the Chinese never saw that email—but if they had, it would have played into their worst fears about Lighthizer. His reputation as a tough negotiator intimidated them. They even held him responsible for Japan's slow-growth woes in the 1990s, believing that the country's troubles could be traced to the "Missile Man."

What really rattled them, of course, was not just Lighthizer, but the fact that for the first time in history, an American president was standing up to Beijing's unfair economic behavior. When China entered the World Trade Organization in 2001 as a "developing economy," it promised to liberalize its economic practices. Yet China had failed to fulfill its promise, even as it gained a larger share of the global market through low-cost goods heavily subsidized by cheap local labor and state investments. By 2018 the US trade deficit with China had ballooned to more than $400 billion annually, up from $83 billion in 2001. At the same time, China forced American companies to disclose their trade and technology secrets as a precondition to doing business in China. In effect, China was stealing our best technology and turning it against us.

By the end of 2018, Trump's confrontation with China had advanced to the point where tariffs covered about 96 percent of all Chinese imports. This raised nearly $40 billion in revenue for the US government in 2018 and 2019 alone. The prevailing wisdom assumed that a Trump-led trade war between the world's two largest economies would tank US markets and threaten a global recession. But Trump didn't buy it. Whenever he imposed new tariffs, the markets got choppy for a few days, but the doom never came, despite the fearful predictions of conventional economists. Even Trump's biggest haters admired his courage to take on a fight that his predecessors had ignored.

After Trump told President Xi in December of 2018 that I would help broker a US-China trade deal, I regrouped with our team. I knew Lighthizer and Treasury secretary Steven Mnuchin had the technical expertise to get the deal across the finish line, so I told them to let me know how I could be helpful to their ongoing negotiations. Wary of Lighthizer, the Chinese preferred to work with Mnuchin, a pragmatic and talented dealmaker who made no attempt to hide his desire for strong and stable economic markets. The Chinese began talking almost exclusively to Mnuchin, attempting to sidestep Lighthizer altogether.

Nothing could change the fact that Lighthizer was America's lead negotiator on trade. He had earned the president's respect and trust. Having seen Lighthizer's effectiveness in our negotiations with Canada and Mexico, I was convinced that the Chinese needed to treat him as Mnuchin's equal to reach an agreement. At the request of Lighthizer and Mnuchin, I called John Thornton, the former president of Goldman Sachs, who in 1997 had helped China Telecom become listed as a publicly traded company in a groundbreaking deal. Thornton had high-level contacts in China, and I found him to be thoughtful and constructive.

"The Chinese are getting nowhere, and it's because they keep trying to play the game on their terms," I said. "That may have worked for them in the past, but it won't under Trump. They need to know that they're never going to get a deal if they don't go through Lighthizer. Please convey to them that he's reasonable and, more importantly, that he has the complete trust of the president. I'll work with him to keep things on track if they engage."

Throughout 2019, the United States and China slapped tariffs on one another, and both countries felt the pain. China's GDP growth rate dropped to a thirty-year low. In the United States, farmers and ranchers saw an important export market close. Not wanting to punish American farmers for China's obstinacy, Trump directed Secretary of Agriculture Sonny Perdue to find a way to provide relief. Through an obscure Depression-era program, Trump redirected revenue from the tens of billions of dollars the United States was now collecting from the tariffs on China to American farmers and ranchers. This administrative masterstroke boosted farmers and gave Trump the leverage and staying power to hold strong in the fight, which he believed was hurting China far more than the United States. Unless the situation turned dire, I knew that Trump was unlikely to fold.

By September, the Chinese blinked. They signaled a willingness to buy American agricultural products. It was a smart way to show that they were serious. Words no longer mattered: as proof that they were not leading us on, we needed them to sign purchase orders and begin shipping containers. Lighthizer and Mnuchin closed in on a deal that would require the Chinese to purchase up to $50 billion in farm products annually, double the amount they had ever purchased from the United States. When Trump described the size of these purchases to the farmers' lobbyists, they said that they weren't sure that America's farmers could even produce that much product to sell to the Chinese.

"Buy bigger tractors!" Trump replied. "The farmers have stuck with me through this fight, and I'm going to make sure they come out stronger."

The agreement also improved intellectual property protections and prohibited the Chinese from forcing American companies to reveal their trade secrets and technologies. And, for the first time in history, China agreed to an enforcement mechanism that would hold them accountable. If they broke the agreement, tariffs would go into effect to offset the economic damage. It was an unprecedented concession for the Chinese to make.

In exchange for these commitments, the Chinese asked the United States to cancel certain tariffs and to reduce others. Lighthizer and Mnuchin disagreed on some of the details: Mnuchin wanted to lift more tariffs than Lighthizer believed was necessary. So we arranged to meet with the president. Trump would decide how to manage the final stage of the negotiation and secure a massive win.

On a Saturday evening in October, Ivanka and I joined Mnuchin, Ross, Lighthizer, Perdue, Navarro, and Kudlow in the Executive Residence for dinner. Mnuchin and Lighthizer made their cases. Trump was relaxed and happy to have an agreement in sight. Sensing that China wanted to make a deal, Trump opted for Lighthizer's more aggressive proposal. After the dinner broke up, he invited the team to watch *Joker* in the White House movie theater, but Ivanka and I decided to get back to our kids. As we walked out, Lighthizer pulled me aside and asked if I could have Thornton communicate our proposal to the Chinese so that they would know exactly what was needed to get a deal done.

I called Thornton, who agreed to carry the message to the Chinese, but also recommended calling Chinese ambassador Cui Tiankai to communicate our position.

Knowing that my call with Cui would be analyzed back in China, I spoke very deliberately. When I first walked the ambassador through the terms, he balked. China wanted more tariff reduction, he said.

I cautioned him not to think in terms of percentage, but to think in terms of what would happen if they didn't make a deal. I said I was fairly confident that Trump would do the deal I had outlined, but warned that dynamics could change if they delayed. If they said yes, it would pause the trade war and create space for the next round of negotiations to occur. If they didn't accept the offer, however, Trump would likely escalate. "My father once told me that no one ever sold him a building because they liked his tie—they sold it to him because he paid the highest price," I said. This was the price they needed to pay to make a deal and avoid a further increase in tariffs. While it was uncomfortable, they would look back and be glad that they did it.

My inflexibility seemed to get through to the ambassador. He assured me that he would relay my message to Beijing and that they would begin drafting a formal offer.

* * *

As this high-wire act of trade talks with the Chinese progressed, I had to confront an unexpected and frightening personal problem. On the morning that I traveled to Texas to attend the opening of a Louis Vuitton factory, White House physician Sean Conley pulled me into the medical cabin on Air Force One. "Your test results came back from Walter Reed," he said. "It looks like you have cancer. We need to schedule a surgery right away."

Before he could say more, I put my hand on his shoulder. "Listen, Doc, let's pretend you didn't just say that and get through the next twenty-four hours," I said. "Come to my office tomorrow morning. Please don't tell anyone—especially my wife or my father-in-law."

The next morning, I told Ivanka what I knew. With as much confidence as I could conjure, I told her not to be concerned. Whatever this was, we would find a way to work through it. She joined me for the meeting with Dr. Conley, as did Avi. Ivanka and Avi graciously offered to find the best specialist in the country. Dr. Thomas Fahey of New York-Presbyterian Hospital concluded that I needed surgery to remove an unusual growth in my thyroid, and we scheduled the operation for the Friday before Thanksgiving. That way, I would miss the least amount of time in the office. My absence might even go unnoticed. That's how I wanted it. This was a personal problem and not for public consumption. With the exception of Ivanka, Avi, Cassidy, and Mulvaney, I didn't tell anyone at the White House—including the president.

I threw myself into my work and tried not to think about the upcoming surgery or the unwanted growth in my body. When I did think about it, I reminded myself that it was in the hands of God and the doctors, and that whatever happened was out of my control. At moments, I caught myself wondering whether I would need extensive treatment. I thought about the many simple things I took for granted that the doctor warned could be different—or even vanish. Every night, before I went to bed, I lingered for a few extra moments in my children's rooms. I watched them sleep without a care in the world. I felt guilty that I had been so distracted and absent over the previous few years. I was always at work or taking phone calls when they wanted to spend time with their dad. I missed plays and sporting events. I had promised myself that when my service in the White House ended, I'd make up for lost time. Now I was forced to confront that possibility that my time might be up. I prayed that the surgery would be successful.

The day before the surgery, Trump called me into the Oval Office and motioned for his team to close the door.

"Are you nervous about the surgery?" he asked.

"How do you know about it?"

"I'm the president," he said. "I know everything. I understand that you want to keep these things quiet. I like to keep things like this to myself as well. You'll be just fine. Don't worry about anything with work. We have everything covered here."

I hadn't wanted him to know because I felt he didn't need another problem to worry about, but now I was glad he did. At the White House, I tried to have his back. Now he had mine, and I was grateful for it.

Thanks to the skill of Dr. Fahey, the operation went well. He had removed a substantial part of my thyroid. When I woke several hours later, Dr. Fahey was standing over me. "Please tell my wife I am okay," I said. The biopsy results arrived a few days later, revealing that the nodule indeed was cancerous. Thank God we caught it early. Before surgery, the doctors had warned me that the procedure could alter my voice, and it could take weeks or months to return to normal. Luckily, the impact was minimal. Several hours later, while I was still in the hospital recovering, I got a phone call from Thornton. The Chinese were ready to make a formal offer based on our proposal.

We discussed a few details, including exactly how much relief China would get on the tariffs. I suggested a compromise that tracked closer to Lighthizer's position than Mnuchin's, and said that if the Chinese agreed to it, I'd take it straight to the president.

After I hung up with Thornton, I called Lighthizer on my secure phone and told him what had happened.

"That is very close to what we wanted," he said. "If they really make that offer, that would be a great deal."

The Chinese had agreed to what would soon become known as the "Phase One" trade deal, a massive victory for the United States. To keep President Trump from further escalating tariffs, China had agreed to an unprecedented series of trade concessions. They consented to keeping the $250 billion in existing tariffs on Chinese imports in place, without retaliating further. This completely reset the US trade relationship with China, raising the cost of their imports, while protecting American workers and netting tens of billions of dollars in annual revenue to the federal government. Through Phase One, the Chinese agreed to make systemic changes in their treatment of intellectual property and in their agricultural and financial services sectors, balancing the competitive playing field between American and Chinese companies. They also agreed to make significant purchases of US agricultural products. Finally, the agreement was enforceable: if China failed to follow through, the United States could impose sanctions—and possibly tariffs. This alone was a major breakthrough for the United States.

Excited by these developments, I called the president.

"That's great," Trump said. "Get it done."

Finally, I called Ambassador Cui and told him the news.

That was to be my last call in the hospital: Ivanka stepped into the room, gave me a kiss, and took away all three of my phones.

38

Soleimani

On January 2, 2020, as Trump met with his campaign team at Mar-a-Lago, National Security Adviser Robert O'Brien entered the room. Trump had recently hired the successful hostage negotiator and foreign policy expert to replace Bolton.

"Mr. President, it's time," said O'Brien.

Trump stood up and began to follow O'Brien out of the room. Before exiting, he turned around.

"Wait here, fellas. I'll be back."

We were in a small and dimly lit room off the library at Mar-a-Lago. Called the Monkey Room because of the intricate monkey carvings on the walls, it exuded a vintage, clubby feel that hearkened back to the Roaring Twenties, when the resort was built. Trump was reviewing options for a television ad to air during the Super Bowl, which would reach an estimated 80 percent of voters. Brad Parscale, Larry Weitzner, Dan Scavino, and I were going over the two spots with him when O'Brien came in.

"I don't expect him to return for a while," I said, after Trump left.

Over drinks the night before, Senator Lindsey Graham had suggested that something big was on the horizon: "What POTUS is thinking about doing tomorrow is courageous," he said, cryptically. "It comes with a risk, but it's going to be a game-changer." I was intrigued by Graham's comment, but I was totally unaware of what was about to come.

As the minutes ticked by, the others looked around restlessly, wondering how long they should wait around and whether the president would come back at all. Sooner than I had expected, the president returned to the Monkey Room.

"Can you play the Alice Johnson spot one more time?" he asked.

As we resumed our discussion, I noticed that Scavino was scouring Twitter. He knew exactly which journalists to follow for breaking news around the world, and he often flagged international events for the president and senior staff long before we received intelligence from officials at the CIA or others elsewhere. Ten minutes passed, then Scavino spoke up.

"There are images of an explosion in Iraq. People are saying it was by the airport."

"Dan," said the president, "follow that closely and tell me if anything interesting comes up."

Five more minutes passed.

"You all have got to see this!" said Scavino.

A journalist in Iran had tweeted a photograph of a severed, ash-covered hand adorned by a ring with a large blood-red stone.[49] Alongside this image, for comparison, was a recent photo of top Iranian general Qasem Soleimani, stroking his beard. On his hand, he wore the exact same ring.

As the news broke, Trump remained coolly engaged in our discussion. It was as if nothing out of the ordinary had happened. This was one of the traits I admired most about Trump. He was one hundred percent focused on the task at hand. The higher the stakes, the calmer and more engaged he became. Many of his critics assumed that he was erratic and undisciplined, especially because of his tweets. This perception missed something fundamental about the president: when making consequential foreign policy decisions, he was careful and deliberate. He always understood the gravity of the moment, and he never wanted to endanger American lives if he could avoid it.

The world soon learned that Trump had ordered the strike that killed the world's top terrorist, General Qasem Soleimani of Iran. If Supreme Leader Ayatollah Khamenei was Iran's head, then Soleimani was its clenched fist. He commanded Iran's Quds Force, an elite unit of twenty thousand soldiers that worked clandestinely to destabilize the Middle

East through Iran's terrorist proxies. He had supplied the roadside bombs that America's enemies used to kill and maim thousands of US troops in Iraq and Afghanistan. In Syria, dictator Bashar al-Assad had given Soleimani free rein to command militias that had access to Syria's borders with Israel, Lebanon, and Iraq.

As Soleimani's military grip on the region tightened, his popularity in Iran and his fame across the Middle East rose to unprecedented heights. Former CIA analyst Kenneth Pollack profiled Soleimani for *Time* magazine's 100 Most Influential People list in 2017: "To Middle Eastern Shi'ites, he is James Bond, Erwin Rommel, and Lady Gaga rolled into one."

In the months leading up to the president's strike, the Iranians had escalated their attacks against America and our allies in the Middle East. On June 20, the Iranians shot down an American drone flying in international airspace over the Strait of Hormuz. Trump initially approved a retaliatory strike, but reversed course just minutes before it was carried out. He tweeted about his decision: "We were cocked & loaded to retaliate last night on 3 different sights when I asked, how many will die. 150 people, sir, was the answer from a General. 10 minutes before the strike I stopped it. Not proportionate to shooting down an unmanned drone. I am in no hurry."

On December 27, Iran-backed Shiite militias fired several rockets into a joint Iraqi-American airbase, killing a US military contractor and injuring four US soldiers. The president decided he had shown restraint for long enough. Trump knew that taking out Soleimani would degrade Iran's military capability and send the strongest possible message that there would be no safe harbor for those who aim to kill Americans.

On January 2, Trump called in the strike. Soleimani had landed at Baghdad International Airport, just twelve miles from the US embassy, unaware that he had only moments to live. He climbed into a sedan and departed the airport. In an extraordinary twist of fate, he was joined by an unexpected passenger: Abu Mahdi al-Muhandis, one of the most dangerous but seemingly untouchable terrorist masterminds in the world. For years, Muhandis had been at the top of America's target list. In 1983 he orchestrated the bombing of the US and French embassies in Kuwait, killing five civilians, before fleeing to Iran and developing a close relationship with Soleimani, who was establishing the Quds Force. In 2003 he shifted his operations to Iraq, creating a sophisticated web of highly trained terrorists known as the Hezbollah Brigade, which killed hundreds of American soldiers. The United States might have killed Muhandis years earlier, but in 2014 he was appointed to an official role within Iraq's government, and the US didn't want to damage its relationship with Iraq as it navigated the volatility in the region. When he placed himself in that vehicle with Soleimani, however, Muhandis unknowingly signed his own death warrant.

As the two killers and their entourage traveled along an airport access road with light traffic, a Reaper drone circling far overhead launched a Hellfire missile. It's unlikely that Soleimani or Muhandis heard the whistling sound of the missile for more than two seconds before it left them in a smoldering pile of ash and steel on the airport access road. They were dead instantly.

Soleimani was a dangerous target. His military influence in the region and close relationship with the Ayatollah meant that killing him risked war. Military leaders who had served in the Middle East understood the implications. "It is impossible to overstate the significance of this action," General David Petraeus said on the public radio program *The World*. "This is much more substantial than the killing of Osama bin Laden. It's even more substantial than the killing of Baghdadi," the leader of ISIS that the military had killed at Trump's direction several months earlier.

As the world reacted to the president's decision, Trump dined with House minority leader Kevin McCarthy on the patio at Mar-a-Lago. I sat at a nearby table with Graham, Scavino, and O'Brien and his wife Lo-Mari. Throughout dinner, O'Brien kept excusing himself and disappearing into a top-secret facility to take phone calls. When O'Brien returned to the table after one of his calls, I asked if he was preparing a statement for the president. Other than tweeting a picture of an American flag shortly after the strike, the president had refrained from commenting publicly. I thought Trump needed to send a clear and strong message about his reasoning for the strike and the consequences Iran would suffer if it retaliated. To my surprise, O'Brien said that he and Secretary of State Mike Pompeo felt that the strike should speak for itself.

The next morning, January 3, I paid a rare visit to Trump's bedroom. He asked how the news was playing, and I said that it was getting massive attention from the press, and that many world leaders were calling O'Brien to express appreciation for the bold move, but they were afraid to say so publicly. When I asked if he was going to make a statement, Trump said that Pompeo had advised against it because it would draw unwanted attention to the strike and escalate the situation.

"That ship has sailed," I said. "This is dominating the news."

Three years earlier, when the president ordered the strike on Syria, I had kept my thoughts mostly to myself because I didn't have confidence in my point of view. But now I knew more about the region. This time I had a strong conviction, and the stakes were even higher.

"Iran is vowing to retaliate," I said. "You have an intended audience of one—Ayatollah Khamenei. He has to know that if the Iranians kill one American, you will unleash fury. Right now he is probably sitting in his version of the Situation Room with his top experts, discussing options. It is important that you explain that this was not a preemptive strike—it was retaliation for all of the murders and maiming of American soldiers that Soleimani had caused. If you don't make a statement, we will be at greater risk of Iran hitting back at American troops in the region."

Trump thought for a minute, and then asked me to put together a set of remarks. As I worked with the speechwriting team, Pompeo called the president to check in, and Trump told him that he was now considering making a public statement and asked the secretary to discuss it with me. When we spoke, Pompeo was initially resistant and a bit annoyed. In fairness, the Department of State, the Department of Defense, the intelligence agencies, and the National Security Council had developed a thorough plan for all aspects of the Soleimani strike, and the consensus was to recommend against presidential remarks.

By the end of our discussion, Pompeo hedged. "There's a fifty-fifty chance that your strategy is right."

"Let me send you the remarks," I said. "See if you like them, and let me know if you have any changes."

A few minutes later, Pompeo called back. "I have no problem with these remarks," he said. "I see what you are trying to do here. If the president is going to say something, this is the right thing to say."

I called the president and updated him on my conversation with Pompeo.

"I want to give the speech today," he said.

Shortly after 3:00 p.m., right before he took off on Marine One for an event at a church in Miami to launch one of his most important campaign coalitions, Evangelicals for Trump, the president walked into Mar-a-Lago's temporary press briefing room and addressed the nation.

"As president, my highest and most solemn duty is the defense of our nation and its citizens. Last night, at my direction, the United States military successfully executed a flawless precision strike that killed the number-one terrorist anywhere in the world, Qasem Soleimani. . . . Under my leadership, America's policy is unambiguous: To terrorists who harm or intend to harm any American, we will find you; we will eliminate you. We will always protect our diplomats, service members, all Americans, and our allies. . . . We took action last night to stop a war. We did not take action to start a war."

On Sunday, January 5, while I was on a run with Ivanka, UK prime minister Boris Johnson called my cell. He had put in a formal request to speak to the president but hadn't heard back.

Johnson and I had been friendly since the transition in 2016, when he was foreign secretary and came to meet with me in New York City. We stayed in touch after he resigned from government in July of 2018 and continued our friendship when he returned to government and became prime minister. I always found him to be accessible, engaged, and imaginative.

I immediately called the president's military aide, who handed the phone to Trump. When I told him that Boris Johnson had requested a call, he was frustrated that no one had told him about it and asked to speak to the prime minister immediately. The conversation went well, and Johnson issued a supportive statement: "Given the leading role [Soleimani] has played in actions that have led to the deaths of thousands of innocent civilians and Western personnel, we will not lament his death."

Johnson's positive message stood in contrast to the tepid responses of other European leaders, some of whom criticized the strike. These leaders knew that Soleimani was an architect of chaos, repression, and terrorism, but they were too scared to admit publicly that Trump had taken the right course of action.

In Tehran, Ayatollah Khamenei called for three days of mourning and openly wept at Soleimani's funeral—and he vowed "severe revenge" against the United States. Seeing this threat, Trump fired off a warning over Twitter: "Let this serve as a WARNING that if Iran strikes any Americans, or American assets, we have targeted 52 Iranian sites (representing the 52 American hostages taken by Iran many years ago), some at a very high level & important to Iran & the Iranian culture, and those targets, and Iran itself, WILL BE HIT VERY FAST AND VERY HARD. The USA wants no more threats! They attacked us, & we hit back."

Trump's bombast on Twitter belied his cool and calm demeanor behind the scenes. He used the platform to wage a psychological battle against our adversaries—and on numerous occasions, his tweets helped deescalate foreign conflicts.

On the evening of January 7, the president was back at the White House when O'Brien reported that Iran had struck installations at two Iraqi airbases where American troops were stationed. The military was still assessing the damage, but early indications were that no American service members had died—they'd been prepared for the attack and were sheltered in bunkers at the time of impact. Meanwhile, the Iranian media was falsely claiming that they had killed many American soldiers. Several hours later, the Iranians conveyed a message through a Swiss intermediary: if we were finished, they were too.

After a tense week, this was a moment of relief. Through Trump's strong, decisive, and unpredictable action, he had knocked Iran's queen off the chessboard, and they hadn't even taken a pawn.

Months earlier, Gold Star husband and Army Green Beret veteran Joe Kent came to the White House and asked to meet with me. His wife, Senior Chief Petty Officer Shannon Kent, had served in the Navy for sixteen years before a terrorist suicide bomber killed her in Syria. She left behind not only Joe but also their two young children. Joe could have been bitter that his wife was gone—no one would have blamed him—but he instead chose to devote his life to raising his two sons and honoring his wife's legacy by fighting to prevent deaths like hers from happening in the future. He reached out to me after Soleimani's death and shared his belief that when President Obama was trying to negotiate the Iran deal, the military operated under a protocol that if Iran struck, we wouldn't hit back. He felt that Iranians knew this, and they kept shooting at our bases and at American soldiers with no repercussions. "Every time Iran killed one of our service members, they faced no consequence," Joe told me. "We were sitting there with our hands tied behind our back for years just being tortured by the Iranians. And Soleimani was the mastermind."

Joe told me about meeting the president at Dover Air Force Base on the worst day of his life—the day that his wife's body was returned in a flag-draped casket. He recalled that Trump spoke his name, shook his hand, and grasped his shoulder.

"I'm so sorry for your loss," Trump told him. "Shannon was an amazing woman and warrior. . . . We are lucky to have people like her willing to go out there and face evil for us."

Joe gave me a bracelet with his wife's name and ID number on it, which I kept on my desk until the day I left the White House. It served as an ever-present reminder of the brave men and women risking their lives on the front lines every day.

39

Bank Shot

Politics rarely provides perfect moments for anything, but by January of 2020, I thought the time was finally as good as it would ever be to release the president's peace plan. If we waited much longer, the noise of the upcoming presidential campaign could overwhelm our efforts.

I sat down in a chair in front of the Resolute Desk, along with Avi, CIA director Gina Haspel, Secretary Mike Pompeo, and National Security Adviser Robert O'Brien. Ambassador David Friedman joined on a secure conference line. Seated on the couches behind us were Mick Mulvaney and Marc Short, the vice president's chief of staff.

"We think now is the right time to release your peace plan," I said, to kick off the meeting on January 13. For more than an hour, Friedman and I walked the president through each aspect of the plan—the parts that would be controversial, the extensive feedback we'd received from both Arab and Israeli leaders, how we expected each country in the region to react, and how we planned to respond to the potential criticisms.

"Both Bibi and Gantz have agreed to endorse your proposal," I said, referring to Prime Minister Netanyahu and his political rival Benny Gantz, the Israeli minister of defense. "This is a huge win, since they are locked in a contentious political campaign and at odds on nearly every other issue, and their joint endorsement will show a united Israeli position."

"So both the Israelis and the Palestinians have agreed to this?" asked Trump.

"No," I said. "We designed it as a 'heads you win, tails they lose' deal. If the Palestinians agree to it as a starting point for negotiation, that's a huge win. I left enough meat on the bone in the plan for the Palestinians to leave a negotiation as winners. But if they don't—which is the much more likely outcome—the Arab world will see that the Palestinians are unwilling to even come to the table to consider a plan with real compromises, including a path to a Palestinian state, and they will likely be more open to normalizing relations with Israel."

I explained that this was the most detailed plan ever released, and the first time Israel had agreed to negotiate on the basis of a detailed map. It was also the first time Israel had made a meaningful commitment to ensuring that Muslims would have permanent access to the al-Aqsa Mosque. If the Palestinians opposed this plan, it would bring Israel and other Muslim countries closer together, which would only increase pressure on future Palestinian leaders, and create the conditions for Arab countries to normalize with Israel.

"I have a lot of issues going on right now," said Trump. "And this is not my top priority. I don't want to do anything if Abbas says no. Set up a call with him. I'll be able to tell by his tone if there's a chance. Otherwise, let's wait to release the plan at a later date and not waste our time."

This was a surprise. Trump's desire to solicit Palestinian president Mahmoud Abbas's approval before we released the plan slammed the brakes on our strategy and flipped it in reverse. Back in my office, Avi collapsed into a chair, exasperated. Friedman called me in alarm.

"It's over now," the ambassador said. "Our plan is never going to see the light of day, and our whole effort was for nothing."

"Let's keep going and see what happens," I said.

Avi looked at me like I was crazy.

Even if the president's call to Abbas somehow failed to derail the plan, we still had many steps to complete before its release. I put together a matrix of every prominent foreign leader I thought might support our proposal, or at a minimum take a neutral posture. Pompeo, Haspel, O'Brien, Avi, and I met at the State Department to divide up the countries according to who had the best relationships, and we started making calls.

At the top of my list was British prime minister Boris Johnson. We had discussed the Israel-Palestinian conflict many times, dating back to the transition in 2016. He had even sent me a letter expressing confidence in my efforts and encouraging me to strike a deal that required courageous compromises from both the Israelis and Palestinians.

Johnson asked if his team could review the text of the plan. As our new special envoy for international negotiations, Avi traveled to London to meet with UK foreign secretary Dominic Raab and Richard Moore, a devoted civil servant who had risen through the ranks and would soon become the chief of the Secret Intelligence Service, known as MI6. Raab and Moore had proven to be trustworthy allies throughout my time in government, and they were glad that our plan included a path to an independent Palestinian state.

Meanwhile, I traveled with Trump to Davos, Switzerland, for his second address to the World Economic Forum. As we flew over the powder-white Alps, the president reviewed a draft of his much-anticipated keynote address.

"What is this trillion trees bullshit?" he asked.

He had come across a line in the speech that pledged America's support for an initiative to plant one trillion trees globally by 2050. I had been working on this initiative privately for several months with Salesforce CEO Marc Benioff, and thought it was a science-based approach to improving the environment without increasing burdensome regulations.

"Are you trying to push more liberal shit on me?" he asked.

"No, it's a smart idea," I responded. "It costs zero dollars right now and conservatives like Kevin McCarthy love it. You always say you agree with the environmentalists in wanting clean air and clean water. The quality of both has actually improved under your presidency, but you never take or get any credit for it."

"Fine. I'll leave it in," the president huffed.

That evening, Trump attended a dinner Ivanka and Larry Kudlow had organized with top international CEOs. Trump clearly felt at ease among his former colleagues. He was a business guy first, politician second, and he would always be one of them. He also never missed an opportunity to recruit jobs and manufacturing to the United States, and he made a strong pitch to these leaders, touting America's improved business climate and the unparalleled talent of America's workers.

"Can we take a group picture?" one of them asked before the event concluded.

As the executives crowded around him, the president spoke up.

"I just want to say: all my life, I have followed you guys. You are the biggest, and I have respected all of you. I've seen you on the covers of magazines. I've read about you. I've done business with some of you. When I built a great building, someone else would build a bigger building. When I made a lot of money, one of you would make more money. I thought to myself, 'I can't compete with these guys. What can I do that these guys can't do?' So, I decided I should become president."

The group erupted in laughter, and Trump grinned from ear to ear. The president didn't take himself too seriously, and I always admired that about him. It made him far more relatable than he often appeared when he was sparring with pundits on television.

From Davos, I had originally planned to make stops in Saudi Arabia and Israel. I was hoping to work with the Saudis to finalize a statement urging the Israelis and the Palestinians to negotiate on the basis of our peace plan, and I needed to resolve a few outstanding issues with Bibi and Gantz. Just before we boarded Marine One for the thirty-minute flight from Davos to the Zurich airport, where Air Force One was holding, the president's military aide announced a bad weather call. We were going to have to drive for three hours to the airport. As we wove through the Alps on icy, narrow roads, I began to reconsider my trip to the Middle East. There had been an unexpected but fortunate development: when a U.S. government official reached out to the Palestinian Security Forces to request a call between Abbas and Trump, amazingly, Abbas declined the call and conveyed that he would only speak to Trump after we released the peace plan. If Abbas had simply agreed to the call, he likely would have derailed our proposal.

We were now just six days out from the date we'd targeted for releasing the plan: Tuesday, January 28. Friedman assured me that there was no need to travel to Israel. The two Israeli leaders were still on board. During a trip that week to Jerusalem, Vice President Pence had met with both Bibi and Gantz and delivered invitations for a White House ceremony. After speaking to Friedman, I called Avi and Brian Hook. They both thought that the Saudi statement was in good shape and that a phone call could bring it the rest of the way. My biggest unknown variable was the president, so I decided to scratch my trip and fly back with him. If he agreed to release the plan, I wanted to be near the Oval Office in the days that followed in case someone tried to change his mind and disrupt the launch.

As Air Force One climbed to cruising altitude en route to Joint Base Andrews, I went up to the president's cabin. He was reading documents and watching the coverage of the opening arguments from his impeachment trial in the Senate. The screen flashed between scenes of Pat Cipollone and Jay Sekulow presenting their case on the Senate floor.

"We did as you asked, and Abbas said he would potentially agree to a call after the plan is released," I said, placing a glossy printout of the proposal on his desk. "Both Bibi and Gantz are ready to come to Washington to support this plan, and many countries have agreed to put out positive statements. I think the time is now."

After an extended discussion, the president finally looked at me and consented.

"I trust you," he said. "I'm not going to nitpick you on the details. Israel can be a combustible file. You've taken responsibility and haven't gotten me in any trouble. At least not yet. If you think this is the right thing to do, let's do it."

Trump was giving me latitude, but was also making clear that he would hold me accountable if anything went poorly. This was all I needed to hear.

Empowered by his approval, I jumped into action. With the assistance of the Air Force One switchboard operator, I called the vice president, the secretary of state, the secretary of defense, and the CIA director to let them know that the president was ready to move forward. We would need to choreograph the plan's rollout. We had to finalize the peace plan document, coordinate dozens of statements of support by foreign leaders, orchestrate the visits of Bibi and Gantz, and alert the appropriate US officials to make security preparations at embassies in the Middle East in case of violence.

I still hadn't decided on how we were going to let the press know that we were releasing our long-anticipated peace plan, but the president was one step ahead. The next day, as he flew to Florida, Trump walked to the back of the plane and spoke to the traveling press pool, as he often did. Unbeknownst to me, he announced that he intended to release the plan within a week. He had taken care of the media strategy himself. I sensed that Trump was floating a trial balloon to gauge people's reactions. When the news broke that both Bibi and Gantz were coming to Washington for the announcement, the press recognized the significance of getting Israel's political rivals to support our plan. Just as we had hoped, the coverage was more positive than usual.

Friedman called a few hours later with a problem: Gantz was apparently saying that he was no longer coming. He had heard that Bibi would speak, and he didn't want to sit in the audience while his political opponent took the podium at the White House. After multiple phone calls, Gantz agreed to come to Washington to announce his support of our plan, as long as he could have a full meeting with the president prior to the rollout.

Gantz would not agree to attend the event, however. This wasn't ideal, but it was better than him not showing up at all—or rejecting the plan altogether. I respected that every time Gantz had to make a decision between what was better for the State of Israel or for himself politically, he always chose his country. When Bibi heard about Gantz's meeting, he insisted on having one as well. Fortunately, Trump agreed to meet with both leaders. As we edged closer and closer to the announcement, we had averted yet another crisis.

40

Chaos and Peace

Two days before the release of the peace plan, while I was on my way to the White House to prepare for a meeting with Israeli prime minister Bibi Netanyahu, Brad Parscale called.

"What do you make of the latest *New York Times* story?" asked Trump's campaign manager. It was late Sunday afternoon. The article was slated for the next day's newspaper, but it was already online.

Political correspondent Maggie Haberman had obtained several excerpts from John Bolton's forthcoming book, in which Bolton claimed that Trump had directly tied Ukraine's foreign aid to the investigation of Hunter Biden. The man whom the president had fired just a few months earlier was contradicting Trump's defense that he had never linked the two. It bolstered the Democrats' accusation of a quid pro quo. To further complicate matters, Bolton had made it clear that he was willing to testify before Congress if subpoenaed.

It was clear to me that Bolton was trying to whip up media speculation to promote his book. Based on this report, the Democrats were already requesting a new round of hearings to investigate the matter. But when it came down to the legal case for impeachment, there was nothing new. Once again, this was not a legal issue but a messaging battle.

When I arrived at the White House, I headed straight to the residence. I entered the Yellow Oval Room to find Trump sitting with his impeachment lawyers, Pat Cipollone and Jay Sekulow. Halfway through the twenty-day Senate impeachment trial, both had dark circles under their eyes and were looking uncharacteristically beaten down. They had withstood round after round of questions from Democrats who were determined to destroy their case and remove the president from office. Their performance had been stellar, reflecting countless hours of careful preparation and the inherent strength of their legal case. Defending a president in an impeachment is a once-in-a-lifetime opportunity for a lawyer—one that defines their career. But the stakes were even higher for the president, and Cipollone and Sekulow both knew it. When I walked into the room, they were in a heated debate with the president about how to respond to Bolton's claim.

"We have a big problem," Sekulow said. "We had the Senate in a perfect place. They were not going to call witnesses. They were going to vote this week. The trial was about to end. This is going to change everything."

"Why is that?" I asked.

"I spoke to Senator Lindsey Graham," Sekulow responded, "and his sense from speaking to others is that they will want to hear what Bolton has to say. Graham knows that Bolton has an ax to grind, but if Bolton is willing to say something under oath, enough senators will feel like it's their duty to hear him out."

"I don't agree," I said. "Unless there is a bombshell that we don't know about, I don't think his testimony changes anything. This is an easy one. There is literally nothing new here, and that has to be our position. If we act panicked, this will be a big deal. If we stand firm and confident, we can make it through this one. We need to get out a statement that pushes back on the Bolton narrative and makes clear that the facts haven't changed."

Trump either liked the fact that I was presenting with confidence, or he understood my strategy. "Jared's right," he said. "Pat and Jay, go work with Jared on a statement and bring it up to me. I want to get it out fast."

As we walked back to the West Wing, I got an earful from Sekulow on how dangerous it was for me to give legal advice on such a sensitive matter. "You don't know what Bolton has written in the book," he warned. "If the president puts out a statement that is incorrect, we are dead."

Sekulow wasn't wrong, but I sensed that he was wound up pretty tight and that we weren't going to reach a consensus. I suggested that he and Cipollone draft a statement, while I worked on a separate draft. They agreed and disappeared into their office. I walked down the hall past the offices of the chief of staff and the vice president to the office of the national security adviser in the far corner of the West Wing.

I found O'Brien at his desk, waiting for me so we could walk across the street to the Blair House, where we were scheduled to meet with Bibi that evening. "I have a small problem to deal with first," I said. "Do you have that Bolton manuscript?" Bolton had been required to submit an initial manuscript to the National Security Council for review to ensure that it did not disclose any classified material.

"I have it locked in our safe," he said. "No one has seen it other than me and the career official reviewing it for classified information."

I passed him a printed copy of the *New York Times* story. "I need to know what Bolton says in his manuscript about the Ukraine aid and whether there is anything explosive or new."

"I'm not going to show you the manuscript," O'Brien said as he reviewed the draft, "but in this section, he does something really interesting. Throughout the book, Bolton constantly quotes the president verbatim, but in this instance, he doesn't. Instead, he implies that it was *his understanding* that the president wanted him to withhold the aid until Ukraine opened an investigation."

"So is it safe for me to operate under the assumption that he does not directly quote the president in a way that contradicts our defense thus far?" I asked.

"Correct," O'Brien replied.

I rushed back to my desk and drafted a statement, then called the White House operator and asked him to connect me to the president. I read the draft statement to Trump, and after he dictated some edits, I printed an updated version and walked back into O'Brien's office.

"If I said something like the following—'I never told John Bolton that the aid to Ukraine was tied to investigations into Democrats'—would that be contradicted?"

"You're on safe ground there," said O'Brien. "Nothing in the book contradicts that."

I took the statement to the residence. Trump carefully edited it. I ran back to my office and printed a revised version. As the president continued to refine the statement, I made several more trips back and forth from the residence to the West Wing. I eventually looked down at my watch: it was past 10:00 p.m.

"I really need to go," I told the president. "Bibi has been waiting for over an hour."

"Bibi can wait," said Trump. "This comes first."

I told my team to let the Israeli prime minister know that I would be delayed further, and I took the statement to Cipollone and Sekulow for their review.

"POTUS has signed off on this statement, and I think this will get us to where we need to be," I told the lawyers as I handed them the draft.

"Don't talk to my client!" shouted Sekulow. "You're going to mess up our attorney-client privilege!"

"Jay, calm down," I said. "This is not a big deal. Everything's going to be fine . . ."

"I AM CALM!" Sekulow yelled.

Cipollone and I burst into laughter, and Sekulow cracked a smile too. I walked over and gave Sekulow a big hug. He was clearly feeling the pressure of a tremendous burden. We agreed on a compromise shortly thereafter.

At 12:18 a.m., Trump released his statement in a series of three tweets: "I NEVER told John Bolton that the aid to Ukraine was tied to investigations into Democrats, including the Bidens. In fact, he never complained about this at the time of his very public termination. If John Bolton said this, it was only to sell a book. With that being said, the . . ." "transcripts of my calls with President Zelensky are all the proof that is needed, in addition to the fact that President Zelensky & the Foreign Minister of Ukraine said there was no pressure and no problems. Additionally, I met with President Zelensky at the United Nations . . ." "(Democrats said I never met) and released the military aid to Ukraine without any conditions or investigations—and far ahead of schedule. I also allowed Ukraine to purchase Javelin anti-tank missiles. My Administration has done far more than the previous Administration."

Meanwhile, I raced back to the Roosevelt Room, where Avi, O'Brien, Friedman, and Brian Hook had been waiting since 8:00 p.m. Just after midnight, we walked over to the Blair House. Bibi was gracious and didn't complain about my delay. He understood firsthand the pressure of investigations. He did, however, make another big request: he wanted the media to be present for his bilateral meeting with Trump. This was not part of the agenda we had already negotiated. With the Israeli elections just a month away, we had orchestrated the visit to avoid showing partiality to either Bibi or Gantz. Both leaders would get a photo with the president—no media, no remarks, no major production. It was to be a simple meeting. But nothing was ever simple with Israel. My team called Gantz, who conceded to Bibi's request and expressed that he just wanted what was best for Israel.

Bibi and I ran through the final version of the peace plan. As we finished, Bibi remarked that he could live with it.

"You won't live with it. You'll thrive with it," I shot back with a smile.

This was typical of the veteran prime minister. We had spent two years haggling over every line, and we had created a thoughtful plan that Bibi believed could actually work. In twelve hours, the right-wing prime minister, who had campaigned for decades against giving the Palestinians a state, was going to endorse a plan calling for a two-state solution. Bibi was careful to make sure that not a single word of the plan would put any Israeli at risk and was understandably nervous about how it could affect the upcoming election. To his credit, he recognized that the plan was reasonable, and the best compromise to solve the Israeli-Palestinian conflict.

We worked until nearly 2:00 a.m. When I checked my phone, I had several missed calls from Senator Graham. We had spoken earlier, and he had warned that the Bolton news could spell disaster for the president's impeachment defense. When I called Graham back, he said that the reaction to the president's statement had been surprisingly positive, and that he thought we had a chance at keeping the Senate Republicans united against calling witnesses.

On Monday, Bibi and Gantz had their separate meetings with the president. Trump was impressed with Gantz, who expressed a desire to try and reach a deal with the Palestinians.

Later on, Trump told me what he thought of Gantz: "I like this guy."

41

A Vision for Peace

At noon on Tuesday, January 28, 2020, we prepared to reveal to the world our proposal for peace between Israel and the Palestinian people.

The morning was a blur of briefings, calls, and last-minute tasks before the rollout event. By 8:00 a.m., the White House had given a handful of reporters a background briefing on the plan, hoping that the added context would result in fair and accurate coverage. I spoke with members of Congress. I wanted them to see that we were proposing a balanced two-state solution. Even though Trump's usual sparring partners were likely to politicize our effort, I hoped the Democrats would consider its merits before issuing their denunciations.

It seemed like each time I checked my phone, I received another positive update. British foreign secretary Dominic Raab said that his government was preparing a supportive statement. Kirill Dmitriev touched base to say that Russia was in a good place on the announcement.

Then Avi called with unbelievable news. That morning, UAE ambassador Yousef Al Otaiba had asked if it was too late to RSVP for the ceremony. This was a major development. Emirati and Israeli officials virtually never appeared together in public. We had invited a wide swath of Arab dignitaries, but we didn't expect them to attend an event with Israel's prime minister. After Yousef confirmed that he would come, Avi immediately called our closest allies within the Arab diplomatic corps, urging them to join Yousef. We didn't know it at the time, but Yousef was also calling his fellow ambassadors with a similar encouragement. To our surprise, the ambassadors from Bahrain and Oman agreed to appear. This meant that three Arab countries with no current diplomatic ties to Israel were prepared to show public support for our plan as the new framework for peace negotiations. This would send a strong signal that the Middle East was ripe for normalization.

As good as it was, it could have been even better. After the event, Egypt's ambassador called me and asked why he had not been invited to the ceremony. He said he gladly would have joined. I was mortified by the thought that we had overlooked him by mistake. When I checked with my team, however, I learned that we indeed had invited the Egyptian ambassador. It turned out that his staff had assumed that he would not want to attend. Despite the snafu, the rollout event was shaping up to be better than we had dared to hope.

Around 11:00 a.m., I went up to Trump's quarters in the residence with the draft speech. Trump was reviewing documents, looking sharp and ready to go.

"Good job last night on the Bolton statement," he said. "This morning, we will completely focus on Israel. Is this going to lead to peace?"

"This is a critical step," I said. "You are going to enjoy it. We've prepared a very special speech."

Just outside his bedroom, we sat down across from Claude Monet's *Morning on the Seine, Good Weather*, the painting former First Lady Jackie Kennedy had donated to the White House in memory of her husband. I handed Trump a manila folder, with the speech inside just the way he liked it: sixteen-point font and unstapled. He took it out and held it up high.

"Five pages? Why is this so long? I told the speechwriters never more than two pages for East Room events."

Except for major events such as the State of the Union address, Trump thought that short and punchy speeches were more effective than longer ones, which were often too wonky and less interesting for listeners. His insistence on brevity forced the writers to refine ideas down to their essence, and his speeches were clear and direct as a result.

"You can't do Middle East peace in two pages," I replied.

He reviewed the draft, making fewer changes than normal. After finishing each page, he handed it to me with his edits, and I coordinated with White House staff secretary Derek Lyons to make sure the changes were loaded into the teleprompter.

"This really is a good speech," he said. "You are right. We have done a lot. I don't even know how we have done so much, since we have had to spend most of our time fighting off phony witch hunt investigations and impeachments."

He paused, looked up from the draft, and quipped: "The being president part of the job is easy; fighting off the crazies is the hard part. Just imagine what we could get done if I could spend all of my energy on issues like this."

At noon in the East Room, as three hundred guests stood up from their chairs and the press hovered in the back, the military aide announced: "Ladies and gentlemen, the president of the United States and the prime minister of the State of Israel."

I took my seat next to Ivanka, and the president began his remarks.

"Today Israel takes a big step towards peace," said Trump.

The room erupted in applause. I exhaled. It was finally happening, and it was off to a good start. I couldn't have been prouder watching the culmination of three years of dedicated effort and careful planning.

As I watched history unfold just a few feet away from me—televised for the world to see—Bibi made a surprising statement. "Mr. President," he said, "I believe that down the decades, and perhaps down the centuries, we will also remember January twenty-eighth, 2020, because on this day, you became the first world leader to recognize Israel's sovereignty over areas in Judea and Samaria that are vital to our security and central to our heritage."

This was not what we had negotiated. Under our plan, we would eventually recognize Israel's sovereignty over agreed-upon areas if Israel took steps to advance Palestinian statehood within the territory we outlined. The two hinged on each other, and it would take time to flesh out the details. The prime minister then repeated the point, going one step further.

"Israel will apply its laws to the Jordan Valley, to all the Jewish communities in Judea and Samaria, and to other areas that your plan designates as part of Israel and which the United States has agreed to recognize as part of Israel."

He had implied that our plan would allow Israel to immediately annex the Jordan Valley and portions of the West Bank. While Bibi had to navigate a difficult political environment at home, this was a step too far.

I grabbed my chair so intensely that my knuckles turned white, as if my grip could make Bibi stop. I had explicitly asked Israeli ambassador Ron Dermer to make sure Bibi kept his remarks brief and above the politics of the day. In both tone and substance, the speech was way off the mark. It contained nothing magnanimous or conciliatory toward the Palestinians. It was essentially a campaign speech for his domestic political audience, and it misrepresented our plan.

As the prime minister approached the twenty-minute mark, I could tell that Trump was becoming uncomfortable. He was pursing his lips, swaying side to side, and periodically glancing down at Bibi's prepared speech to see how many pages were left.

I looked over at the three Arab ambassadors, and thought about our friends and partners in the region, whose trust I'd spent three years building. I had walked them through the peace proposal and given them my word that Trump would present a dignified and balanced proposal—one that required compromises on both sides. But that certainly wasn't the deal Bibi was describing.

Had the rollout gone according to plan, it would have put Abbas in an impossible position. Reacting harshly against a credible proposal would further alienate him while exposing the hollowness of his position. But the Israeli prime minister had given Abbas exactly the kind of opening he needed to reject our plan and potentially to persuade the rest of the major players in the region to side with him. I had expected to spend the afternoon on offense, selling the plan through the media. Now I was worried about damage control.

42

The "Misunderstanding"

A s I walked with the president along the colonnade back to the Oval Office right after the announcement concluded, Trump turned to me with noticeable disappointment on his face and said, "Bibi gave a campaign speech. I feel dirty." Neither of us wanted the plan to become political, which was why we had waited more than a year to release it and had invited Bibi's political rival to the rollout. This transcended politics. This was about making peace.

As it turned out, Ambassador David Friedman had assured Bibi that he would get the White House to support annexation more immediately. He had not conveyed this to me or anyone on my team. Shortly after the president's announcement, he told reporters that Israel "does not have to wait at all" on the annexations and that the only limiting factor was "the time it takes for them to obtain internal approvals."

When I confronted Friedman, he told me that he had accurately represented the plan. Our conversation got heated, and I pulled out the plan from the folder on my desk.

"Where does it say that in here?" I asked. "It doesn't say that in here. You're one of the best lawyers in the world. You know that's not what we agreed to."

Realizing he was losing the argument, Friedman tried to turn on the charm. "What's the big deal?" he said. "Why don't we just stay ambiguous and let Bibi say what he wants and let it play out?"

"You haven't spoken to a single person from a country outside of Israel," I shot back. "You don't have to deal with the Brits, you don't have to deal with the Moroccans, and you don't have to deal with the Saudis or the Emiratis, who are all trusting my word and putting out statements. I have to deal with the fallout of this. You don't."

Friedman now saw that Bibi's words posed a big problem and indicated that he was ready to back down. I asked him to speak with Bibi to clarify our position and to let the prime minister know that while I was going to try to minimize the glaring gap between our two positions, we weren't going to back him on this one.

"Tell him," I said, "that if we're lucky, this hasn't completely killed my credibility with other countries, and I will still be able to get the statements of support I have teed up."

To his credit, Friedman cleaned up the misunderstanding with the Israelis and the media.

My own afternoon was filled with one television interview after the next. As I worked to shape the coverage and defend our plan, Avi was busy behind the scenes calling our closest partners and clarifying our position. He found that while some leaders were confused, they were still ready to move forward as long as we were not pushing immediate annexation. Avi assured them that we were not.

The trust we had built was holding up, and our partners were preparing to release their positive statements as originally planned. No one wanted to go first, however. I called Dominic Raab, who said that he had paused his statement until he was able to confirm that the United States was not going to support immediate annexation. I gave him my word, and the United Kingdom published the statement as planned: "This is clearly a serious proposal, reflecting extensive time and effort," it said. The rest of our partners followed suit. Within twenty-four hours, over a dozen countries released statements of support. Saudi Arabia noted the king's appreciation for our efforts and encouraged the "start of direct peace negotiations . . . under the auspices of the United States." The United Arab Emirates called the plan "a serious initiative that addresses many issues raised over the years." Bahrain commended the US for "its determined efforts to advance the peace process." The Moroccans praised the plan and expressed their "wishes that a constructive peace process be launched." Egypt thanked the United States for its persistent work to "achieve a comprehensive and just settlement of the Palestinian issue, thereby contributing to the stability and security of the Middle East."

Diplomacy is commonly an exercise of words. People guard against new terms and sentiments. No one gets fired for sticking to the old talking points. This was why these statements of support were improbable and unprecedented. For nearly two decades, every Arab nation had held up the 2002 Arab Peace Initiative as the appropriate framework for negotiations. Now the most influential Arab nations were praising our plan as a starting point for the next round of talks. Importantly, the European Union and the United Nations refrained from denouncing our plan, and instead called for both sides to begin negotiations. Given how negative both bodies had been toward Israel in the past, we viewed their neutrality as a major step in the right direction.

In interviews, I made clear that the Trump administration did not support immediate annexation, and I tried to minimize the gap between our position and Bibi's pledge. Behind the scenes, our relationship with the Israeli government had reached its lowest point to date. I felt like I was trying to move the Israelis forward and build partnerships with the broader world while they were stuck on internal politics.

Israeli ambassador Ron Dermer, who was usually a constructive force, came to see me several days after the rollout. I was expecting him to apologize on behalf of Bibi or to propose some kind of compromise. Instead, he said that Bibi needed to move forward with annexation immediately.

I couldn't believe it. Trump was still fuming over Bibi's speech. In fact, he had asked me whether he should take the unusual step of endorsing the prime minister's political rival, Benny Gantz. Had I walked twenty feet down the hall to the Oval and asked Trump to go forward with annexation, the president would have thrown me out.

Although the immediate response to our announcement had been positive, the African Union and the Arab League used Bibi's statements as grounds to condemn our plan. We had enough allies in the European Union to block the EU's top foreign policy official, Josep Borrell, from making an official statement, but he was so upset that he broke from protocol and issued a scathing personal statement rejecting our plan and condemning annexation. Russia also began walking back its initial support.

We had done so much to strengthen America's alliance with Israel—moving our embassy to Jerusalem and recognizing Israel's sovereignty over the Golan Heights, withdrawing from the Iran deal, and waiting a year to put out the peace plan to accommodate the turbulent Israeli elections. Now they wanted even more. Dermer said that if we didn't support the immediate annexation, Israel would no longer be able to trust the administration.

"Don't take us for granted," I warned. "We worked our asses off for three years to get to this point. For the first time, Israel has the moral high ground. You're offering the Palestinians a state and a map that Arab countries actually support as a starting point for negotiations. But now it's all screwed up. You guys think you have been so effective with this administration. I hate to break the reality to you, but we didn't do any of these things because you convinced us to. We did them because we believe they were the right things to do."

Dermer saw that he had gone too far. He apologized and left soon after, knowing that it was up to them to clean up the political mess that Bibi had created.

Between Friedman's conversation with Bibi and my altercation with Dermer, the Israelis got the message. After three years of policies that had strengthened the US-Israel relationship, Trump's popularity was so high in Israel that Bibi couldn't afford to go against him. The prime minister walked back his statement about the Jordan Valley, and the Israelis canceled their plans to begin moving forward with immediate annexation. In private, Bibi continued talking tough with us, threatening to recognize the Jordan Valley within weeks, but I knew that he was bluffing. It would be political suicide to move forward without the backing of their closest ally and supporter.

As I tried to think about how to keep advancing our goal despite this setback, I took comfort in a lesson from Lawrence Wright's account of the 1978 Camp David conference in *Thirteen Days in September*, one of my favorite books on Middle East peace.[50] It describes how a profound misunderstanding led to progress. Egyptian president Anwar Sadat's closest adviser, Hassan el-Tohamy, an astrologer and Sufi mystic, told Sadat that he had learned through back channels that if Sadat traveled to Jerusalem and gave a speech before the Knesset, Israeli prime minister Menachem Begin would transfer control of the Sinai Peninsula back to Egypt. As the story goes, Sadat made the brave and historic journey to Jerusalem. After his speech, he met privately with Begin and asked how Israel wanted to proceed with the Sinai exchange. Begin said he had no idea what Sadat was talking about, and Sadat left Jerusalem empty-handed and disappointed.

Despite this misunderstanding, Sadat's visit shattered a barrier and changed the world's outlook on the Middle East conflict. It showed that peace with the Arabs did not have to run through the Palestinians, and that separate, bilateral peace deals were possible. This set off a chain of events that led directly to a breakthrough at Camp David: the first peace agreement between an Arab country and Israel in modern times.

Although it was hard to appreciate in the moment, something similar would happen with our efforts. This proved to be the greatest paradox of peace: Bibi's annexation threat, and the tension and urgency it created, ultimately led to the breakthrough that became the Abraham Accords.

43

Battle at the United Nations

M adam Speaker, Mr. Vice President, members of Congress, the First Lady of the United States, and my fellow citizens," said the president as he rehearsed for his State of the Union address.

"Do I really have to say Madam Speaker?" Trump asked. "That crazy woman just impeached me over nothing. Maybe I can just leave her out and see if anyone notices."

Trump's sarcastic hypotheticals were famous among his friends and family, and we always got a kick out of how they landed with those who didn't know him well enough to realize that he was joking. When he did it in public, his supporters appreciated his sense of humor. His critics, on the other hand, didn't try to understand it. Writer Salena Zito best summed up the dynamic during the 2016 campaign: "The press takes him literally, but not seriously; his supporters take him seriously, but not literally."

It was Tuesday, February 4, 2020, and Trump was standing in the corner of the Map Room, a small, wood-paneled parlor on the ground floor of the White House residence. Franklin D. Roosevelt had once used the room as a top-secret communications hub, a sort of precursor to the modern-day Situation Room, where he could track the latest military developments on large maps during World War II. Ceremonial versions of these maps still hang in frames around the room, giving it a sense of history. I faced Trump behind a table, along with Vice President Pence, Stephen Miller, Dan Scavino, Derek Lyons, and speechwriters Vince Haley and Ross Worthington. As he went through the draft, Trump stopped every few lines to insert an idea, tweak a phrase, or add his signature flair.

Some of my favorite moments in government came during the State of the Union address. There was always a temptation to load the speech with wonky policy proposals geared toward Washington special interests and political allies. But Trump's speechwriters labored to keep the speech focused on a few core policy goals, while also using the world's biggest stage to demonstrate how Trump's pro-American policies were changing lives and restoring hope in our nation. Speechwriter Brittany Baldwin, who kept a running list of ideas generated throughout the year, drafted the stories of the "gallery guests"—the cast of heroes whose lives of courage, grace, and patriotism created some of the most unforgettable moments. Ivanka and I sat with these remarkable individuals in the gallery of the House of Representatives each year, and it moved us to see their faces light up with pride as the president honored them. I will never forget standing next to former inmates Alice Johnson and Matthew Charles in 2019 as America celebrated their redemption stories, or joining with the entire chamber to sing "Happy Birthday" to Holocaust survivor and Tree of Life congregant Judah Samet, or watching D-Day hero Herman Zeitchik share an embrace with Joshua Kaufman, a Holocaust survivor whom he had liberated from Dachau, a Nazi prison camp.

These speeches had personal significance to Trump, who would make changes down to the minute he departed for the Capitol. I typically blocked off the entire day on my schedule so I could help him prepare—and 2020 was no exception. As we entered the final hours before the address, I received a note that Avi needed to speak to me. He knew I was in speech prep, so I had a feeling it was urgent.

"I just got off the phone with Dermer," Avi said when I called him. "The Tunisians are circulating a UN Security Council resolution condemning our peace plan as a violation of international law."

This was indeed urgent. That year, Tunisia was the Arab League's rotating representative on the UN Security Council. In the week since we'd released the plan, the Palestinians had waged an all-out public-relations assault against it—and they were gaining momentum at the United Nations.

"Call the Tunisian ambassador to the White House immediately and ask him why, after all America does for Tunisia, they are prioritizing their relationship with the Palestinians over America," I said to Avi. I had learned from our previous experience that the delegations at the UN complex in New York weren't always in sync with their leaders at home.

Twice before I'd fought and lost battles at the UN. As we geared up for our third test, I knew the survival of our plan was at stake. If the UN denounced our effort, it would validate the Palestinian intransigence and effectively preclude our plan from being a credible basis for peace talks. I decided to make an emergency trip to New York to address the Security Council directly.

Avi reached out to the Tunisian ambassador at the embassy in Washington, DC. When they met at the White House, Avi expressed our consternation about the resolution. The ambassador turned pale and apologized profusely. There had been a miscommunication, he claimed. Avi requested a call between me and the prime minister of Tunisia, and the ambassador quickly agreed.

When the prime minister called on February 6, I had just arrived in New York and was en route to a UN Security Council meeting. I expected the standard diplomatic runaround about his difficult political situation or international position. But to my great surprise, the prime minister explained that his UN representative had gone rogue. Tunisia was scrapping the resolution, and the prime minister had relieved the diplomat.

When the news broke that Tunisia had recalled its representative, it sent shock waves through the sleepy corridors of the UN. This development caught everyone off guard and showcased the progress we had made over the past three years.

At noon that day, I stepped off an elevator onto the penthouse floor of the US Mission to the UN, a recently renovated event space with thirteen-foot ceilings and an imposing panoramic view overlooking the East River and the United Nations complex.

As the fifteen Security Council representatives took their seats around an oversize square table, they were greeted by two large documents that

I had placed at each setting: a copy of the peace plan and a PowerPoint presentation. I projected the presentation on several large screens and launched into the merits of our plan as if I were talking to a corporate board of directors. My first chart illustrated the irrationality of sticking to the failed approaches of the past. Since 1993, there have been nine rounds of peace talks between Israel and the Palestinians. Each time the negotiations broke down, Israel's settlement activity increased, and the Palestinian Authority received more money from the international community.

DEFINITION OF INSANITY:
DOING THE SAME THING AND EXPECTING A DIFFERENT RESULT

- *Assistance to Palestinians Since 1993 is over $50 billion cumulative*
- *Neither Party Incentivized to Compromise*

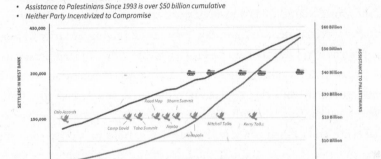

"For twenty years, much of the peace process has been a charade," I said. The UN had adopted nearly seven hundred resolutions with regard to the Israeli-Palestinian conflict. Yet nothing had changed. Why would either side ever have reason to compromise so long as we continued along this path? Meanwhile, the quality of life for those in the West Bank and Gaza had deteriorated. The jihadists were manipulating the conflict to radicalize young Muslim boys and promulgate a false narrative that they needed to take violent measures to reclaim the al-Aqsa Mosque.

"I keep getting urged to play by the old rules, but the old rules don't work," I argued.

Next, I walked them through the practical elements of the new offer on the table: a realistic two-state solution that maintained Israeli security and improved the lives of the Palestinian people. I reminded them that this plan was the first time that Israel had conceded a path to Palestinian statehood. Following my presentation, the representatives asked me questions for more than an hour. The tenor of our discussion was collegial and productive, and I departed the meeting hopeful that I had broken through.

For decades, Abbas had dominated at the UN. He was accustomed to winning every fight he started. But now it was unclear whether his Security Council resolution would get the unanimous support it needed to isolate the United States, and anything less would signify that his previously impregnable position at the UN was beginning to crumble.

After a brief period of uncertainty, the Palestinians recruited Indonesia, the world's largest Muslim-majority nation, to carry the resolution. A draft was circulated on Saturday, February 8, giving Security Council representatives twenty-four hours to provide edits—a step known in UN parlance as "breaking the silence." Everyone assumed that we were simply going to veto the resolution. When Avi and I read the draft, however, we had another idea. Rather than vetoing it, what if we "broke the silence" and offered constructive edits? We could turn the resolution into a positive statement about the importance of pursuing new ideas and improving the lives of the Palestinian people. This sleight of hand would transform the resolution from a condemnation of our plan into an endorsement of our general approach. It would force Security Council representatives to decide whether they were really against the constructive path forward we were proposing. With the help of our new UN ambassador, Kelly Craft, who had succeeded Nikki Haley, we refined the new proposal over the weekend and called our contacts at each of the UN Security Council member countries. By Monday, Germany, France, the United Kingdom, and the Dominican Republic voiced support for our language. China, Vietnam, and Niger indicated that they would abstain. Left to choose between failure and retreat, the Palestinians chose retreat. Indonesia announced that it was delaying the resolution indefinitely.

We had won. We beat the Palestinians at their own game. Abbas arrived in New York on Tuesday, February 11, planning to formally address the Security Council after what he assumed would be another loss for the United States. Instead of taking a victory lap, he delivered a meandering speech before the Security Council, where he vehemently rejected our proposed state as "Swiss cheese." He browbeat the representatives with familiar and hollow words, threatening that if they didn't act quickly, "the situation could implode at any moment."

Everyone who tried to talk to the Palestinian leader on our behalf came away frustrated and hopeless. Even the Arab leaders were losing faith in Abbas. Their own people were growing tired of a Palestinian cause that was tethered to the past. One leader privately shared a common Arabic saying to sum up his feelings toward the Palestinian president: "It's better to have a smart enemy than a dumb ally."

* * *

The same day that I had made my presentation before the Security Council, the president was in the East Room of the White House, celebrating his acquittal in the Senate. I was sad to miss the special moment. Trump had weathered an historic onslaught of attacks and come out stronger than before. His approval rating jumped ten points. Heading into an election year, the Republican party was united and brimming with energy.

Soon after Trump returned from his February trip to India, the president asked for his acting chief of staff. He wanted Mulvaney's opinion on a pressing policy question, but Mulvaney was nowhere to be found. Trump soon learned that he had left on a personal trip. This was not the first issue that had arisen with Mulvaney. Deeply displeased, Trump called me down to the Oval Office, where he was sitting with Dan Scavino.

"It's time to make a change," he said.

"Mick's actually had a discussion with me about this," I said. "He told me that if you ever wanted him to resign, he'd be willing to do so on good terms. He's not going to be a problem, and he's already identified a job that he wants."

Surprised but also relieved, Trump said, "Okay, well, it's time to call him. Tell him that I want to be on good terms."

Then he asked us who we thought would be the right replacement. Without hesitation, both Scavino and I said, "Meadows."

Trump nodded in agreement and asked me to find out whether Mark Meadows, the North Carolina congressman and Freedom Caucus chairman, would accept an offer.

Trump had previously considered Meadows for the job. The two men had forged a strong mutual respect through the fire of the Russia investigation and impeachment. Yet the congressman was worried about the legal fees and controversy that seemed to follow everyone who had served in Trump's orbit.

When I called Meadows to discuss the role, he was reluctant. "Mick's my friend," he said.

I filled him in on where the president was headed: a change was inevitable.

"This year is high stakes," said Meadows. "You know I love the president. If he's going to make a change anyway, I'd be honored to serve."

My next call was to Mulvaney, who accepted Trump's decision. That evening, Friday, March 6, Trump tweeted: "I am pleased to announce that Congressman Mark Meadows will become White House Chief of Staff. I have long known and worked with Mark, and the relationship is a very good one. . . . I want to thank Acting Chief Mick Mulvaney for having served the Administration so well. He will become the United States Special Envoy for Northern Ireland. Thank you!"

As Meadows prepared to enter the West Wing, I decided not to hold back on offering advice. Having so strongly supported his hiring, I felt like I had a responsibility.

"I've now worked with three chiefs of staff, four national security advisers, and more than thirty cabinet members," I said. "I have seen people take over government organizations. Some do it well, and others fail miserably. My advice to you is to come in and empower many of the great staffers we already have. Each chief of staff before you brought in their own people from their previous job, and this inevitably led to a culture clash. Reince brought in the RNC, Kelly brought in his DHS team, and Mick brought in his OMB staff. Each time it frustrated the existing team members, who'd been working hard for Trump and were uncomfortable reporting to a new cast of characters who didn't understand the president and hadn't paid their dues.

"The chief of staff has an impossible job—you need to manage the staff in the building while also being fully available and attentive to the president. You'll only be able to do this job well if you have a strong team around you."

I advocated for Chris Liddell as deputy chief of staff for policy to keep the trains running on time, Derek Lyons as a problem-preventing super lawyer, and Hope Hicks as a strategic communicator. I also made the case for Dan Scavino as another deputy chief of staff. "Dan is the most important staffer in the White House," I told Meadows. "He's often with the president ten hours a day. When you go home, and the president is in his office working late into the night, Scavino will be sitting right next to him as he reads his materials, expresses his private opinions, and talks about all kinds of things. If something is going wrong, you want Dan to feel ownership over it and then work with you to correct it."

Meadows thanked me for the advice, and for the most part, he took it.

As spring approached, the president's prospects for the 2020 election had never looked more promising. Impeachment had finally come to an end. The American economy was humming. Since the 2016 election, the Trump economy had created 7 million new jobs, lifted 10 million people off welfare, brought 3.5 million working-age people off the sidelines and into the labor force, and raised wages for low-income workers. The unemployment rates for Hispanic Americans, African Americans, and Asian Americans had reached their lowest levels in history. Congress had just passed legislation approving the new USMCA trade deal, the final step in making it permanent. We had a trade deal with China. The president's approval rating had soared to a personal best of 49 percent, a number we always thought underrepresented Trump's supporters. An astounding 94 percent of Republicans favored him.

In Meadows, we had a new chief of staff who had Trump's confidence and felt like the right fit for the job. By nearly every indicator, Trump was positioned to sail toward reelection.

Then everything changed.

44

Code Red

Y ou should come to this meeting in the Oval," deputy chief of staff Chris Liddell whispered into my ear. It was Wednesday morning, March 11, 2020, and I was midway through a meeting in the Roosevelt Room with a bipartisan group of lawmakers and business executives to discuss accelerating the Trillion Trees initiative, which Trump had given me the green light to coordinate after his Davos endorsement.

"The president's considering closing down travel from Europe," Liddell said. "This is a pretty major decision, and you should be there."

Up until that moment, I had not been involved in the White House response to COVID-19. Before Trump banned travel to and from China on January 31, Health and Human Services (HHS) secretary Alex Azar and deputy national security adviser Matt Pottinger had been running a coronavirus task force—a team of federal officials to monitor the spread of the virus and oversee the administration's response. As the virus spread, the nation's top doctors and health-care experts began visiting the White House on a daily basis, following a predictable flight pattern that started down in the Situation Room, before moving up to the chief of staff's office, over to the Oval Office, and then back down to the Situation Room. It was impossible not to notice the buzz of activity—or the mounting worries that it represented. And it certainly caught my attention when Trump put the vice president in charge of the task force on February 26, in response to growing concerns about testing shortages nationwide.

The *New York Times* later cited an unsubstantiated source to claim that I downplayed the virus internally. This was false. I was told that the virus was a serious threat and that the government's medical and public health experts had the response to the public health emergency under control.

On Monday, March 9, two days before Liddell pulled me into the Oval Office, the Dow plummeted two thousand points, the largest-ever drop during intraday trading. As the stock tickers descended deeper into red, the television screen carried real-time footage of the *Grand Princess* cruise ship docking in the San Francisco Bay and more than three thousand people entering quarantine. One elderly passenger had already died from the virus while the ship was at sea. At 1:45 p.m., the vice president's chief of staff, Marc Short, came to my office and beseeched me to help them.

"We're having a big problem with the task force," he said. "We're not getting support from White House comms or the Domestic Policy Council. The vice president's office is a pretty slim operation, and for this to work, we need more support from the rest of the White House. But they're refusing to work with us. Could you help bridge the gap?" I told Short that I'd try to resolve the issue.

Now, as I dismissed myself from the Roosevelt Room, I sensed that COVID-19 was about to become a crisis far beyond the scope of a typical public health emergency. The meeting in the Oval Office was already in motion, so I snuck in, slipped into one of the yellow chairs at the back of the room, and listened as the discussion unfolded. It was a large group: Vice President Pence, Steven Mnuchin, Alex Azar, Dr. Anthony Fauci, Dr. Robert Redfield, Dr. Deborah Birx, Robert O'Brien, Matt Pottinger, Larry Kudlow, Ivanka, Hope Hicks, and Stephanie Grisham. They were debating whether to block travel from Europe, and Mnuchin and Kudlow were explaining the devastating impact that the decision could have on the economy and global markets.

The president listened intently, weighing the magnitude of the decision and considering all the variables. He seemed to be siding with his national security and public health teams, who wanted to impose a ban immediately.

"We don't know what we don't know," Fauci argued. "Taking this step could end up being a really big deal."

The situation in Italy, as portrayed on television, appeared borderline apocalyptic. Hospitals were running out of ICU beds. Patients lined hallways and field hospitals as overwhelmed doctors triaged the sick and were forced to make life-or-death decisions about who would receive care. A travel ban could help prevent this from happening in America.

As I listened to the debate, I was struck by its abstractions. The two sides were discussing the idea of a ban in principle, but it wasn't clear if they had developed a concrete proposal or implementation plan. These details should have been fleshed out through the White House policy process, which was finally operating at a high level of professionalism. This was the best way to present Trump with clear options, informed by stakeholders. But due to the urgency of COVID-19, the topic had bypassed the policy process.

I sensed that the president might appreciate a bit of time to think about such a consequential decision. Stopping travel from our closest partners and allies would be unprecedented. After sitting quietly through the meeting, I suggested that the staff recess for two hours and come back with a tactical plan for the president to consider.

The team assembled in the Cabinet Room, and I started to ask questions. What would a travel ban mean for trade and commerce? What would be required of returning American citizens? Would we attach an expiration date to it, or leave it open-ended? How quickly would we begin enforcing it?

We worked through these questions and went back to the president several hours later. Trump was ready to make his decision.

"Let's do it," he said. "This is a big step to take, and I'm going to get a lot of blowback from our allies, but we have to do it. If this is a mistake, the Europeans will do more complaining, corporations will lose some money, and travel plans will be delayed. If we don't do it, and this threat is as real as it looks, people are going to die."

I was proud of the president's decisiveness. It was a strong move that would help keep America safe and show the country that he was willing to go to great lengths to deal with the virus.

We agreed that a presidential address from the Oval Office was the best way to explain the decision and calm the public. The president would show Americans that he was steady, in charge, treating the matter with concern, and taking definitive action.

Over the next few hours, we scrambled to put together a speech. Our speechwriting team was top-notch, but because the policy was not fully fleshed out, we struggled to get the input we needed from the key experts to make sure we struck the right message and proper tone. I huddled in Stephen Miller's office on the second floor of the West Wing with Pence and staff secretary Derek Lyons, trying to write the perfect speech. The topic was new to us, and we were hopelessly pressed for time.

At 9:00 p.m., the president began his second-ever address to the nation live from the Oval Office. "My fellow Americans: Tonight, I want to speak with you about our nation's unprecedented response to the coronavirus outbreak that started in China and is now spreading throughout the world."

In a ten-minute address, he announced the Europe travel ban, framing it as the latest installment in a series of bold actions the administration had taken to keep Americans safe. He had closed travel from China, declared a national public health emergency, and activated a mandatory quarantine for the first time in more than fifty years. In the middle of his speech, as he began to describe the practical aspects of the Europe ban, the president misread his speech, adding a word that was not in the script. The travel restrictions "will not *only* apply to the tremendous amount of trade and cargo," he said. In reality, the ban did not apply to trade or cargo, but the inclusion of the word *only* reversed the meaning of the line. I made a mental note of the mix-up, but the president recovered and finished strong. The speech wasn't a masterpiece, but it provided critical information to the nation on the severity of the threat, the reasons for the European travel ban, and Trump's plan moving forward.

Immediately after the speech, the White House released a statement clarifying that trade and cargo were excepted from the travel ban. It was critical to get the message out quickly: the US and EU exchange $700 billion in goods on an annual basis, and stopping this flow would disrupt our economies.[51]

The rest of that evening, and throughout the following day, the media covered the speech like they would cover a scandal, with an initial round of criticism followed by a series of process stories ascribing blame for the missteps. They found a familiar target. "The speech was largely written by Kushner and senior policy adviser Stephen Miller," reported the *Washington Post*.

On the night of the speech, Pence stopped by my office. "Thank you for what you did today," he said. "Can you get involved and help me with the task force? This is a big challenge, and if we are going to be successful for the president and for the country, I need the muscle of the full White House and the entire federal government."

It wasn't an assignment I had invited, and I knew it would draw criticism, but through an intense three years in government I had learned how to navigate the federal bureaucracy and deliver results. And after seeing the task force in action that afternoon, I was concerned about the state of the federal response. Many of the task force members had frozen like deer in headlights. I felt a responsibility to the president and the country to help where needed. I told the vice president that I would clear my schedule for the next thirty days and work at his direction: "I'm all in."

* * *

On my way to the White House early the next morning, March 12, my brother Josh called from New York City. He described the worrisome signs: the city had canceled its annual Saint Patrick's Day parade, thousands of people were self-quarantining, and millions more were leaving the city. When I told him that I was asked to jump into the response, he made a suggestion: "You should call Adam."

Adam Boehler was the CEO of the International Development Finance Corporation, a powerful new $60 billion foreign investment agency within the federal government. I'd known him since the summer of 2001, when we had roomed together in a quad unit in New York University's student housing. Boehler went on to start four successful health-care companies, including Landmark Health, the nation's largest in-home health-care provider. And we had remained good friends over the years.

In the spring of 2018, Seema Verma, administrator of the Centers for Medicare and Medicaid Services (CMS), asked me to help recruit Boehler into her agency. I invited him and his wife, Shira, to the Navy Mess for lunch. At the time I was at one of my lowest, most politically toxic moments. Recent polling showed that I had the lowest approval rating in the Trump family: 10 percent. "That means that even though I rarely speak publicly, thirty-three million Americans appreciate the job I'm doing," I joked.

Boehler and his wife laughed, and I leveled with them about what public service entailed. "It's hard to get attacked—by people who don't know you—for giving up your business and comfortable life to do what you think is right," I said. "Don't come to Washington for appreciation, but if you want to have an impact on the country and millions of people's lives, there's no better place to work."

After a prolific eighteen-month stint at CMS, Boehler was tapped to lead the Development Finance Corporation, and the Senate confirmed him unanimously in September of 2019. Boehler was the perfect person to help us with the federal government's COVID response, especially because he had the skills to overcome the fierce rivalries among the administration's health-care team. I called him, and after failing to get him on the phone, I messaged him at 7:47 that morning: "Come to White House."

A few minutes later I walked across the street to the Eisenhower Executive Office Building for an 8:00 a.m. meeting in Matt Pottinger's office with Chris Liddell and Dr. Deborah Birx, who was the vice president's hand-selected coronavirus task force coordinator. A well-regarded physician, Birx was a retired Army colonel and had successfully led the federal government's global HIV/AIDS response for years. About a dozen people were scattered around a conference table and sitting on the couches and chairs that dotted the office. Seven minutes into the meeting, Boehler arrived.

Birx led us through a sprawling discussion that touched on nearly every aspect of the COVID response. She grew animated when discussing her frustrations with the bureaucracy and her inability to get people to move with the urgency she had been feeling over the previous weeks. Two points stood out: we had fallen far behind on testing, and supplies of critical materials like face masks, gloves, and gowns would soon become scarce.

After the meeting, Boehler and I huddled in my office and began sketching out how we could help with testing and supplies. To get additional support, we called our mutual friend and successful health-care entrepreneur Nat Turner. We also reached out to three of the very best public servants at HHS: Brad Smith, head of the CMS innovation office; Brett Giroir, head of the Public Health Service Commissioned Corps; and Secretary Azar's deputy chief of staff, Paul Mango.

That afternoon, in the Situation Room, a man whom I had just seen on television approached me. "Thank you for what you did yesterday," said Dr. Anthony Fauci, the top infectious disease official at the National Institutes of Health. "It's really not fair how the press is beating you up. You made a very positive contribution. If you'd like me to say that to the press, I would be happy to."

"Thank you for saying that, Doctor," I said. "I have come to accept that when I step into a problem situation, I tend to become an irresistible target."

With Pence sitting at the head of the table, Birx updated the group on the latest COVID case data from across the country. Her charts showed grim developments. New York cases were beginning to swell, and she expressed concern that the virus had proliferated in New Orleans during Mardi Gras several weeks earlier. CDC director Robert Redfield promised that the CDC and FDA were looking at all options to ramp up testing. Admiral Brett Giroir, a gifted public health official with impressive medical credentials, reported that we had completed only 30,000 tests to date, well short of where we needed to be.[52] By contrast, the South Koreans had already completed more than 230,000.[53] The discussion exposed deep acrimony between HHS leadership and CDC about who bore responsibility for the debacle. The agencies were playing the blame game. They were clearly more focused on explaining why it wasn't their fault than on mapping out a concrete plan to fix the situation moving forward. Nowhere is the expression "failure is an orphan" truer than in politics.

As we wrapped up the two-hour meeting, the vice president ran through his talking points for that evening's task force press conference. He went around the table to get final thoughts and asked for my perspective about the message they were planning to convey from the podium.

"I'm the new guy at this task force, so take what I say only as my blunt assessment with the limited perspective I have," I said. "Right now, we have a *facts* problem, not a *messaging* problem. The public won't be satisfied until we can describe a concrete plan for fixing this testing mess we are dealing with. Once we have a plan in place to fix this testing nightmare, then we will be able to communicate better."

Back in my office, I challenged my team to think strategically about how we could accelerate the distribution of tests and improve public access. We examined South Korea's system of drive-through testing for potential best practices and brainstormed about how we could implement a similar system in America.

"What is in every community in America?" Boehler asked.

I paused and thought for a moment before guessing: "Walmart?"

Partnering with large pharmacies like Walmart, CVS, Rite Aid, and Walgreens would play to the great strength of America's private and not-for-profit sectors. Unlike the federal government, which was not equipped to roll out testing at scale, these large pharmacies were efficient operators that already provided millions of flu shots each year. With their collaboration, we could distribute COVID-19 testing as quickly and widely as possible.

We jumped into action, calling the CEOs of the companies to pitch the idea. Without hesitation, they agreed to devote resources to explore the possibility further.

That evening Boehler, Smith, Giroir, Turner, and I joined a conference call with the pharmacy CEOs and their teams. The heads of the large testing companies Labcorp, Quest Diagnostics, Roche, and Thermo Fischer joined as well. Our initial concept involved patients entering the stores for testing. As we sketched out a plan, a CEO piped up.

"I don't think it would be too good for our businesses to bring sick people into our stores. It would be much better if we could use our parking lots."

It was a fair point, so we adjusted, and the CEOs agreed to work through the night so that we could finalize details when we met the next day at the White House. As we hung up the phone, my team looked at each other with a surge of hope. We had stepped into the middle of a mess, but our crazy long-shot idea might just work. If we could harness the power and imagination of America's private sector, we might have a chance at turning testing around.

We hunkered down and fleshed out the drive-through testing plan. There were so many variables: we needed to know how many tests the US had, where exactly they were located, how many we could acquire immediately, which communities needed them the most, and the best way to distribute them. We heard that Verily, a sister company of Google, was setting up a pilot program in California to help connect people to testing services. Turner called its CEO, Andy Conrad, to see about setting up something similar on a national scale.

By the time we broke for the night, it was 4:00 a.m.

Early in the morning on Friday, March 13, after we prodded the FDA for an expedited review, the agency announced that it had granted approval for a coronavirus test developed by Roche—a significant breakthrough for the testing effort and a necessary step for our plan to work. The new test could be processed ten times faster than the existing tests. We had promised Roche that the FDA would approve the test in record time, as long as its data was accurate. And the company had taken us at our word, prepositioning systems to process the tests throughout the United States.

At around 11:00 a.m., the pharmacy and testing company CEOs convened around the conference-room table in the Roosevelt Room to continue our planning from the previous evening. We outlined the rough parameters of a public-private partnership between their companies and the federal government to deploy four hundred testing locations in communities across America. Everyone was energized and excited to help. I've rarely seen such a powerful mix of altruism and collaboration from the private sector. There was only one word for it: patriotism.

Afterward, I stopped by the Oval Office to update the president on our work. "Can we make an announcement today?" he asked.

Trump always had a keen sense of public sentiment, and he felt that people were anxious to see the government taking decisive action at a time when testing delays continued to dominate the headlines, and the markets were headed for another rotten day.

"We could, but we hadn't planned on that," I said. "Not everything is fully fleshed out, but it's promising."

Trump decided to make an announcement that afternoon. This presented a contrast in our management techniques. I preferred to be methodical and never wanted to make an announcement until I had painstakingly mapped out the potential scenarios, next steps, and contingencies for when things went off course. This took time, and it was difficult to resist the public pressure to share information. Trump, on the other hand, was much more willing to make a bold announcement and trust his team to live up to it. I was hesitant, but deferred to his instinct as our nation's leader to make the call.

Back in the office, where our team had assembled, I updated them on Trump's request. Their jaws dropped. We'd been working on this plan for less than twenty-four hours. Everyone rushed into action to finalize the key outstanding details and to make sure our stakeholders knew what was coming. Thankfully, they all stayed on board.

In the Rose Garden at 3:30 p.m., flanked by officials and CEOs, Trump spoke: "Today, we're announcing a new partnership with the private sector to vastly increase and accelerate our capacity to test the coronavirus."

As the press conference continued and the president described our plan, aided by Birx, Fauci, and the CEOs, the markets began to rally.

"Google is helping to develop a website," the president went on to say. "It's going to be very quickly done, unlike websites of the past, to determine whether a test is warranted and to facilitate testing at a nearby convenient location. . . . Google has seventeen hundred engineers working on this right now."

By the market's close, twenty minutes after the start of the press conference, the Dow had rallied fourteen hundred points—a 6 percent jump and the first positive economic news in days. We weren't trying to juice the market, of course, but we saw this result as immediate positive feedback.

Bad news came that evening around 5:30 p.m. Google and its sister company Verily released a statement that scaled back their commitment to the drive-through testing effort, announcing that the website at its outset would serve only the San Francisco Bay Area.[54] That was not what Andy Conrad had promised. In fact, before the president delivered his speech, Boehler had specifically read him the lines describing Verily's involvement.

I was in the Oval Office when Sundar Pichai called me. I motioned for Boehler to follow me into the president's study, and we put the CEO of Alphabet, the parent company of Google and Verily, on speakerphone.

"What happened?" I asked.

"Andy gets ahead of himself sometimes," said Pichai. He blamed the misunderstanding on an internal miscommunication.

"Sundar, this website will help a lot of people, regardless of the misunderstanding. Can you get it up and running?" I asked.

"Let me look into what's possible," he said. "We're here to help, and I don't want to let the country down."

Eager to make the president look bad, the media had a field day with the mix-up. "Trump Oversold a Google Site to Fight Coronavirus," gloated the *New York Times*.

The experience provided an important lesson in the early days of the crisis. It was a reminder that in this extremely difficult situation, even the minor mistakes we made would be broadcast in real time.

On Saturday I convened a conference call with the CEOs of the companies supporting our drive-through testing plan and encountered new and unexpected headwinds, perhaps caused by the misunderstanding with Verily. Several of the executives on the call expressed reservations about the legal liability posed to their companies by a drive-through testing system. They were growing hesitant, and I worried that they would back out. The tide was turning against us.

Then Walmart's CEO, Doug McMillon, interjected: "Guys, if we don't do this, who's going to do it? Our country needs us right now. Walmart is willing to take the risk."

By the end, the CEOs had redoubled their commitment to help. It made me proud to be an American as we rushed headlong into the fight of our lives.

45

Battle Rhythm

W hatever happened between Wednesday night and Friday afternoon at the White House, let's please have more of it," wrote the *Wall Street Journal* editorial board in a March 13 op-ed that boosted morale among my team.

But suddenly, we didn't have enough cotton swabs in our country.

I learned about the problem on Sunday afternoon, as we congregated in an office on the seventh floor of the Health and Human Services headquarters. Brad Smith described the problem, which he had just discovered: "We only have one point two million cotton swabs in the entire Strategic National Stockpile."

I knew the federal government kept a strategic stockpile of basic medical supplies. It hadn't occurred to me that cotton swabs were among them, but of course they were—and each COVID test required at least one cotton swab. We were short on lots of other supplies as well, from gloves and gowns to masks and ventilators. The H1N1 flu pandemic in 2009 had seriously run down the stockpile, and for some inexplicable reason, nobody had bothered to build it back up.

This was a major kick in the stomach. Smith had obtained this information only after wrestling it from Dr. Bob Kadlec's team in the Office of the Assistant Secretary for Preparedness and Response (ASPR). Prior to the pandemic, all of these supplies were low-cost and readily available on the market. ASPR could easily have purchased tremendous amounts to fill the stockpile. Now that we were in the middle of a pandemic, however, the supplies were nearly impossible to find and procure. Private citizens, businesses, and hospitals were buying up everything. Was the lowest-cost item really going to be our bottleneck? How could the world's most powerful nation not have enough testing supplies for a single city, let alone the entire country? How were we so unprepared, on basically every front? As much as I wanted to understand ASPR's failure, these questions would have to wait. We were in triage mode, and we needed every spare second to stop the bleeding.

We had to find millions more swabs in short order.

As we dealt with the shortage of cotton swabs and other supplies, we faced another problem: the need to develop public health guidelines. Given that people across the country were confused and concerned, Birx and Fauci had been discussing the need for a unified set of federal standards to help Americans understand what they should do to keep themselves safe and slow the spread of the virus. They insisted that these guidelines would help prevent hospitals from becoming overwhelmed. Despite all the talk over the past week, no one had taken steps to produce a document. When Nat Turner flagged the issue, I asked him to coordinate with Derek Lyons to produce a draft and encouraged him to call Dr. Scott Gottlieb, the former head of the FDA and a renowned public health expert. I had been trying to persuade Gottlieb to come back into government for a short-term stint to help us better organize our response and support our effort to develop a vaccine.

When we called Gottlieb, he was grateful that we were preparing guidelines. "They should go a little bit further than you are comfortable with," he said. "When you feel like you are doing more than you should, that is a sign that you are doing them right."

That evening, I received an unexpected call from Governor Andrew Cuomo of New York. I had known Cuomo for years. He had reached out after my father's arrest back in 2004, which my family never forgot. "I've had highs and lows as well. You'll be back," he told my dad at the time.

On the phone with me, his typically confident voice was shaky with alarm. "Jared, this is getting really bad, and I fear we are soon going to run out of ICU beds," he said. "We only have three thousand ICU beds in the city, and at this rate, we could need another hundred and fifty thousand in the coming weeks. I'm pleading to you and the president as fellow New Yorkers. I need your help. I need the help of the federal government to get through this."

He said that he was looking at retrofitting college dormitories and buildings to create space for additional beds. "From my time as secretary of housing and urban development," he said, "I know that if you want to build fast and money is no object, the Army Corps of Engineers is the best. Can you send them up here immediately to help me start converting facilities?" The governor went on to express his fear about how uncontrollable the spread of the virus could potentially be, especially among New York's elderly population. "For nursing homes, this could be like fire through dry grass," he said.

I promised that I would do everything I could to lend the federal government's support, and that I would be available to him 24/7.

"I want you to be prepared," Cuomo said. "There's going to be a lot of things we can't solve, but let's just acknowledge on the front end that we'll do everything we can do. And then let's show people that we're leading, because right now, people need to feel that their leaders are working together and leading. People are so freaked out at this point, you almost can't make a decision that's too extreme. Indecision is the only bad decision you can make. You have to be decisive."

After a thirty-minute discussion, I asked him who on his team I should contact to coordinate our response. "Just deal with me directly," he said. For the next sixty-odd days, we worked on a daily basis to ensure that we quickly addressed any concern raised by New York.

The panic in Cuomo's voice and his dire predictions, compounded by the exhaustion I felt from four straight days of trying to improve the testing situation, hit me hard. We were miles behind where we needed to be, and I felt powerless to improve our outlook. The worst-case scenarios flashed through my mind: nurses and doctors without protective equipment, overflowing hospitals with no beds for patients, ventilator shortages forcing doctors to choose who would live and who would die, limited ability to detect new outbreaks due to the testing supply shortage, and tens of millions of Americans stuck in their homes, growing more and more anxious by the hour.

I looked at my watch. It was past 9:00 p.m. I walked downstairs, through a mostly empty West Wing, to see how the team was doing, and found Turner, Smith, and Giroir huddled around Derek Lyons's computer in the staff secretary's office. They were working on the draft guidelines, which they had titled "15 Days to Slow the Spread." They were running on fumes, too, but they were determined to get a draft ready to present to the president the following morning.

"You guys have done great work over the past ninety-six hours, and I have no doubt that what we're doing will help save lives," I said.

I didn't want the team to sense the fear I felt, but my voice dropped down a bit, betraying my lack of confidence in our ability to avoid a disaster. I struggled to speak. "There is a chance that the challenge we are about to face is bigger than we thought. Maybe there are problems that are just too big to solve. I hate to say this to you guys, but right now, it feels to me like we are on a beach working frantically trying to build a protective hut made of sand and leaves, while a massive tsunami is coming."

As the team looked at me, not sure what to say, I regained my composure. "We're all exhausted. After you finish this revision, go home and get some rest so that we can be ready for the fight we have ahead of us."

On the short ride home, I sat silently in the back of the Secret Service SUV, replaying my conversation with Cuomo in my head. As I walked through the front door, Ivanka sensed my mood and asked how it was going. I recounted my last twelve hours.

"Right now is the calm before the storm. Nothing else matters anymore," I told her. "This is bigger than politics, bigger than every other problem we've had to solve, combined. Bigger than immigration, trade deals, and prison reform."

Ivanka looked at me with concern. "This is the first time I have ever seen you wear your worry," she said, wrapping her arms around me. I realized that not since my father's arrest had I faced a challenge so out of my control that I let fear and helplessness overtake me.

As we talked, the vice president called to compare notes from the day.

"We made good progress, but I fear it's too little too late. This is going to be really tough," I said. "I'm not confident we are going to be able to meet the demand for supplies. This could be a horror show."

I will never forget the vice president's calming response: "Jared, I was a governor. At the federal level, we will absolutely do our best," he said, steady as always. "But we won't have to solve this alone. Governors have resources, teams, and their own ingenuity. And in times of crisis, the American people step up and figure it out."

I don't know if he felt as confident as he sounded, or if he was showing the strength of leadership I had failed to show to my team earlier that night, but it was exactly what I needed to hear to jolt me out of my discouragement. More importantly, it was what I needed to believe as I prepared for the battle ahead. Pence lifted me up when I needed it the most, and his words looped in my head as I fell asleep.

*   *   *

The next morning, Monday, March 16, I woke up at 5:00 a.m. as usual, but that day I had a renewed sense of resolve. As I rode to the office, I thought about my message to the team the previous evening. I had made a mistake by showing them the cracks in my confidence. If we were actually going to pull off miracles, I needed to show them that I believed we would pull off miracles.

"This is going to be the hardest thing that we've ever done," I said to them that morning. "For whatever reason, God put us here. The only judgment we should care about, when this is all over, is being able to look ourselves in the mirror and say that we did everything possible to make the greatest difference. And if we come up short, we come up short, but we're going to give it everything we've got."

Turner and the team had finished the first draft of the guidelines. We walked the document to Pence's office for a final review with Birx, Fauci, and Redfield. They offered a few constructive changes, but overall they were very positive on the document we had produced.

"You think the president will support these?" Fauci asked.

"I don't know," I said. "But we're going to try."

Alone in my office, I called the president to preview the guidelines and give him a chance to react honestly, without fear of the doctors leaking about his response to the press.

"Vice President Pence and the doctors are going to come to you with strong public health guidelines," I said. "They may seem draconian, but we think they could save tens of thousands of lives. We are critically low on supplies, and it will take us several weeks to track down more. Asking people to take these precautions will slow down the spread of the virus, reduce the number of new cases, and buy us much-needed time."

He understood. That afternoon, we took the guidelines to Trump. Birx, Fauci, and Redfield made their case.

"That's it?" Trump said. "I thought you were going to ask me to call in the military to make people stay in their homes. We can't do this forever, but people will tolerate this for a few weeks."

At a press conference a few hours later, Trump announced "15 Days to Slow the Spread," which urged all Americans to work and attend school from home, to avoid gatherings of more than ten people, to postpone travel, to avoid eating or drinking in restaurants and bars, and to refrain from visiting nursing homes and retirement centers.[55] Health experts later estimated that the guidelines helped save millions of lives.

As Americans hunkered down for fifteen days, we ramped up our efforts to wartime-level operations. Dr. Bob Kadlec was running point on the operational aspects of the coronavirus response, from the repatriation of passengers aboard cruise ships to the management of the stockpile. Kadlec seemed overwhelmed by the responsibility of it all. According to ASPR's estimates, we would need at least 3.5 billion masks to confront the pandemic. We had one percent of the masks we needed, and our current supply would expire within weeks.

"Don't worry," Kadlec said. "I ordered six hundred million masks."

"Great." I exhaled. "When will they be delivered?"

"The first shipment comes in June."

"Are you fucking serious?" I threw my pen against the wall. It was one of the few instances in which I lost my composure. "We are in March! We could run out of masks in a week. We could all be dead by June!"

Kadlec was a nice man and had a reputation for being a hard worker, but he clearly needed help. On March 18, we transitioned the COVID response from Kadlec's office at HHS to FEMA's National Response Co-ordination Center (NRCC), an interagency operation designed to function on a 24/7 basis during national emergencies. With all of its research and intellectual capacity, HHS runs more akin to an academic institution, while FEMA has a completely different orientation. It is built to move fast, make decisions, and handle hurricanes, wildfires, floods, blizzards, and other natural disasters. With offices around America, it serves as a key federal interface with the governors.

For weeks, Birx and the NSC's top disaster response staffer, Brian Cavanaugh, had pushed to activate the NRCC, but Azar fought the idea. The secretary had lost much of his control when Pence took over the task force, and moving the center of operations from HHS to FEMA would further loosen his grip on the response. I knew Azar wouldn't like it, but I went to the vice president and told him that we needed to make the change immediately. Activating the NRCC would give the governors a system they understood for receiving, adjudicating, approving, and shipping requests.

I knew immediately that it was the best decision we could have made. FEMA administrator Pete Gaynor was a former Marine Corps lieutenant colonel with the mentality of a wartime planner. "We need to establish a battle rhythm," he told me when we first met at FEMA headquarters. "Right now we are spending too much time in meetings discussing high-level topics. When principals are in these meetings, they can't be running their departments. We need to identify objectives, create a chain of command, and then start making this happen." It was music to my ears.

The following day, at our request, the Joint Chiefs of Staff dispatched Rear Admiral John Polowczyk, one of the military's top logistics experts, to FEMA to run point on procurement and distribution of supplies. Working together, Gaynor and Polowczyk brought structure and credibility to the management of the stockpile, and directed the process for managing incoming requests and shipping materials to states within twenty-four hours.

Around that time I was sitting in Gaynor's office at the NRCC when New York senator and Democratic minority leader Chuck Schumer called to plead for supplies for his state. I was in constant contact with New York's state and local decision-makers. They called me regularly— usually to express gratitude for the targeted flow of supplies we were sending. Schumer apparently wasn't in touch with them, and complained that we weren't sending supplies. I reached for a folder that contained the latest data on the supplies we had sent to New York and rattled off the extensive list of supplies that were en route. Then I told Schumer that we would even work with him on an announcement so that he could take credit for the delivery.

Gaynor had listened to this conversation, and when I hung up, he gave me a wary look. "We need to be meticulous on all of this," he said in his thick Rhode Island accent. "Once we get through this crisis, every single contract, every single delivery, is going to be investigated. I'm going to be called before Congress, and I'm going to have to answer questions."

"You're telling me that there's an unprecedented natural disaster, for which we were theoretically prepared but not actually prepared, and while everyone is running away and trying to avoid blame, you run into the disaster, use every bit of ingenuity and whatever else you can think of to save lives, and then your reward for doing all of this is that you get hauled before Congress and harassed with subpoenas to answer questions about the small percentage of things that went wrong?"

"Precisely," Gaynor said with a wry smile.

"And you volunteered for this job?"

"We're a sick bunch in the emergency response community," said Gaynor. "We're gluttons for punishment, but at least the pay sucks."

46

Project Airbridge

This place is a black box," said Adam Boehler. "It isn't designed for a global pandemic."

Boehler was calling from FEMA headquarters, where he and a dozen private equity volunteers he had recruited set up a makeshift office in the basement and were urgently dialing around the world for supplies. To our surprise, we were locating more equipment than we expected, but FEMA's procurement system wasn't allowing us to make rapid purchases. Even though the agency often responded to fast-striking natural disasters, it typically drew from stockpiles it had built during long periods between national crises. Now, in a time of global crisis, we needed FEMA to buy millions of items at breakneck speed. Brad Smith forwarded me an itemized list of supplies they had found but couldn't get approval to purchase: in total 160 million masks, 223 million gloves, and 1.3 million gowns, among other critical items.

"Ask the FEMA leadership team to meet me," I told Boehler. "I'll be over there in thirty minutes." I called White House counsel Pat Cipollone and budget director Russ Vought and asked them to join me.

More than two million civilians work directly for the federal government, but after spending five minutes at FEMA with Admiral Gaynor that afternoon, March 18, there was one person whom I desperately wanted to meet: Bobby McCane. As FEMA's chief procurement officer, he was in a unique position to buy the medical equipment and supplies we needed to fight the spread of the virus. But like many bureaucrats, he wouldn't be empowered to act unless he was given prior direction from leadership.

I asked Gaynor's team to get McCane. When he appeared in our room, my team's eyes lit up: here was the guy they'd been trying to find for several days.

"Bobby, right now you are the most important person in the entire federal government," I said. "My team is finding badly needed equipment from all over the world, and we're at risk of losing it if we can't contract fast. People's lives depend on it."

I asked him to solve our problem by creating a form that listed the criteria we needed to provide so that he could approve a purchase order quickly. I motioned toward Pat Cipollone and Russ Vought. "I've brought the top White House lawyer, and the top government funder. If you need extra authority or money, they will solve it for you. We will do whatever due diligence we need to do on the front end, but we need you to be able to sign the order and wire the money within ten minutes," I said.

I sensed that McCane was excited by the prospect. Having worked in the federal government for many years, he had learned to perform his job within the confines of seemingly irrational mandates. We were offering to cut through the red tape. It was a procurement officer's dream.

I wrote my cell phone number on a piece of paper and gave it to McCane. "You have two hours," I said. "Call me if you need anything."

About seventy-five minutes later, when I was back at the White House, my phone rang. It was McCane. He'd gotten it done.

The system we established at FEMA unleashed a global procurement effort not seen since World War II. Boehler, Smith, and the FEMA team leaped into action, calling every major medical supplier around the globe in a race to purchase millions of masks, gowns, gloves, testing swabs, and other critical supplies. As we sourced supplies from all over the world, we discovered that the factories with the most available supplies were in China. Despite their abundance of product, the Chinese government was blocking supplies from leaving the country. I knew that in time Americans would be able to manufacture much of what we needed, but at this moment we had no time to spare.

We needed to ask the Chinese government if they would allow us to purchase supplies, which meant that we needed to address the growing tension between our two governments. As the coronavirus grew from a localized problem in Wuhan into a global pandemic, the president's rhetoric toward China had grown increasingly antagonistic. He was genuinely upset that China had unleashed the virus, especially because it had tried to cover up the source of the problem and failed to alert the world about the nature and scale of the threat. For example, the Chinese restricted flights from Wuhan to Shanghai and Beijing but didn't stop flights to Milan and Los Angeles.

I went to speak with Trump privately. "We're scrambling to find supplies all over the world," I told him. "Right now, we have enough to get through the next week—maybe two—but after that it could get really ugly really fast. The only way to solve the immediate problem is to get the supplies from China. Would you be willing to speak to President Xi to deescalate the situation?"

"Now is not a time to be proud," said Trump. "I hate that we are in this position, but let's set it up."

I reached out to Chinese ambassador Cui Tiankai and proposed that the two leaders talk. Cui was keen on the idea, and we made it happen.

When they spoke, Xi was quick to describe the steps China had taken to mitigate the virus. Then he expressed concern over Trump referring to COVID-19 as the "China Virus."

Trump agreed to refrain from calling it that for the time being if Xi would give the United States priority over others to ship supplies out of China. Xi promised to cooperate. From that point forward, whenever I called Ambassador Cui with a problem, he sorted it out immediately.

As we worked to source supplies, I was impressed by the spirit of devotion and public service in America's private sector. The executives were willing to put the common good ahead of themselves and their companies. When Boehler and Smith first began hunting for Personal Protective Equipment (PPE), for example, they asked US-based manufacturing companies for production data, such as how many supplies they were making in their factories around the world and what portion of the supplies were coming to America. This data was key to knowing what supplies were available. Nearly every company shared the information, and when we dug into their spreadsheets, we found that most were sending about 70 to 80 percent of their supplies to America. Those sending a lower percentage willingly agreed to step up and increase the allotment for the United States.

One corporation, however, was initially resistant: 3M, a Minnesota-based company and the world's leading manufacturer of masks.

Boehler tried to get ahold of Mike Roman, the CEO of 3M. A few hours later, he received a call back—not from Roman, but from a government-affairs representative. "We understand you want to know about our masks," the representative said, "but we already sent a million to the stockpile, and we're reading media reports that they haven't been distributed. You can speak to Mike, but we need you to tell us what happened with those masks."

Boehler told the 3M representative not to believe everything he read in the press and promised to track down the status of the masks, but insisted on speaking to Roman immediately.

When Roman finally called Boehler, the CEO admitted that of the tens of millions of masks his company was making in China and elsewhere, the United States was receiving only about a quarter.

"We have a factory in the United States that serves the United States," Roman said. "It accounts for twenty-five percent of our global production, and those are the masks you are getting—about thirty million a month. We just announced a big investment that will increase that capacity in a few months. Our factories in the United Kingdom, China, Singapore, and South Korea are serving those areas with about ninety million masks a month, and those masks will stay there."

"We need seventy million of those masks," Boehler pressed.

"I'm not sure that's possible because the masks made in China are made for smaller faces and I'm not sure they'll fit Americans."

Boehler told him that we would work out the sizing issues, and reiterated that we needed the masks within twenty-four hours. The next day,

3M's government liaison, Omar Vargas, followed up with Boehler and refused to provide the number of masks in China. After a heated back-and-forth, Vargas admitted the truth: "We have business relationships in China, and we're not going to break those relationships."

Boehler filled me in on the conversation. Just then, Vice President Pence walked in. He knew we were clashing with 3M.

"Great news!" he said. "The CEO of 3M just called me, and ten million more masks are on the way."

In an apparent attempt to circumvent Boehler and me, Roman had called Pence, promised the ten million masks, and encouraged the vice president to announce it at a task force briefing that afternoon. It was a crafty ploy, but it also confirmed that our pressure was working.

"That's a good start," I told Pence. "But don't let him buy us off that cheaply. If we get more, we can solve our short-term crisis with this deal alone."

Pence agreed and left it to us to close the deal.

To compel 3M to send us the masks, we'd have to invoke the Defense Production Act (DPA). For several weeks, Trump had faced tremendous political pressure to use that heavy-handed authority, a vestige of the Korean War. So far, we hadn't needed it because most US companies were eager to help America in this hour of crisis.

In the Oval Office, we explained the problem with 3M to the president.

"Bring me a DPA," he requested. "I'm dying to use it. It's important to make sure that every American company is pulling its weight."

Trump signed the order and announced it with a tweet: "We hit 3M hard today after seeing what they were doing with their Masks. 'P Act' all the way. Big surprise to many in government as to what they were doing—will have a big price to pay!"

Later, I called Roman and told him that we were sending him a contract for all of 3M's masks in China.

"I can't sell them to you," he said. "The Chinese government has taken over my factory and is controlling my distribution."

"That's not your problem anymore," I said. "It's our problem. Under the DPA, we technically control your company. We're going to send you a contract, and federal law requires you to sign it. You can tell the Chinese that you had no choice."

Within thirty minutes, Roman signed the contract and the masks were ours. Now I had to work with the Chinese to get the masks to America.

"I need your help with an important issue," I said to Ambassador Cui. "We have a contract with an American company for forty-six million masks per month for the next six months. We need them right now. I'm told that there's an issue with the Chinese government holding them. I can't imagine that's the case right now. People in America are very angry at China. If word gets out that the Chinese government is not allowing us to ship masks we contracted from an American company, this could get very ugly."

The ambassador said that he would look into the issue. An hour later, he called back and said that the masks we had contracted for were cleared for travel.

Once Roman saw that we had used diplomacy to work out the situation with the Chinese government, and that we weren't looking to take the rest of his global supply, he became much more agreeable. In the end, he and 3M became great partners in our effort.

* * *

Now that we could rapidly source and procure materials from around the world, we needed to figure out how to get them quickly to our shores. Typically, supplies from overseas were transported by boat and took an estimated forty-five days to cross the ocean. Airlifting the supplies, however, would reduce the transit time to twenty-four hours, if we could just find the planes to carry the tons of cargo.

The military was an obvious choice, but its planes were slow and required refueling stops. I thought that FedEx and UPS might be better options. They had fleets of cargo planes built for carrying massive loads. When we called Fred Smith, the chairman and CEO of FedEx, and David

Abney, the chairman and CEO of UPS, they both immediately agreed to have their companies help. We didn't even have to mention the DPA.

"Consider our planes to be your Air Force," Smith said. "We will do whatever it takes."

These two phone calls commenced a public-private partnership that delivered tons of PPE to our nation's health-care workers on the front lines of the pandemic.

Admiral Polowczyk dubbed the initiative Project Airbridge and meticulously ran the operations of the monumental undertaking. Between March 29 and June 30, Project Airbridge completed 249 flights and delivered approximately 1 billion gloves, 130 million masks, 60 million surgical gowns, and other lifesaving supplies to hospitals, nursing homes, and health-care facilities, right as their shelves were becoming bare. It was nothing short of miraculous. When I ran into Abney several months later, he pulled me aside and said, "Jared, in all of my career, I have never seen anything like Project Airbridge. As fast as we could land our planes, you guys filled them up and turned them around."

In the first days of responding to COVID, there were many factors we couldn't control. We couldn't extinguish the pandemic or instantly invent a vaccine. That would take time. But we could search the globe for lifesaving supplies. My goal was to do everything within our power to give health-care professionals the supplies they needed to save every possible life. I will never forget the innumerable ways that Americans from across the country rose up to serve their fellow citizens. Factory workers and truck drivers worked long hours to produce and deliver supplies. Students brought groceries to elderly neighbors. Communities came together to help families in need. And brave doctors, nurses, and health-care workers risked their own lives to save others. The spirit, strength, and sacrifice of the American people carried our nation through one of the greatest trials of the twenty-first century.

47

Life Support

On the night of March 23, I spoke to Governor Andrew Cuomo of New York and promised that the federal government would send his state 4,400 ventilators from a national stockpile of less than 11,000. As a percentage of the stockpile, this was a big shipment, but New Yorkers desperately needed ventilators. So I was surprised the next day when he attacked us.

"FEMA says, 'We are sending four hundred ventilators,'" he complained at his press conference on the morning of March 24. "Really? What am I going to do with four hundred ventilators when I need thirty thousand? You pick the twenty-six thousand people who are going to die because you only sent four hundred ventilators."

He misstated the number we had sent, and I knew from our call the previous evening that he was unsure how many ventilators he really needed. When I asked him to share how he arrived at his thirty thousand estimate, he couldn't answer. He had no data on how many ventilators he already had, how many were in use, or how many he anticipated needing in the next week.

At that time, the medical experts still believed that ventilators were the most critical medical device available for saving lives. Doctors used them on patients whose virus-ravaged lungs could not supply their bodies with enough oxygen. As cases of COVID-19 skyrocketed, every governor in America demanded the largest possible share of the federal stockpile's diminishing supply. They didn't know how many they would need, but they feared that the stockpile would run out, so they requested as many as they thought they could get from us.

Amid the flood of competing requests, we needed to create a process to allocate this scarce resource. Nat Turner recruited Blythe Adamson from Turner's former company, Flatiron Health, to help our team estimate how many ventilators, ICU beds, and other critical medical supplies America would need. A brilliant PhD epidemiologist and economist, Adamson had a colorful background: she was raised by hippies and grew up in a tree house in Washington State. Adamson initially planned to help us for a few weeks until we'd built reliable models, but she decided to extend her stay. She was inspired by the military service members at FEMA, who regularly went on long deployments away from their families. If the service members could be gone for six months at a time to keep Americans safe, Adamson wanted to stay longer to serve our country in this medical crisis.

Five hours after Cuomo's comments, I headed from the White House to FEMA to get the first draft of Adamson's ventilator projections. In a windowless conference room, Adamson briefed Pence, Azar, Gaynor, Boehler, Smith, Turner, and me. She handed out a one-page chart forecasting the expected ventilator shortages. I looked at her sheet in shock.

"So you're saying that Cuomo's estimate is actually right, and we will be thirty thousand ventilators short within a week?" I asked.

"Yes," said Adamson. "That's what the current data projects, assuming the spread of the virus continues to accelerate at this rate. The best-case scenario is that due to mitigation efforts now in place, the rate of transmission will slow, ventilator demand will drop, and we will have more time to source ventilators. But things also could get worse."

Based on the current trajectory, her numbers also showed that we would need 130,000 ventilators in two weeks. I shuddered at the possibility. Until that moment, I thought the worst of the supply crisis was behind us. All the PPE in the world wouldn't matter much if we ran out of ventilators for critically ill patients.

I couldn't bring myself to look at Azar. I was livid that the secretary had not done more to prevent the shortage. Maybe it was unfair of me

to feel this way, but it was his department's job to anticipate and prepare for this kind of problem. There was no chance we could procure or
manufacture anywhere close to 130,000 ventilators in two weeks. We
were staring at the possibility of two football stadiums full of prevent-
able deaths. In Italy, people were dying on gurneys in hospital hallways
because they couldn't get ventilators. We could not let that happen in
America, a country that prides itself on having the most advanced and
innovative health-care system in the world.

"This is way worse than the swab shortage," I said to the vice president
as the meeting at FEMA broke up. "People are going to die if we don't
figure something out."

Sensing my worry, Pence invited me to ride back with him to the
White House. The sidewalks and streets of our nation's capital were eerily
empty and matched the bleakness I felt.

"Jared, all we can do is our best," Pence said as the lights and si-
rens blared outside the vice presidential limousine. "We'll find a way
through it."

Once again, I appreciated the vice president's optimism. I didn't yet
know whether it was possible to prevent a ventilator catastrophe, but I
was absolutely determined to try.

* * *

As much as I respected Adamson, I had just met her, and I wanted a
second opinion about her projections. I knew enough from my career in
business that predictive models are a sort of science fiction. Their projections are only as good as their assumptions, which can vary wildly. In a
novel pandemic defined by variability and uncertainty, it would be nearly
impossible for anyone to make assumptions that would lead to accurate
predictions.

I called Kevin Hassett, an accomplished economist and the former
chairman of the Council of Economic Advisers. I had recruited him back
to the White House the week before to strengthen our data operation.

In the Roosevelt Room, Hassett, CEA economist Tyler Goodspeed,

Birx, and others combed through Adamson's data and assumptions. They grilled her on everything, and she offered quick, concrete, and confident answers. By the end of the session, Hassett and Birx believed that Adamson's methodology was credible and that her projections could be accurate. I steeled myself for our most critical fight yet, hoping that the "15 Days to Slow the Spread" guidance would reduce hospitalizations and buy us enough time to distribute the ventilators we had and to find more.

FEMA was receiving increasingly panicked calls from governors requesting ventilators. In addition to Cuomo's demand, John Bel Edwards of Louisiana sought 5,000, Phil Murphy from New Jersey asked for 2,300, and Gretchen Whitmer of Michigan and Ned Lamont of Connecticut wanted thousands as well. Put together, these requests far exceeded the number still in the national stockpile.

Everyone was terrified. White House chief of staff Mark Meadows got a call from a hospital CEO in his former congressional district who requested 150 ventilators. At that time, there were no reported COVID-19 cases within a thirteen-county radius of the hospital. Meadows asked why the ventilators were needed. "We're just scared," the CEO admitted. It was one of many examples of panic-induced hoarding, which exacerbated the supply shortages.

That night, while I worked with Adamson and Hassett to analyze the data we had collected from the states, Boehler, Smith, and Avi joined Colonel Pat Work, a hypercompetent Army officer whom I had met three years earlier in Iraq. They called every major ventilator manufacturer in the United States, asking each company how many ventilators they had on the shelves now and how many they could produce in the weeks and months ahead. Working with Bobby McCane at FEMA, they sent letters of intent to purchase all of the American-made inventory.

I also called Cuomo and told him that we needed to know how many ventilators New York had, how many were being utilized, and how many they projected they would need over the next seven days. We would send ventilators based on data, not on guesses or intimidation.

"We aren't going to send them to you just because you bash us in the press," I said. "You need to get the information to us."

Cuomo complained that he couldn't get data on New York City from Mayor Bill de Blasio. They were barely on speaking terms. Their deep-seated rivalry had reached toxic levels through the pandemic, and I worried that it could cost lives. Trying to find a way to work around the feud, I phoned Jessie Tisch, a close friend from college who served as the city's chief information officer.

I had called Jessie a week earlier, after seeing an alarmed New York City nurse tell CNN that her hospital was out of masks. I asked Jessie to find out what the hospital needed. Before long, she organized a conference call with the CEOs of every hospital in New York, so that I could better understand their PPE shortages, and what they needed from the federal government. One by one, we went through their needs and mapped out a plan to get them the requisite supplies.

Now, I explained the ventilator situation to Jessie and asked if she could help get the data from the hospitals. She got on it immediately, helping me navigate the dysfunctional relationship between Cuomo and de Blasio. Her efforts proved invaluable and helped ensure that the federal government's supplies got to New Yorkers in need.

The next several days were like trying to steer a ship in a violent storm. FEMA initially resisted my data-driven approach. The agency was accustomed to taking governors' requests at face value and approving them upon demand, and they were not used to the intense public pressure. At one point, top FEMA officials wanted to send every ventilator in the stockpile to New York.

Knowing that once we sent them out, we would never get them back, I asked FEMA administrator Pete Gaynor how long it took to ship a ventilator. He said twenty-four to thirty-six hours. We agreed to position a thousand ventilators in New Jersey, where we could deliver them to the New York metropolitan area in just four hours if needed. Fearing there would be shortages across the country, we required governors to report the real number of ventilators needed based on the facts on the ground, and we were not going to be swayed by political or media pressures.

In the first week of April, Jessie called me with an update: based on her data, New York City was six days from running out of ventilators. The previous week, we had shipped 4,400 ventilators from the stockpile to New York. I was told that Cuomo had funneled 2,000 of them to a state-run warehouse, where they were not being used, rather than sending all of them to New York City, which was the epicenter of the outbreak in the United States. Jessie's estimate did not account for the 2,000 sitting in Cuomo's warehouse. It included only the 2,400 the city had received—and nearly all of these were already in use. I had to confront Cuomo.

"We did not send the ventilators from the federal stockpile to sit unused in New York's stockpile," I said. "Please send the two thousand ventilators to New York City before people die."

While I was trying to break the impasse, Boehler rushed in with an urgent problem: the vice president had authorized sending an additional tranche of ventilators to New York City. I raced down the hall to Pence's office and explained that Cuomo was sitting on two thousand unused ventilators. Based on our projections, we still had seventy-two hours before the situation in New York City turned dire, and I wanted to use every available second on the clock to ensure that we didn't distribute ventilators to a place that didn't absolutely need them. I told him that we had put additional ventilators in a federal facility in New Jersey, so that we could deliver them to New York City within four hours if Cuomo remained obstinate. Thankfully, the governor relented the next day and sent ventilators to New York City.

That same week George Helmy, the chief of staff to New Jersey governor Phil Murphy, called with a request for five hundred ventilators. A talented and affable former management consultant, Helmy had a precise answer to all my questions about New Jersey's usage rates. He sent me a spreadsheet calculating that New Jersey's ventilator supply would run out in three days. We sent five hundred right away. We agreed to speak every day at 7:00 a.m., and I promised that as long as we had the supplies and he had the data, we would stay twenty-four to forty-eight hours ahead of New Jersey's needs. Several months later, Helmy was among the few brave Democrats to defend our efforts against partisan attacks: "From the president on through the highest levels of the administration, we always felt we were a priority to the administration," he said in *Newsweek*.

At the same time, in Louisiana, local officials warned that hospitalizations were rising, and New Orleans was running out of ventilators and PPE. Governor John Bel Edwards implored us to send four hundred ventilators, but he hadn't submitted the data. Boehler called the CEOs of the two largest hospital systems in Louisiana. They said that they were prepared for the wave and had ventilators in reserve. They added that they needed gowns, which we sent the next day. We were later told that Edwards, a competent and gracious governor for the most part, had reamed out the CEOs for undermining his request. He was doing his job to fight for the people of Louisiana, but our job was to see through the smoke signals and make sure we matched our limited ventilator supply to real demand. Our approach of working directly with hospitals to get the data was not the typical government protocol, but had we stuck to the normal processes we almost certainly would have failed to get hospitals the supplies they needed immediately.

Not surprisingly, whenever we denied a request, or shipped fewer than the desired number, the governors aired their grievances in the media, generating headlines such as this one in the *Washington Post*: "Governors Plead for Medical Equipment from Federal Stockpile Plagued by Shortages and Confusion."

On April 2, as we briefed the president before his daily press conference, he brought up the issue: "Why are you not sending out the ventilators to the states?" he asked. "I'm getting killed on this."

"They don't need them yet," I said with uncharacteristic force, allowing my frustration to show after weeks of hardly any sleep and balancing life-and-death situations. "Governors want them preventively. They are worried about what *could* happen. Once we send them out, we're not getting them back. We have a small chance to meet the real demands, but only if we are as precise as possible."

Sensing that I was confident in my approach and had the situation under control, Trump responded with a jab. "You're a hoarder. You're hoarding the ventilators."

"I promise you that no one is going to die because I am holding back on sending them out," I said. "We may not have enough to get through the next two weeks, but when there is a real need, we will send them out within twenty-four hours. I am willing to take the blame if I am wrong."

"Okay, then you're speaking in the briefing today, and you're going to explain to the press why we aren't sending them all out," the president said.

Less than thirty minutes later, on the evening of April 2, I stood behind the podium in the White House press briefing room, looking thin and pale from hardly eating or sleeping for three weeks.

"The notion of the federal stockpile was it's supposed to be our stockpile. It's not supposed to be states' stockpiles that they then use," I said in response to a question about our supply management strategy. It was not my most eloquent moment, but I thought this was a pretty obvious point: we wanted states to use the supplies we sent, not to stash them away in warehouses.

The next morning, my brother Josh called to see if I was okay. When I asked him why, he said that I was getting destroyed in the press. Soon I saw the headlines: "Jared Kushner's Coronavirus Briefing Debut Sparks Outcry, Confusion," wrote the *Washington Post*. "Heaven help us, we're at the mercy of the Slim Suit crowd," wrote *New York Times* columnist Maureen Dowd.

The intensity and volume of the media's vitriolic outrage caught me off guard. But I didn't have time to dwell on anything but the crisis at hand. Requests continued to pour in. At our lowest point, we had just twelve hundred ventilators in the stockpile. The only good news was that our "15 Days to Slow the Spread" guidelines were making a difference. The growth in hospital usage rates was slowing, and our efforts to purchase every available ventilator we could find were beginning to pay off.

48

On the Brink of
Economic Collapse

O n March 27, President Trump signed the single largest government spending package in history: a $2.3 trillion economic
stimulus package. The CARES Act came together in less than
two weeks. We were shooting bullets into a cloud of smoke and hoping
that enough of them would hit their targets to save an economy veering
toward collapse.

The CARES Act only passed because Congress worked with the White
House in a way that I always hoped it would. Republicans and Democrats both agreed on what needed to be accomplished. They ultimately
drafted a plan for getting $1,200 cash payments directly to middle- and
low-income Americans. Equally important was the Paycheck Protection
Program (PPP), which would provide hundreds of billions of dollars in
federal loans to small- and medium-sized businesses.

The mechanics of the legislation were complicated. For the PPP loans
to work, America's banks needed to participate in the program voluntarily, and applicants needed to request funds through a hastily created
online SBA portal. Ivanka and SBA administrator Jovita Carranza called
the CEOs of every major bank as well as many local banks and urged
them to participate. The last time the federal government had rolled out
a new web-based program at this scale was the Obama administration's
catastrophic rollout of healthcare.gov, and our team wanted to avoid a

similar fate. Ivanka paid special care to ensure that funding was accessible to minority communities. Within the first fourteen days of the program, PPP processed fourteen years' worth of loans.

Because of these timely efforts, the country staved off a new economic depression. PPP alone saved as many as 17.3 million jobs through loans to small businesses. Despite this success, the closure of restaurants and catering companies disrupted America's food supply chain. Many farmers had no place to sell their fresh food, while at the same time thousands of newly unemployed people were lining up at food banks.

Not wanting food to go to waste while Americans went hungry, Ivanka reached out to Secretary of Agriculture Sonny Perdue to see if they could work together to help solve the quandary. Within days, they launched the Farmers to Families Food Box Program, which purchased food from farmers and distributed it to Americans in need. The program helped faith-based and community groups deliver more than 173 million boxes of fresh meat, dairy, and produce to families over the next twelve months. Ivanka worked around the clock to stand up the program and traveled across the country to help distribute boxes to families. Her passion for helping others and her core belief in the goodness of people were on full display.

*   *   *

On April 9, the first full day of Passover, I was looking forward to a special Seder dinner with Ivanka, Avi, and the kids—my first family meal in weeks. On the way home, I got a phone call from John Hess, the CEO of the Hess Corporation. He was an old friend from when I lived in New York, as well as the commissioner of my former fantasy football league, a hobby I had to drop when I came to Washington.

"The industry is out of oil storage tankers," Hess said. "We have nowhere to store the oil coming out of the ground. This could break the American oil and gas industry. The president has to get involved."

I'd received similar reports earlier in the week from other leaders in the sector, including Vicki Hollub from Occidental Petroleum and Harold Hamm from Continental Resources. If oil prices remained at $20 a barrel, energy companies would be forced to lay off millions of American workers, and our country's energy independence would be in jeopardy.

"You need to call the president directly," I told Hess. "He likes cheap oil. And I can't do anything on this unless he directs me to."

Half an hour later, my phone rang again.

"Jared, I never thought I'd be asking you to make a deal to raise oil prices," said Trump. "This is getting really bad. Call the Saudis and the Russians and work with them to make a deal."

I dialed Secretary of Energy Dan Brouillette to get the lay of the land: he had been immersed in negotiations between Saudi Arabia and Russia for months. He explained that the two countries had been close to an agreement one month earlier. When Russian president Vladimir Putin walked away from that deal, the Saudis cut their oil prices in response, leading to the current crisis. Brouillette was working closely with the negotiators on the two sides to broker a compromise. They were close to an agreement, but several significant issues remained unresolved.

I called MBS. The Saudi crown prince described his frustration with Russia. He thought they were playing games with the international oil supply and trying to force Saudi Arabia to cut production. Then I dialed Kirill Dmitriev, the powerful Russian financier and Putin confidant who had been helpful with the Middle East peace plan. We agreed that this was an opportune moment for Russia and the United States to work together on a matter of global importance.

When I arrived home, I sat down with Ivanka, Avi, and the kids for Seder dinner. As we practiced the sacred rituals and partook in the Passover meal, it almost felt as if life was normal again. Never mind the weeks of sleepless nights, missed family moments, and returning home long after the kids had gone to bed. I was savoring every second. As we sang my favorite Passover song, "Vehi Sheamda," a prayer about God's promise to deliver each generation of the Jewish people from their oppressors, the familiar sound of my phone broke the serenity: it was MBS, and I had to take it.

Ivanka nodded knowingly, of course, but I couldn't help but notice the kids' disappointed faces as I walked out of the room.

That night I was on the phone back and forth between MBS and Kirill, and our calls continued throughout the next day. Thirty-six hours later, we had nearly finalized an agreement to reduce production by around ten million barrels per day, which would be the largest cut in history.

"I think we got to the right number," I told Trump, who agreed to speak with Saudi Arabia's King Salman bin Abdulaziz and Putin to close the deal.

On Sunday, April 12, the three leaders spoke.

As Trump congratulated them for reaching a deal, King Salman interrupted: "Well, we don't have a deal yet. We need Mexico to reduce its production by four hundred thousand barrels per day."[56]

Mexico was part of an extended oil compact called OPEC+ that included the thirteen OPEC nations and ten non-OPEC countries, including Russia. Salman explained that if Mexico refused to cut production, any agreement would fail because the other OPEC+ countries would resent Mexico's free riding.

Trump passed me a note: "CALL MEXICO ASAP."

When I spoke to Alfonso Romo, chief of staff to President López Obrador of Mexico, he said they were trying to lower production but hadn't reached a decision.

Secretary Brouillette explained what was really going on. "It's the Hacienda Hedge," he said.

Mexico had nationalized its oil industry in the early 1900s and relied on oil production for a significant portion of its government revenue. To protect against dips in the oil market, it made an annual billion-dollar hedge on Wall Street. If prices fell below Mexico's hedged position, the country reaped billions of dollars, offsetting the losses in oil production revenue caused by the reduced price of oil. The more the price dropped, the more Mexico made from its "Hacienda Hedge."

"The Mexicans are currently hedged at fifty-five dollars per barrel," said Brouillette. "They're indifferent to the low prices. Why would they agree to reduce a single barrel of production?"

A few hours later, Romo came back and said that as a concession to Trump, Mexico would cut production by a hundred thousand barrels per day. That was not nearly enough. When I updated Trump, however, the president was surprisingly upbeat.

"That's great," he said. "Tell the Saudis that we'll make up the three-hundred-thousand-barrel difference."

"But we don't control our oil markets," I said, not sure what he was thinking. The United States could not order its oil companies to halt their drilling.

"Just tell them we're doing it," said Trump. "We've got to get this deal done—I have an idea."

Instead of focusing on the obstacles, Trump identified an opportunity. He recognized that US oil production was already coming offline. American producers couldn't make a profit unless the price per barrel was more than $40, so as prices fell to $38 and even lower, they naturally reduced production. This reduction far exceeded 300,000 barrels per day. The president realized that we could credit the amount to Mexico and strike a deal. I floated his idea past MBS, who agreed to consider it. The negotiations were nonstop, and everyone was exhausted, but they included moments of brinkmanship and jousting. During a conversation with Putin, Trump pivoted to an entirely different topic.

"Aren't you concerned about China's buildup on your southern border?" he asked. "That's where a lot of your country's wealth is—aren't you concerned at some point they may get a bit more aggressive and look to expand?"

Without skipping a beat, Putin shot back that he wasn't the one building a wall on his southern border.

The three leaders eventually reached an agreement, with OPEC approving a reduction of 9.7 million barrels per day. The deal saved millions of American jobs in the oil and gas sectors.

The Pulitzer Prize–winning energy writer Daniel Yergin praised the agreement as Trump's "biggest and most complex" deal ever. The *Wall Street Journal*'s editorial board echoed the sentiment, giving "credit to

Mr. Trump for using US global influence to mitigate the mayhem"—high praise from a source frequently critical of Trump's international economic policy.

*　*　*

On April 15, Trump called me to the Oval Office and said that he wanted to end the COVID-19 lockdown and reopen the economy the following day. While he believed that the federal guidance to slow the spread was justified to flatten the curve and build up lifesaving supplies, it was supposed to be temporary, and he believed that the doctors wanted it to go on indefinitely. As he fielded calls from business leaders, economists, and members of Congress, it was clear that the unemployment rate would soon jump to 30 percent. He told me that he wanted to make an announcement immediately. I implored him to give me a few more days, explaining that the governors had asked for clear reopening guidelines and that Dr. Birx was in the process of formulating a plan that Trump's medical and economic teams could support. I cautioned him that if he moved forward before a plan was finalized, his own advisers would distance themselves from the decision and Americans would lose confidence in the federal response. "If we can have consensus on a plan, it will be much better," I said. Trump ultimately agreed to give me twenty-four hours to achieve a consensus on reopening.

In a meeting with the president the next day, April 16, Fauci strongly advised against a full reopening. Continued lockdowns would save lives, he argued, and we should keep them as long as possible.

"I'm not going to preside over the funeral of the greatest country in the world," Trump declared.

"I understand," said Fauci meekly. "I just do medical advice. I don't think about things like the economy and the secondary impacts. I'm just an infectious diseases doctor. Your job as president is to take everything else into consideration."

Fauci was a shrewd politician and smooth communicator. Nobody rises to the top of a bureaucracy like the National Institutes of Health and survives six presidential administrations over three and a half decades without knowing how to self-promote, outmaneuver, and curry favor with the powerful.

Early in the pandemic, Fauci was sitting in my office when his phone rang. We both glanced down and saw the caller's name: Jim Acosta, the president's chief antagonist on the generally hostile news network CNN. Neither of us acknowledged the awkward moment, but it stuck in my mind. Members of our task force resented that Fauci would participate in these meetings, and then criticize the federal government's response as if he was not involved with it.

That very week he told the Associated Press that "we're not there yet" on testing, and that "we have to have something in place that is efficient and that we can rely on." The comments demoralized staffers who were working twenty-hour days while Fauci was chatting with his friends in the media. His statement also struck me as odd. It came at the end of a seven-day period in which we'd conducted a million tests. We were rapidly scaling, and we finally had a reliable system in place. Rather than highlighting this progress to build confidence, he focused on the negative. As a full member of the task force, Fauci attended all the meetings and knew what we were doing. Yet he continually distanced himself from the White House when discussing the effort publicly.

"Is he a sportscaster or is he a member of the team?" asked one task force member. "He knows the challenges we face and everything we are doing to solve them. If he has recommendations, he should give them to us!"

One day, after Fauci gave another doom-and-gloom interview, Trump tried to convince him to change his approach: "Anthony, you've got to be more positive. We need to give people hope."

Fauci pushed back: "My advice in situations like this is that we should make people feel as bad as possible. We want to explain the worst possible scenario. If it comes true, we were right. If it doesn't, then we did a better job than people expected."

"I'm not like that," Trump said. "I take the opposite approach. I am like a coach who believes in the team even if they are down to give them a reason to keep fighting. We can't let people give up. People are losing their jobs. They are drinking and doing drugs; they are depressed, suicides are going up. That is not America. We will get through this, but we have to stay positive; we have to give people a reason to keep their businesses open so that our country can bounce back."

"Fine, I'll be a little more positive," Fauci said, but he never made good on this commitment.

Fauci wasn't the only one beating us up on testing. Cuomo also attacked. On one of our calls, I confronted him about his public criticism: "What more do you want us to do?"

"This is not a scientific answer," said the governor. "What's enough testing? No one knows. Once I say there's enough testing, the media narrative against you guys will stop. Why don't I come to the White House, and we'll come to an agreement?"

Trump approved the meeting, and so Cuomo came to the West Wing. He said that New York was administering twenty thousand tests a day. He thought fifty thousand would be enough.

"Done," I said.

Cuomo was surprised. Unaware of the progress we had made on ramping up supplies, he had proposed a number he didn't think we could meet. Afterward, he spoke to the press in front of the White House and called his visit "very functional and effective." As we met New York's testing demand, he stopped his attacks on testing. When he did, the media's narrative petered out. Learning from this approach, the team had similar discussions with nearly every governor. Some pundits pushed for hundreds of millions of tests per day. Others wanted us to track the movements of Americans and conduct aggressive contact tracing. But we were not going to let America become a surveillance state.

During the pandemic, it would have been much easier if every challenge had one clear, scientific answer, but that was not the case. As I used to tell my team, three factors went into solving big problems:

imagination, money, and gravity. We had the first two. We just could not change the laws of gravity. We could only manufacture products so fast. We did, however, pair the power and resources of the federal government with the nimbleness and creativity of the private sector to confront the biggest challenge of our lives.

49

Operation Warp Speed

On the morning of April 15, Health and Human Services secretary Alex Azar and his deputy chief of staff, Paul Mango, came to my office with a proposal. Mango handed me a PowerPoint deck, and Azar pitched a plan to develop and deploy a coronavirus vaccine within six months.

I looked up from the presentation quizzically. The fastest vaccine to ever come to market was the mumps shot, and that had taken four years. It didn't have to take that long, said Azar. If the federal government could work with vaccine developers to streamline the regulatory approvals and fund the early production of the vaccines, we could dramatically truncate the timeline.

Before joining the administration, Azar had served as president of Eli Lilly, one of America's foremost pharmaceutical companies, renowned for commercializing both the polio vaccine and insulin. This was his wheelhouse. Growing visibly excited, the secretary explained that the FDA approval process was typically a huge choke point in the production of vaccines. On top of that, vaccines could be expensive to manufacture and store, so pharmaceutical companies usually waited for FDA approval before producing the doses at scale. If the federal government offered to underwrite the production costs, however, the companies could begin manufacturing vaccines as they entered clinical trials, the FDA's three-stage process for validating the safety and efficacy of a vaccine. Conducting these processes concurrently would shave months off the vaccine rollout timeline without compromising safety. Once the FDA approved a COVID-19 vaccine, we could begin shipping it to Americans the next day.

The cost of failure was high: if a vaccine didn't work or the FDA rejected it for safety reasons, the federal government would be stuck with warehouses full of useless doses and a big bill.

It was clear that Azar was coming to me for two reasons. First, he needed a bulldozer to keep the bureaucracy of the task force and anyone else out of his way. Second, and more importantly, he knew that if the project with its massive price tag went sideways, people would look for someone to blame. I was willing to accept this risk, because I knew a successful vaccine could potentially save millions of lives, while helping the country get back to normal.

Behind the ambitious vaccine plan was Peter Marks, an MD and PhD who led leukemia research at Yale University before joining the FDA to run its Center for Biologics Evaluation and Research. He rejected his field's fatalism about long timelines for the development of vaccines.

At a meeting with pharmaceutical executives at the White House on March 2, Stéphane Bancel, the CEO of Moderna, revealed that his company had already developed a vaccine and that they were waiting for regulatory approval to move forward with clinical trials.

"So, you're talking over the next few months, you think you could have a vaccine," said Trump.

Fauci cut in: "You won't have a vaccine. You'll have a vaccine to go into testing."

Later in the event, Trump again suggested that a vaccine could be ready within months instead of years. With the press cameras rolling, an exasperated Fauci declared that a vaccine would be available to the public "at the earliest, a year to a year and a half, no matter how fast you go."

Yet Marks identified ways to accelerate the production of a new vaccine. He knew that both Pfizer and Moderna had spent years testing a new mRNA vaccine technology that could work against COVID-19. Further, the ubiquity of the virus would help. For most vaccines, the trials dragged on for years as pharmaceutical companies searched for volunteers to enroll. That wouldn't be a problem with COVID-19. Marks also anticipated that the FDA would review the vaccines under a special emergency process. He calculated that with a good effort, we could deliver a vaccine before the end of the year. He called his initiative Project Warp Speed.

As Azar and Mango described the Warp Speed concept, my mind flashed to a call I had received a month earlier from Ken Griffin, one of America's most successful business visionaries. "You have to start mass-producing vaccines while you are still doing phase three safety trials," he said. "You may lose money on a few, but if one hits, it will go down as the best investment ever made. If you invest a few billion now, you will spare the economy trillions in damage."

I loved the idea. It was critically important to deliver a safe and effective vaccine as fast as possible. Through the drive-through testing program and Project Airbridge, I had seen the effectiveness of a well-run partnership between the government and the private sector. I had also seen the price of turf wars and government incompetence. We needed to nail the execution. There was no margin for bureaucratic missteps, power struggles, or needless delays.

"Let's keep this out of the task force," I said. "Let's run it out of HHS, with logistics support from DOD. Set up a meeting at the Pentagon, and I'll represent the White House."

Later that day, I described Operation Warp Speed to the president.

It would cost $2 billion to mass-produce each vaccine candidate, and we were looking to take a portfolio approach involving four to six promising candidates, costing upward of $14 billion.

"That's a lot of money to risk, and vaccines are only partially effective," said Trump. "What are we doing about therapeutics? I think people would prefer to know that they can be cured if they are hospitalized."

Trump's question drew from his personal experience. COVID-19 had killed several of his close friends, including Stanley Chera, a fellow real estate developer in New York. He felt like the federal government's response should include treatments that could save the lives of the infected, but he gave the go-ahead to move forward with Operation Warp Speed.

Our discussion had convinced him that the potential benefits to public health, safety, and the economy greatly outweighed the financial risks.

"To do this right, we need to cut through all of the bureaucracy," I said. "This approach will make some enemies. I need your permission to take liberties to do whatever is necessary to get it done."

"Do it," said Trump. "Anything that gets in your way, come back to me. But do what you need to do so that nothing slows you down."

Azar, Boehler, and I met at the Pentagon with defense secretary Mark Esper, deputy secretary David Norquist, and chairman of the Joint Chiefs of Staff Mark Milley to ask for the military's partnership in the project. Esper suggested that four-star Army general Gustave Perna would be the right man to lead the logistics of Operation Warp Speed. Perna had served for more than forty years in the Army and was weeks away from retiring. He patriotically accepted one final mission.

We needed to recruit a coleader with vaccine production experience to serve alongside Perna and manage the vaccine development. We interviewed four high-caliber candidates, but in the end, one stood out: Dr. Moncef Slaoui. The Moroccan-born scientist was smart, humble, and exacting. As head of vaccines for GlaxoSmithKline, he had helped develop fourteen vaccines in ten years. Smooth and confident, he came to his interview in a leather jacket with a T-shirt underneath. "I delegate," he said. "I take credit for all the failures and give credit for all of the successes." That resonated with me. It would help him survive a government bureaucracy plagued by resentment and leaks. Slaoui had another important trait: he was the only candidate who believed we could bring a vaccine to market in less than a year. He was our unanimous choice.

On May 15, with America's foremost scientists already hard at work to develop a vaccine, Trump walked into the Rose Garden with Perna and Slaoui.

"I want to update you on the next stage of the momentous medical initiative," said the president. "It's called Operation Warp Speed."

50

Turmoil

The murder of George Floyd under the knee of a Minneapolis police officer was an injustice that shocked, saddened, and outraged every decent American. The day after the president saw the video footage, he remarked to reporters: "I feel very, very badly. That's a very shocking sight."

In Minneapolis, the streets filled with protesters. During the day, the crowds stayed mostly peaceful. As night fell, however, looters violently smashed glass storefronts, robbed local businesses, and burned down buildings, including a police precinct. Living through my father's prosecution and incarceration had exposed me to the helplessness that many families feel when they are on the other side of the criminal justice system. Furthermore, through our criminal justice reform efforts, I had met families who had suffered unjustly at the hands of a few bad actors in law enforcement. I understood the hurt that many people were feeling. Americans have a right to peacefully protest, but there is no excuse for violence.

Within a matter of days dozens of people were shot and hundreds more injured across the nation. As the scenes of chaos and disorder blanketed the television screen, Trump grew more troubled. Peaceful protests were understandable, but this murderous mayhem was not. The president believed in federalism and respecting the jurisdictions of local authorities, but he also felt that the riots had become a national concern. He called Attorney General Bill Barr and asked for advice on what he could do to quell the violence. Barr said that Trump could activate the National Guard and send in the military, but that doing so without the express consent of the governor would set a dangerous precedent. Trump spoke to Minnesota governor Tim Walz and offered to send in the National Guard, but the Democratic governor declined the assistance, allowing the situation to escalate.

After midnight on Thursday, May 28, Trump released his frustration with a post on Twitter and Facebook: "These THUGS are dishonoring the memory of George Floyd, and I won't let that happen. Just spoke to Governor Tim Walz and told him that the Military is with him all the way. Any difficulty and we will assume control but, when the looting starts, the shooting starts. Thank you!"

By the time I woke up early that morning, Twitter had hidden the president's message behind a warning label claiming that it "violated the Twitter rules about glorifying violence." This was the first time a social media platform had censored the president of the United States. Twitter's censorship of Trump actually called more attention to the words the platform sought to suppress. Democrats and the media seized upon the censored tweet as a chance to redirect the rage of millions of Americans toward the president, who until that point was not a major focus of the national discussion.

Since 2016, Democrats tried to label Trump a racist, in what I learned was a common tactic used against Republicans. Before Trump ran for office, he was a pop culture icon embraced by the entertainment business and leaders in the Black community. But in August of 2017, in the wake of the horrific attack in Charlottesville, the Democrats and the media took Trump's words—"very fine people on both sides"—out of context. Trump was referring to peaceful protesters, some of whom supported, and others of whom opposed, tearing down the monument to Robert E. Lee. Time and time again, Trump had forcefully denounced the heinous violence of neo-Nazis and white supremacists, but the media seized upon every opening to call Trump a racist. I knew from my personal relationship with the president that the charge was nonsense. Trump's commitment to the rights and advancement of African Americans was fully apparent from the policies and priorities of his administration. In addition to passing the largest criminal justice reform in recent history—which mostly benefited Black males who had been unfairly sentenced through the justice system—Trump increased funding for Historically Black Colleges and Universities (HBCUs) and made the funding permanent so that these schools would no longer have to lobby Congress every year. Prior to the pandemic, Black unemployment reached a record low, Black youth unemployment reached an all-time low, and wages for Black workers were rising at an historic rate. By early 2020 Trump was consistently polling at over 20 percent with the Black community, a potentially game-changing twelve-point gain from the 2016 election. Democratic leaders were desperate to reclaim these Black voters, and they were willing to resort to reckless and unfounded accusations of racism.

By June 1, the riots had spread beyond Minneapolis. In New York, protesters threw bricks at police officers. In Atlanta, hundreds of people vandalized buildings and businesses. In Indianapolis, two people were shot and killed. In Washington, DC, protesters gathered in front of the White House and chanted curses at the president, burned American flags, broke through barricades, threw bricks at Secret Service officers, set off fireworks, and started a fire in St. John's, an historic church across Lafayette Square from the White House that every president since James Madison has visited. More than sixty Secret Service officers sustained multiple injuries defending the White House. What had started with a call for justice in response to the death of George Floyd devolved into an excuse for violence, theft, and anarchy.

It was time for the president to address the nation and assure Americans that their commander in chief was committed to restoring safety and peace in their communities. "My fellow Americans, my first and highest duty as president is to defend our great country and the American people," Trump said in the Rose Garden on the evening of June 1. "All Americans were rightly sickened and revolted by the brutal death of George Floyd. My administration is fully committed that for George and his family, justice will be served. He will not have died in vain, but we cannot allow the righteous cries and peaceful protesters to be drowned out by an angry mob. The biggest victims of the rioting are peace-loving citizens in our poorest communities. And as their president, I will fight to keep them safe. I will fight to protect you. I am your president of law and order, and an ally of all peaceful protesters."

Trump strongly urged governors to deploy the National Guard. He said that if they refused to do so and the violence continued, he would use the military to restore order.

"I take these actions today with firm resolve and with a true and passionate love for our country. By far, our greatest days lie ahead," he said. "Now I'm going to pay my respects to a very, very special place."

Trump was referring to St. John's Church, the historic place of worship that had been set on fire the night before. Minutes later, when Trump walked through the empty Lafayette Square, which the US Park Police had cleared, he was surprised to find the church boarded up. He had planned to go inside and say a prayer, so he improvised by holding up his Bible in front of the church. The press alleged that the president had emptied the square expressly for his visit. But that was not the case—the Park Police had cleared it as part of a preexisting plan to create a safer perimeter around the White House. A year later, under the Biden administration, the Interior Department's inspector general released an official report that stated the following: "The evidence we obtained did not support a finding that the [Park Police] cleared the park to allow the President to survey the damage and walk to St. John's Church. Instead, the evidence we reviewed showed that the [Park Police] cleared the park to allow the contractor to safely install the antiscale fencing in response to destruction of property and injury to officers occurring on May 30 and 31."

In a show of support for the protesters, Democrats in Congress were calling to defund the police. The White House needed to respond with a constructive alternative. For several weeks, I had been working behind the scenes with criminal justice reform advocate Jessica Jackson and other police reform advocates to listen to their concerns, and to see if we could find common ground with law enforcement to prevent grave injustices in the future.

On June 4, I convened a meeting in the Roosevelt Room with Attorney General Bill Barr, Ja'Ron Smith, Jessica Jackson, and the leaders of all the major police associations. Around the conference table filled with law enforcement leaders, Jackson floated two concepts that were important to police reform advocates: a new national mandate to require police officers to use force only as a last resort, and the elimination of qualified immunity, a legal protection for police officers that keeps them from being held personally liable for actions they take in the line of duty, unless they are clearly in violation of a court precedent or law. The police groups understood the urgency for reform, but they immediately dismissed the idea of eliminating qualified immunity because it would put officers at risk of being sued simply for doing their jobs.

"What problem are we solving for?" asked Liz Lombardo, a White House Fellow from the New York Police Department's legal bureau whom Chris Liddell suggested join the group. "There's a question we always ask at the NYPD: 'Will this new rule keep an officer from getting out of the car?' There is no law that compels a police officer to get out of the car when they see a crime occurring. They do it out of a selfless love for their communities. It requires incredible courage. If we create confusion about the standards, while eliminating their legal protections, officers are not going to get out of the car, and our communities will be less safe."

Lombardo and Ja'Ron Smith explained that the killings of George Floyd, Breonna Taylor, Ahmaud Arbery, Philando Castile, and others all had something in common. The officers who killed them were from police departments with outdated and unaccredited use-of-force policies.[57] These agencies were failing to prepare their officers to make split-second decisions related to deadly force. Staff secretary Derek Lyons and deputy counsel Pat Philbin argued that rather than mandating a federal standard, as some reform advocates were suggesting, we could create incentives for the departments to seek accreditation voluntarily, and we could ensure that the accreditation standards expressly forbid use of the choke hold—the policing tactic that killed George Floyd. Everyone seemed amenable to the idea, including the representatives of the police groups. I became convinced that a deal was possible and that we could build consensus on meaningful reforms.

Ja'Ron Smith and I discussed our ideas with Senator Tim Scott, a Republican from South Carolina who had recently offered his own reform bills. We drafted an executive order for the president's review. By the second week of June, after working closely with law enforcement and police reform advocates, we had a document ready for the president's signature, backed by both law enforcement and the victims of police brutality. Scott agreed to draft legislation that would expand upon these reforms and codify them into law.

On June 16, the president walked into the State Dining Room of the White House, where police officers sat next to families who had lost a loved one due to police misconduct. I had worked with Jackson, Desiree Perez of Roc Nation, and a lawyer named Lee Merritt to invite the families. In a closed-press meeting, the president asked each family to tell their story. African American moms and dads spoke of how their sons and daughters had suffered at the hands of police. The officers in the room expressed their sorrow for what a few bad cops had done to hurt their loved ones. A sheriff put his arm around a mother who had choked up while telling the story about how her son was wrongfully killed by a police officer. I wished that the nation could have witnessed the sincere dialogue between the families and the officers, and the profound compassion the officers had for the families. It was a powerful moment of healing and unity that I will never forget. During the meeting, which lasted well over an hour, the president listened intently and treated each family with compassion and understanding. Barr took careful notes and promised to follow up with each family on their case.

After the private roundtable, Trump honored the families and officers in the Rose Garden and pledged to "fight for justice for all of our people." He announced that he was signing an executive order encouraging police departments to adopt the "highest professional standards." The order would provide funding to police departments that received accreditation for their training policies, eliminated the choke hold, shared information about police misbehavior, and addressed mental illness and addiction.

"It strikes a great balance between the vital need for public and officer safety, and the equally vital need for lasting, meaningful, and enforceable police reform," said Fraternal Order of Police president Pat Yoes.

Even Laurie Robinson, who cochaired Obama's task force on policing, commended our reforms. "One of the things we'd hoped for is presidential leadership, and that's why his stepping out on these issues may be really helpful," she said.

Though Trump took bold action to make America's justice system fairer for all citizens, his rhetoric didn't resonate with some Americans. When asked about Trump's tough language, Senator Tim Scott gave the best explanation: "The president's love language has never been words of encouragement. I like to think of his love language as acts of service. And that's one of the reasons why I focus on the policy positions that we take that produce the type of change that will be necessary for a healthier, stronger middle class in the African American community."

While this was one of the most challenging periods of my service, I tried to stay focused on finding a constructive resolution that brought the country together. By December of 2021, hundreds of law enforcement agencies had adopted the reforms promoted by the executive order. In contrast, only eighteen police departments had adopted the reforms suggested by the task force on policing that Obama formed in response to the Ferguson riots in 2014. Trump's action was the most significant step taken by the federal government in recent memory to improve policing in America. Even in a starkly divided country, there are always opportunities to build bridges.

Three months of constant negative news focused on the COVID-19 pandemic and civil unrest caused Trump's poll numbers to sink sharply. At the same time, the president's reelection campaign struggled to adjust its fundraising efforts to the pandemic, which would require hosting virtual events. The RNC and the campaign did not hold a single presidential fundraiser from March through most of June, and while they still had more than $150 million in the bank, they failed to meet our finance goals for the period.

By July, Trump wanted to make a change in his campaign's leadership.

Deputy campaign manager Bill Stepien had proven to be a smart and capable team player, both on the 2016 campaign and during his tenure running the White House political affairs team. In Stepien, the president would gain a savvy, low-profile campaign manager with a decade of experience running high-profile campaigns. Trump asked me to speak with campaign manager Brad Parscale, who agreed to stay on and run the digital and marketing operations. The new structure allowed the campaign to build momentum, and it raised nearly $250 million in the third quarter alone. Down by a big margin and with a lot of macro trends working against us, we needed a strong comeback, and with Stepien at the helm, the campaign began working with renewed energy and unity.

51

Suicide Squeeze

On June 24, 2020, I spoke to the UAE's de facto ruler, Crown Prince Mohammed bin Zayed (MBZ). It felt like five years had passed since the two of us had talked, but it had only been five months. So much had happened.

MBZ warned me that if Israel annexed areas of the West Bank, it would reverse the progress we had made to bring Israel closer to its Arab neighbors. This was not a threat but rather a caution, as MBZ was hopeful about our progress. He reiterated that he would still maintain the UAE's partnership with the United States either way and added that he believed that if peace was going to happen, it would happen through our team.

It was a remarkable and humbling endorsement, but the best was yet to come: MBZ said that the UAE was ready to normalize relations with Israel—and that he wanted to begin during our administration.

My immediate concern was that we were about to disappoint him. A few weeks earlier, Ambassador David Friedman had emailed me Israel's proposed map for settling the territorial disputes in Jerusalem and the West Bank. It was the product of four months of painstaking work by a mapping committee we had established in February to draw the street-by-street and neighborhood-by-neighborhood maps envisioned by the president's peace plan. Getting Israel to agree to a map was a crucial step in the peace process. When I first met one-on-one with President Abbas in June of 2017, the Palestinian leader had assured me that if I could get a map out of Israel, "we will be flexible and everything else will be easy." But the map revealed an issue: we were struggling to convince Bibi, a master negotiator, to agree to a compromise that would give tangible life improvements to the Palestinians.

Friedman began calling me repeatedly, asking for the president's sign-off on the maps. Having gone through the diligence of drawing them, the Israelis were pushing hard to declare sovereignty right away. Each time I spoke with the ambassador, I pressed him for an update on Israel's concessions. Our conversations intensified. He became impatient, and I grew exasperated. "I know that this is your top issue right now," I said, "but I have a million issues to work on with the president, and this is not in his top hundred. After Bibi's speech in January, this hasn't been his favorite topic."

I told Friedman that I wasn't going to bring the annexation issue to the president unless we had a fair proposal that advanced our peace plan.

Several days later, on June 11, Yousef Al Otaiba called Avi. The pragmatic Emirati ambassador who had attended the rollout of our peace plan said that he was planning to publish an op-ed the next day based on a conversation with the renowned Jewish philanthropist Haim Saban. The op-ed, which Yousef had written in Hebrew, appeared in *Yedioth Ahronoth*, a prominent Israeli newspaper. "Israeli plans for annexation and talk of normalization are a contradiction," wrote Yousef. "A unilateral and deliberate act, annexation is the illegal seizure of Palestinian land. It defies the Arab—and indeed the international—consensus on the Palestinian right to self-determination."

It was a bold play. Yousef was sending a public warning to Israel: if Bibi moved forward with annexation, it would kill the possibility of normalization with the UAE and other Arab nations.

It also complicated our mapping negotiations with the Israelis. I had hoped that reaching a resolution on the annexation issue would move the peace process past an impossible sticking point and bring us closer to a deal. The compromise I envisioned would create a framework for resolving the land dispute between Israel and the Palestinians, and it would freeze expansion of settlements beyond the predetermined borders. But MBZ's call and Yousef's op-ed were forcing me to rethink whether reaching a normalization breakthrough between Israel and the UAE should be a higher and more immediate priority than our relentless pursuit of an Israeli-Palestinian peace deal. Unless the Emiratis were bluffing, annexation would reverse the progress we'd steadily made toward normalization.

I didn't think they were bluffing, but it was difficult to imagine that normalization would actually come to fruition. Only two Arab nations had taken the step to normalize relations with Israel: Jordan in 1994 and Egypt in 1978 as part of the Camp David Accords. Whenever I spoke with colleagues and confidants about the possibility of additional Arab nations normalizing with Israel, they thought the concept was impossible without first resolving the Israel-Palestinian conflict. They presupposed it would never happen, and that the Arabs would never follow through.

As I prepared for our meeting with the president, Friedman stopped by my office to compare notes. The two options—annexation or normalization—weighed heavily in my mind as we talked. My team had invested more than three years into our peace plan, and we were on the precipice of entering a critical new phase. But was it worth setting aside a long-shot chance at normalization?

As we sat down in the Oval Office, Friedman began with an update on the mapping effort and asked the president whether he was ready to support Israeli annexation of areas of the West Bank. Frustrated, Trump cut him off. He'd already done much for Israel, he said, and there were priorities to pursue with other countries.

He went around the room and asked each of us for our opinions on annexation. I told him that I thought we could do it in a way that minimized backlash from the Arab world, but we had to ensure that the Israelis made concessions to materially improve the lives of the Palestinian people.

By the meeting's end, Trump was ambivalent. "Let's be neutral," he said. "Mike, do what you think is best."

This was his way of telling Pompeo and the rest of us that we could move forward, but that if anything went wrong, he would hold us accountable.

As we left the Oval Office, Friedman, an experienced litigator who typically exuded confidence, was sweating bullets.

"Jared, that was close," he said. "I don't know how you do this every day on so many topics. That was really hard! You deserve an award for all you've done."

"I don't need an award, I just want to make progress," I told him.

"That was a fifty-one to forty-nine vote in the Senate," he said.

"No, that was fifty-fifty with Pompeo casting the deciding vote," Avi shot back.

Pompeo was more accustomed to these policy debates in front of Trump. He was happy with the outcome and ready to charge ahead, and once Friedman's shell shock wore off, the ambassador didn't need any convincing to move forward either. Despite my inner turmoil about the implications for normalization, the decision had been made. We were moving forward with annexation, and the president's team was prepared to execute.

On June 25 Avi and Friedman flew to Israel to meet with Bibi. I had sent Avi with one objective: finalize a mapping agreement that would advance our long-term goal of peace between Israel and the Palestinians. Our proposal was eminently fair. The Israelis would annex only those areas where Israeli settlements currently existed. In return, they would grant the Palestinians civil control over some neighborhoods where the Palestinians lived in the West Bank. While this was a blunt action, we believed it would advance an inevitable outcome. We couldn't imagine a peaceful scenario in which either the Israelis or the Palestinians who currently dwelt in these areas would be uprooted and placed somewhere else. Our proposal would acknowledge this reality. It would allow Bibi to claim a win and declare Israeli sovereignty over disputed territory, but I hoped it would also show the Arabs that we

had a plan for breaking through the stalemate of the past and ultimately reaching a resolution that included an independent Palestinian
state. I also hoped that the Emiratis would appreciate the progress we
were making, have a change of heart, and keep their normalization
offer in play.

At 9:30 p.m. on Saturday night, June 27, Friedman and Avi met with
Bibi. The prime minister quickly rejected their proposal. He was unwilling to make the concessions we suggested.

"You're hanging on by a thread with Trump," warned Friedman.

Bibi knew that Friedman was a pro-Israel hawk, so his words rattled
the prime minister. This was one example of Trump's effective management style. His hesitancy to move forward made clear to Friedman that
he should negotiate from a position of strength and should not agree to a
deal that failed to advance America's overall objectives in the region. Our
message resonated: Bibi wasn't getting annexation for free. Israel needed
to give something in return.

Trump was the most popular US politician in Israel, and Friedman
and Avi made clear that if Bibi didn't compromise, there was a good
chance Trump might publicly oppose annexation. Additionally, with
annexation, Bibi risked near-unanimous condemnation at the United
Nations. And if he went forward unilaterally, there was no guarantee
that our administration would block the international sanctions against
Israel that might follow.

After three days of meetings, our team failed to reach a deal that included sufficient Israeli concessions. I couldn't in good faith recommend
that the president endorse the current package. It felt like we had reached
a dead end.

My mind drifted back to my conversation with MBZ the week before.
Since the call, my team had been advancing the annexation proposal and
hadn't raised MBZ's potential normalization offer with the Israelis.

"Maybe now is the right time to bring up the normalization pathway,"
I said to Avi. "Why don't you see if Bibi would pursue it in exchange for
dropping annexation?"

The following morning, on June 30, Avi told the prime minister about my call with MBZ and asked if he would call off his annexation plans if we could get full normalization with the UAE.

Bibi was still skeptical that the offer was real, but said he would be interested if we could deliver.

Yousef called Avi just after he landed back in Washington the next day, July 1. As they discussed the impending Israeli announcement, Yousef floated the idea of the UAE entering into a nonbelligerence agreement with Israel in exchange for dropping annexation. Avi told Yousef he appreciated the offer, but said that he didn't think that would be enough. After further discussion, they both agreed to take to their teams the following proposal for consideration: the UAE would fully normalize relations with Israel if Bibi dropped his push for annexation. Until this phone call, the Emiratis had only said that if Bibi moved forward with annexation, he would ruin any chance of normalization. But they had not specifically offered to go forward with normalization in exchange for Israel suspending annexation. While MBZ had hinted at this during our June call, this was the first direct offer with achievable terms. At midnight that same day, Bibi's self-imposed annexation deadline passed without incident. He had refrained from making the explosive announcement. Normalization was still in play.

At 5:00 p.m. on July 2, Yousef came to my office and described UAE's offer, which he had now vetted with the Emirati leadership. The UAE would fully normalize if Israel would suspend its annexation plans.

This was getting serious. On July 5, Yousef provided a normalization offer in writing. We shared parts of it with Bibi. Intrigued, the prime minister said he was willing to move forward and pursue the offer. He was beginning to appreciate the significance of the opportunity in front of him. He also must have known that annexation was too perilous without US support.

We were on the brink of a breakthrough. Less than twenty-four hours later, however, Avi rushed into my office with a message from Israeli ambassador Ron Dermer: Bibi would not make a deal with only the UAE. "He will only drop annexation if we can get three countries to normalize," said Avi.

I couldn't believe it.

"Please remind him if we can get this deal, it will change the whole global dynamic for Israel and likely lead to other countries normalizing," I said. "I don't blame him for asking, but it will be impossible to keep this a secret if we try to include other countries. And one untimely leak could spark protests across the region and kill the process. Also, remind him that he doesn't have annexation without us."

The next day, we heard back from Dermer. Bibi had agreed to postpone annexation for the time being, but he would not say for how long.

That was a good start, but we still needed Pompeo's sign-off on several critical items. On a visit to the State Department, I showed him the UAE's normalization proposal. Pompeo had been continuously supportive of the effort, but partially to manage my expectations, he outlined some of the hurdles we still faced to close a deal between Israel and the UAE.

"If this happens, it would be game-changing," Pompeo said. "In my experience, the Emiratis are serious people and don't waste time. This is a high-impact but low-probability objective . . . but crazier things have happened."

Over the next several days, my team and I worked around the clock to build out the details of a deal that would be agreeable to the president, the Israelis, and the Emiratis. At no point during these discussions did the Israelis and Emiratis speak directly to each other. Instead, Avi and I served as the interlocutors.

At one point, I suggested to Yousef that, to expedite the process, we should bring the two sides together for direct talks. The Emirati ambassador just shook his head. "I much prefer to work through you and Avi," he said.

A few weeks earlier, Dermer had called Yousef to complain that his op-ed in the Israeli newspaper had unhelpfully contradicted Bibi's very public prediction that annexation would not harm Israel's relationships with the Arabs. Despite this consternation, Yousef's courageous op-ed was the best possible thing for the Israelis. It was a stroke of genius that pushed us one step closer to changing history in the Middle East. It was now up to us to get a deal to the finish line.

52

The Call That Changed
the World

Throughout July, our talks with Israel and the UAE continued to progress, but we knew that negotiations could break down at any point—and several times they nearly did.

Our White House team met with the Emiratis daily to iron out the exact details of the normalization agreement. Avi and Major General Miguel Correa shepherded the negotiations on behalf of the US delegation, supported by National Security Council officials Rob Greenway, Scott Leith, and Mark Vandroff. Our NSC team, which had the full backing of Robert O'Brien, could speak in technical detail about various aspects of the deal and navigate within the federal bureaucracy to deliver results.

By the end of July, we had reached a tentative agreement with Israel and the UAE. Recognizing that numerous problems could still surface, we made plans to announce the deal on August 13. In preparation, we needed to draft a joint statement from the UAE, Israel, and the United States providing the high-level details of the agreement. To avoid telephones, which could be monitored, we relied on personal visits. Avi began cycling between the UAE embassy, the White House, and the Israeli embassy to work out the open issues with Yousef and Dermer, and consulted me on any sticking points. General Correa did double duty with his Emirati contacts to assure them that all open issues would get resolved.

Ten days of around-the-clock shuttle diplomacy produced more than a hundred versions of the document. On multiple occasions, negotiations almost came to a halt. Understanding the magnitude of the agreement, both sides treated every word as a life-or-death issue.

By August 7 the normalization talks were on the verge of breaking down. Avi presented me with the latest draft that he thought was the best possible compromise. Yet problems remained. One of the outstanding issues was that Bibi would say only that he had agreed to "postpone" the annexation rather than "suspend" it, and the Emiratis found this, as well as several other issues, unacceptable.

"In the diplomacy business, words matter," I said. "Tell both sides that we are not in the diplomacy business. We are in the results business."

I reviewed the draft, made edits, and handed it back to Avi.

"This should solve everyone's issues," I said. "Tell them to put their pencils down and that this is now the final version. The shop is closed."

The next day, August 8, Yousef called to say that MBZ had agreed to the joint statement and was ready to move forward with a full peace agreement. That same day, the United Kingdom's ambassador to the United States, Karen Pierce, called to warn that if the United States recognized Israel's annexation, the British government would recognize Palestine as a sovereign state. I had found Pierce to be a thoughtful and talented diplomat, and we were often up-front with her about our impending foreign policy actions. But in this instance, to keep her off our tracks, I made an argument for annexation and explained why it would make sense to proceed as we had planned. The call confirmed that not even our closest allies knew that a peace agreement was about to be announced.

On August 12, one day before the scheduled announcement, the deal nearly died twice. First, the UAE flagged a technical issue, which they said was a deal-breaker. As we scrambled to solve it, Dermer came to my office with his own unwelcome surprise: he said the deal was off because the timing wouldn't work.

A domestic political opponent had introduced legislation effectively barring the prime minister from forming a government, and there was a chance that Israel would have elections the following day.

Sensing that both sides were growing nervous about everything that could go wrong, I tried to be patient and exude confidence. "That's not an option," I said. "I know Bibi will put what's best for Israel before his personal political situation. We've come too far, we're so close. This deal is happening. We're announcing it tomorrow."

As we concluded our discussion, Dermer disclosed a frustration: Throughout these entire negotiations, the Israelis hadn't spoken directly to anyone from the UAE.

"I did ask them if they wanted to coordinate with you directly," I explained, "but they insisted on going through Avi and me. I know your call with Yousef about the op-ed didn't go well. When this is done, we'll get everyone together, and I have no doubt they'll come to see what a special advocate and partner you will be."

Sensing there was no way to change the plan, Dermer promised that he would speak with the prime minister and urge him to move forward with the normalization announcement the following day. He expressed his enthusiasm for what was about to come.

Early that evening, I briefed the president on the final agreement under the assumption that we'd keep the UAE and the Israelis from walking away in the final hours.

"This is going to be a big surprise," he said. "How do you think people will respond to it?"

"This deal is massive," I said. "It will send shock waves throughout the world. You allowed me to do this the unconventional way, and we are about to achieve what diplomats have only dreamt of for decades."

By 7:00 p.m., just sixteen hours before the scheduled announcement, we resolved the outstanding issue with the Emiratis, and Bibi was ready to go forward as well. That night, Ivanka and I walked our dog around the neighborhood, and talked about all of the improbable twists and turns that led to this moment. I went to bed praying that there would be no further problems and hoping that nothing would leak. As I drifted off to sleep, I thought: Tomorrow, the world is going to change.

Early the next morning, August 13, I called my dad from the car—something I did each morning during my time in government. One of my Secret Service agents later said that he was so touched by these morning calls that he adopted the same habit with his own father. "Be on the lookout for positive breaking news at around 11:00 a.m.," I told my dad. There was a limit to what I could say about my White House endeavors.

In my office at 8:00 a.m., my team grilled me on questions for my upcoming press interviews. Then we walked several reporters through the details of the impending announcement, under an embargo, so they could prepare to publish accurate stories as soon as we released the joint statement announcing the deal.

As the president prepared for a phone call with Bibi and MBZ to announce the deal, General Miguel Correa popped into my office. "We should call it the 'Abraham Accords,'" he said.

Until then, we had been so busy ironing out details that we hadn't thought to name the agreement, but "Abraham Accords" immediately struck me as perfect. It would remind everyone of the original Abrahamic roots of brotherhood that united the Arab and Israeli peoples.

At 10:15 a.m. I entered the Oval Office with our whole team, including Avi, David Friedman, Brian Hook, General Correa, Rob Greenway, Scott Leith, and Mark Vandroff. I had called Treasury Secretary Mnuchin the night before and invited him to come to the White House. I didn't tell him why, but I assured him he wouldn't want to miss the meeting. Since the 2016 campaign, Mnuchin had been a rock-solid friend and ally. He had supported me in my early days when I was discouraged, and we had been together for many of the most meaningful moments, from election night to our trip to the demilitarized zone between North and South Korea. He had attended the opening of the embassy in Jerusalem and helped execute the economic conference in Bahrain. I knew he would want to be present.

Sensing that something big was about to happen, more and more people started shuffling into the Oval Office.

Finally, the call began. Trump was on the line with MBZ and Bibi.

"This is very, very historic," said the president. "This is something that is incredible."

MBZ thanked the president for his leadership and emphasized the importance of the agreement for the advancement of peace in the region. He called it a transformative event that would create fresh energy for positive change, economic growth, and a new understanding between the Arab and Israeli people.

Bibi graciously thanked MBZ for his courage and Trump for his leadership. He said it was a "turning point for peace" and the biggest advance in more than a quarter century.

When the call ended, everyone in the Oval Office was silent as we paused to absorb the gravity of what we had just heard and witnessed. Mnuchin stood up and clapped, and one by one, everyone else rose to their feet and applauded. The president watched in amazement and enjoyed the applause. Then he too stood up and joined us all in clapping. We had just struck a peace agreement between Israel and the United Arab Emirates—a deal that no one expected, or even thought possible.

Moments later, Dan Scavino had a tweet teed up.

"Dan, if you're okay with it," I asked, "can Avi press send?"

Dan proudly held out the iPhone toward Avi, who paused for a second and then pressed the blue "tweet" button: "President Donald J. Trump, Prime Minister Benjamin Netanyahu of Israel, and Sheikh Mohammed Bin Zayed—Crown Prince of Abu Dhabi and Deputy Supreme Commander of the United Arab Emirates—spoke today and agreed to the full normalization of relations between Israel and the UAE."

We were live.

The full statement detailed the key elements of the groundbreaking agreement.[58]

The news caught the world by surprise. In a city where there are no secrets, not a single reporter had inquired about the agreement before its announcement—a fact that we considered to be a worthy accomplishment, though of course our real triumph was the deal itself.

White House press secretary Kayleigh McEnany opened the French door to the west colonnade, and the White House press pool burst into the Oval Office. Within seconds the bewildered reporters formed a scrum, wrestling for prime positions in front of the president.

"Just a few moments ago, I hosted a very special call with two friends, Prime Minister Benjamin Netanyahu of Israel and Crown Prince Mohammed bin Zayed of the United Arab Emirates, where they agreed to finalize a historical peace agreement. Everybody said this would be impossible. . . . After forty-nine years, Israel and the United Arab Emirates will fully normalize their diplomatic relations. . . . It will be known as the Abraham Accord."

With his characteristic good humor, Trump added, "I wanted it to be called the Donald J. Trump Accord, but I didn't think the press would understand that."

The entire room erupted in laughter. After the media exited the Oval Office, the mood in the room was triumphant, and we continued to shake hands, hug, and soak in the remarkable moment.

"Jared's a genius," said Trump. "People complain about nepotism— I'm the one who got the steal here."

I smiled at the joke and shot back: "Maybe in the future, more presidents will haze their sons-in-law by tasking them with impossible problems."

To keep the deal a secret, I hadn't previewed it for any other Arab countries, but I had a feeling that others would take a similar step, so long as there was not significant fallout in the region. What I didn't know was how quickly that chance would present itself. That very afternoon, Avi received a phone call from Sheikh Salman bin Khalifa, the finance minister of Bahrain, with whom I had developed a great partnership and friendship while planning the Peace to Prosperity workshop.

He congratulated us and remarked that the first one was big, but the second one would cement the deal. He relayed that under the right conditions, Bahrain could go second.

Later, as I called to thank people who had been helpful throughout our peace efforts, I dialed Rick Gerson, an investor and fellow New Jersey native who had become a good friend over the decade I lived in New

York City. "There is a good chance this agreement wouldn't have happened without your initial connection," I said. "It helped us establish a foundation of trust. Thank you." Back in 2016, during the transition, Gerson had introduced me to several of his longtime close friends in Emirati leadership. Gerson's introduction led to my first meeting with MBZ and his national security adviser, Tahnoun bin Zayed (TBZ), and commenced a constructive dialogue about how to end the endless wars, confront extremism, and pursue a future of greater prosperity and peace. Perhaps in a foreshadowing of what was to come, I was immediately struck by their respect for Israel and their acknowledgment of overlapping interests between the two nations.

Around the same time that we announced the Israel-UAE peace agreement, I received a call from Ric Grenell, special envoy for Serbia and Kosovo peace negotiations. Grenell, who had previously served as ambassador to Germany and acting director of national intelligence, was in the middle of negotiating an unprecedented economic agreement between Serbia and Kosovo—two former adversaries that do not share diplomatic ties. Grenell saw an opportunity through these negotiations to build momentum for the Abraham Accords. He believed that Kosovo, a Muslim-majority country, might be willing to normalize relations with Israel as part of its economic agreement with Serbia.

"If you can get that done, it would be amazing," I told him. But it seemed like a long shot.

Several days later, Grenell followed up with incredible news: Kosovo had agreed to normalize relations with Israel. Further still, both Serbia and Kosovo had decided to place their embassies in Jerusalem. This was a completely unexpected development, a confluence of Grenell's creative diplomacy and the progress we'd made. The Abraham Accords were already starting to reshape the Middle East and the broader Muslim world. The sands were beginning to shift.

Hours after we announced the Israel-UAE peace agreement, Yousef and Avi came to my house for dinner. We were excited, exhausted, and proud. The bonds we had built elevated our friendship and trust to a rare level. We were now partners in changing the world. We exchanged stories on the positive feedback we were hearing from world leaders and talked about the work ahead to get the agreement signed and implemented as soon as possible.

The public response to the diplomatic achievement was overwhelmingly positive. An op-ed by Middle East expert and *New York Times* columnist Thomas Friedman epitomized the uncharacteristically exuberant coverage. "A Geopolitical Earthquake Just Hit the Mideast," his headline read. "For once, I am going to agree with President Trump in his use of his favorite adjective: 'huge,'" Friedman wrote. "The U.A.E. and Israel and the U.S. on Thursday showed—at least for one brief shining moment—that the past does not always have to bury the future, that the haters and dividers don't always have to win."

The next morning, August 14, the president called: "I've never gotten better press coverage in my life," he said. "This is the most positive coverage I've gotten on anything that I've done since I've been president."

The release of our peace plan, along with our unconventional diplomacy, ultimately proved to be an essential step to reaching the Abraham Accords. It offered the Palestinians a pathway to self-determination and a more prosperous future. It showed the Arab public that the decades-old conflict had become more about enriching Abbas and the Palestinian leadership than finding a lasting resolution for the people. It exposed the Palestinian leadership's illogical and outdated positions, even as it proved that Israel was ready to take an unprecedented step forward and agree to a detailed two-state solution. These steps ultimately allowed people to accept that there were in fact two separate conflicts—the Palestinian-Israeli conflict and the Arab-Israeli conflict—and that the cost of linking them was too high. This created the conditions for the beginning of the end of the Arab-Israeli conflict.

When I first came to Washington, almost everyone accepted former secretary of state John Kerry's assessment of peace with Israel: "There will be no advanced and separate peace with the Arab world without the Palestinian process and Palestinian peace." I had questioned this assumption and instead embraced a new approach, based on my belief that countries would engage in new partnerships that offered more promise for their citizens than the status quo.

Over the course of three and a half years, we advanced American interests by uniting our partners in the region against our common threats. Countries in the Middle East were now sharing more of the defense burden, and American troops were coming home. Trump was ending the endless foreign wars, and now he was forging peace in the Middle East. Economic ties were beginning to form that would prevent future conflict. These unprecedented changes would not only improve the lives of millions in the region but also protect countless Americans, especially our brave men and women in uniform.

53

First Flight

On the tarmac at Israel's Ben Gurion International Airport, I stood before an Israeli El Al jet. Painted on the side of the blue-and-white plane was the word *peace* in English, Hebrew, and Arabic. Over the previous three years we had kept a low profile on my travels in the Middle East, but this trip on August 31 was different. News networks carried live coverage as I stepped forward to say a few words: "We are about to board a historic flight: the first commercial flight in history between Israel to a Gulf Arab country. While this is a historic flight, we hope that this will start an even more historic journey for the Middle East and beyond. I prayed yesterday at the Western Wall that Muslims and Arabs from throughout the world will be watching this flight, recognizing that we are all children of God and that the future does not have to be predetermined by the past."

Up to now, the peace agreement between Israel and the UAE had been confined to the words of President Trump, Prime Minister Bibi Netanyahu, and Crown Prince Mohammed bin Zayed. This flight would show what the peace would mean in practice.

A week earlier, Avi had suggested that instead of taking a US military plane on the flight from Israel to Abu Dhabi, we could try to arrange a commercial flight and bring along a delegation of Israeli officials. I immediately embraced the concept, and we called Yousef to discuss it. As long as the Israelis remained constructive, he didn't think it would be a problem. And just like that, we set the plan in motion.

We still needed to solve an important logistical issue. The most direct route was a three-hour flight that passed over Saudi Arabia. Since the kingdom has no diplomatic relations with Israel, it did not typically allow commercial planes flying to or from Israel to travel through its airspace. We would need a special waiver. Flying around Saudi Arabia would add four hours to the trip, and I couldn't see why a journey of three hours should stretch to seven.

I asked Avi to call the deputy defense minister of Saudi Arabia, MBS's brother Khalid bin Salman, with whom we had developed a close working relationship during his stint as ambassador to the United States. He promised to help. It took several more calls, but the night before our flight, the Saudi aviation authority approved the waiver. That authorization was its own major diplomatic achievement. The relationships we had built over the previous three years allowed us to break old conventions, move past bureaucracy, and chart a more constructive path forward.

Just before takeoff, the Israeli pilot, Tal Becker, who had been flying for forty-five years, made an overhead announcement. I pulled out my phone and recorded most of his words: "For the very first time an Israeli-registered aircraft will [fly over] Saudi Arabia, and after a nonstop flight from Israel, land in the United Arab Emirates. The duration of the flight, with the shortened route over Saudi Arabia, will be approximately three hours and twenty minutes instead of what would have been more than seven hours up to now. At the end of this historic nonstop flight, the wheels of the aircraft with the flag of the State of Israel on its tail will touch down on the runway of Abu Dhabi, the capital of the United Emirates. This will be another significant event in our history, just as El Al was when peace was signed between Jerusalem, Cairo, and Amman. We are all excited and look forward to more historic flights that will take us to other capital cities in the region, advancing us all to a more prosperous future. Wishing us all salaam, peace, and shalom. Have a safe flight. Thank you."

We all clapped at his impromptu speech. I sent the video to Ivanka, who was so moved by it that she posted it on Twitter. Millions of people heard the message from the Israeli pilot, who so beautifully captured the sentiment of his fellow citizens.

As the plane sailed through Saudi airspace, the Americans and Israelis passed around their plane tickets and exchanged signatures to commemorate the experience. For many, this flight was the pinnacle of a long career in public service. We all felt the significance of the moment. We were making history. I thought of my grandparents, and wondered what they would think of their grandson leading the delegation on behalf of the president of the United States to make peace between Israel and a leading Arab nation.

The image of the Israeli plane in the United Arab Emirates captured the imagination of millions of people and ignited hope throughout the region. The older generations had accepted the illogical status quo as a given and had grown skeptical of ever seeing a breakthrough in their lifetime. Now many started to wonder: If peace was possible with the UAE, why not with the other Gulf Arab states? Arab observers began to see the enormous benefits of normalization: they could travel to Israel for business, leisure, or religious pilgrimages, opening up new possibilities for commerce and collaboration. Just like that, the unthinkable was now within their grasp.

The trip taught me an important and humbling lesson: despite all of our meticulous work to reach the normalization deal, the flight drew more attention than the deal itself. While I always paid careful attention to the policy details, I often shortchanged the power of effectively communicating our efforts. As the saying goes, a picture is worth a thousand words, and the striking image of the flight made the peace agreement real to people.

When I traveled as a government official, I kept my itinerary tightly focused on the business at hand and avoided tourist activities like visiting historic sites. With this trip, however, I made an exception and agreed to join a cultural event with the Israeli and Emirati delegations. From the airport, we caravanned to the Louvre Abu Dhabi and visited the Gallery of Universal Religions, where a Qur'an, a Bible, and a Torah have long been displayed side by side. The exhibit served as a visual representation of what the Abraham Accords were all about.

That evening, Tahnoun bin Zayed Al Nahyan pulled me aside. The

UAE's national security adviser was one of the first foreign officials I had met, and he had become a trusted friend. He was a deep thinker, and someone whose strategic counsel I often sought when considering our next steps for advancing peace. So I was particularly honored that evening when he presented me with two special gifts: a copy of the official flight authorization that had allowed the first Israeli plane to land in Abu Dhabi, and a copy of a new Emirati law to reverse a boycott of Israel that the UAE had enacted in 1972. He explained that they had to get special sign-off from their parliament to give me a copy of the federal decree since I was a foreigner. I was deeply moved by this presentation, and I cherish these two gifts for what they represent: the ties of friendship and brotherhood that we forged between Israel and the UAE.

Then came an inaugural dinner between the Israeli and Emirati delegations. In keeping with the Emiratis' famous hospitality, the lavish buffet had an entire kosher section, which met the highest standards both of quality and rabbinic supervision. During the introductions and small talk, it struck me that the senior officials present from the two countries had never spoken to one another. I felt like I was facilitating a blind date. At one point, an Emirati official mentioned he was eager to align banking systems so that investments could flow between the countries. Instead of focusing on formalities and celebration, I suggested that we get to work on this right away. Several Israeli financial and Emirati finance officials who were part of the delegations left the dinner immediately to start navigating the hurdles. By 4:00 a.m. the next day, they'd hammered out the details to connect their banking systems.

We were especially eager to ramp up tourism so that Israelis and Emiratis could visit each other's countries and begin to forge friendships, which would build public support for the peace agreement. As Arab visitors made pilgrimages to the al-Aqsa Mosque in Jerusalem and posted photos of visiting the holy site in peace, it would strike at the heart of the inaccurate prejudice that the al-Aqsa Mosque was under siege. Soon after our visit, Israel and the UAE agreed to allow their citizens to travel between the two countries without a visa—a major diplomatic accomplishment.

We also brought Israeli officials from the aerospace, health, and tele-communications departments so they could meet their Emirati counterparts and begin collaborating. The Israeli and Emirati medical teams integrated immediately so they could coordinate more closely on scientific advancements to combat COVID-19. I was surprised to learn that it was impossible to place a call between Israeli and Emirati cell phones. After identifying this issue, we set in place a process to rectify it. I had underestimated how little of a connection there was between the two countries. The trip was more than a symbolic flight. It linked the two countries on a practical level. For the first time, Israeli and Emirati officials dined together, exchanged business cards, and discussed opportunities to work together.

As the Israelis and Emiratis built trust before our eyes, those who opposed progress grew increasingly isolated. Abbas turned to the terrorist group Hamas, convening a meeting to strategize against our efforts, and the Iranian regime issued bombastic statements against the UAE. Iran's supreme leader, Ayatollah Khamenei, even targeted me in a tweet: "The nation of Palestine is under various, severe pressures. Then, the UAE acts in agreement with the Israelis & filthy Zionist agents of the U.S.—such as the Jewish member of Trump's family—with utmost cruelty against the interests of the World of Islam. #UAEStabsMuslims."

In the midst of this predictable opposition from the bad actors in the region, we needed to keep building momentum for peace. From the UAE, I traveled to Bahrain in hopes of bringing a second country into the Abraham Accords.

* * *

In Bahrain, before I made my case for normalizing relations with Israel, I presented King Hamad bin Isa Al Khalifa with an unlikely gift: a Torah scroll.

Just over a year earlier, during the Peace to Prosperity workshop, several Israelis had taken the opportunity to visit the synagogue in Bahrain's capital of Manama. Founded in 1935, the synagogue hadn't held public services in years. During the summit, however, enough Jews visited to have a minyan—a quorum of at least ten men, the number needed to hold a congregational prayer service. It was a profoundly moving experience for those who attended, but they noticed that the synagogue lacked a Torah scroll, which had to be written by hand. Upon hearing this, I personally commissioned one to be made for the synagogue and dedicated it in the king's honor: "For his vision, courage and leadership bringing peace, respect and religious tolerance to the Middle East."

The king was touched by the story. He remarked that we were all sons of Abraham, and he had always believed that Jews, Christians, and Muslims must understand and respect each other.

With that, he gave me and Crown Prince Salman bin Hamad Al Khalifa his blessing to finalize the peace agreement. We spent the next several hours together working through Bahrain's priorities and concerns and came to a framework that we believed would be acceptable for the normalization agreement.

From Bahrain, I flew to Riyadh, Saudi Arabia, to meet with MBS. The Saudis asked that we take extra COVID-19 tests before getting off the plane, so my assistant Charlton Boyd administered a round of tests for our group. The White House Medical Unit had trained him for this exact scenario, so he was prepared to give us the tests. The Saudis adhered to the strictest pandemic protocols we had encountered in the region, which when combined with their royal protocols made for an uncharacteristically formal visit. MBS and I wore masks and sat in chairs placed roughly fifteen feet apart. Our previous meetings had been long and informal, but this meeting was rigid and brief. It was, however, extraordinarily productive.

As we discussed the peace deal with the UAE, I sensed from his tone that MBS was impressed by our progress.

"What about Saudi Arabia?" I asked.

MBS noted that Saudi Arabia shared common interests with Israel, but wanted to continue to let the region process the normalization agreement with the UAE and see if progress could first be made with the Palestinians. He also expressed that he wanted to resolve the rift with Qatar.

Next I decided to take a chance, even though I knew it might push his limits.

"Thank you for permitting our Israeli plane to fly over Saudi airspace to Abu Dhabi. Since we were flying at forty thousand feet, no one seemed to notice," I quipped, before making a serious request: "Can we make that permanent for commercial routes to and from Israel?"

MBS said that he would try to get it done.

I was encouraged by his response. Opening the airspace would demonstrate Saudi Arabia's tacit support for normalization. It would have great practical value, making flights between Israel and the UAE shorter and more affordable for travelers. It would also make it easier for Israeli planes to fly to destinations in Asia. We had been laying the groundwork for this since May of 2017, when the Saudis permitted Air Force One to fly from Riyadh to Tel Aviv on the president's first foreign trip.

I thought the Saudis would take their time before making a decision. To my delight, however, the very next day Saudi foreign minister Faisal bin Farhan Al Saud announced that all Israeli flights going to the United Arab Emirates would be allowed to traverse Saudi airspace. The announcement marked another diplomatic triumph. It caught everyone by surprise, including me.

I was planning to visit Qatar and its emir, Tamim bin Hamad Al Thani, the next day. So before my meeting with MBS ended, I asked him whether he had made any progress with Qatar, the country on Saudi Arabia's eastern flank that had been the focus of a Saudi-led blockade for more than three years.

MBS said that they had prepared a proposal that they were going to send to the Qataris through Kuwait, which was serving as their intermediary.[59]

"I'm flying to Qatar tomorrow morning to see Tamim," I said. "I can bring the proposal with me and save you the postage stamp."

MBS said I'd have it before I boarded the plane.

As we concluded, I asked one more question: "If he wants to talk, would you be open to doing a call with him?" Without hesitation, MBS indicated that he would be. He liked Tamim personally, and he wanted to resolve the issues and move forward.

The next day, before we boarded our plane, Faisal greeted us on the tarmac with a box of Saudi dates as a gift and handed me an envelope with Saudi's proposal for Qatar.

Upon arriving in Qatar, Avi, Brian Hook, General Correa, and I headed to the palace for a meeting with Tamim and several of his trusted advisers. Our relationship had started on difficult terms because of Tillerson's inaccurate suggestion that I was responsible for the Gulf rift. It improved steadily through the years as we met and engaged in strategic dialogues. When I had shared the Peace to Prosperity economic plan with him, Tamim had predicted that we were underselling the plan: If we achieved peace in the region, the explosion in economic activity would be even bigger than we imagined.

As we sat in his royal office on September 2, I updated Tamim on the positive developments between Israel and the UAE, and I asked him if he would consider joining the Abraham Accords.

Tamim expressed openness to doing so at the right time, citing the many areas where Qatar was cooperating constructively with Israel, including helping them to mediate their issues with Hamas. But he wanted to solve the blockade with Saudi Arabia first.

This was the perfect opening.

"I have a proposal from MBS," I said. "I went through it with my team, and while it's not perfect, I think it's a good start."

If they resolved the dispute, the paper wouldn't matter, Tamim said. What mattered was Saudi Arabia's intent. Qatar had invested a great deal of time in trying to reach a compromise, but never seemed to make progress. He asked if I thought the Saudis were truly ready to resolve the conflict.

"Not everyone," I said. "But MBS is ready. You have to trust me when I say that I believe he genuinely wants to resolve it."

I handed Tamim the document, and he started reading through it.

After we discussed some of the outstanding issues, I asked if he would be open to having a quick call with MBS to hear directly from the Saudi crown prince on the sincerity of this offer. Tamim was hesitant, reminding me that the last call between them was pleasant, but then became problematic when both countries published conflicting summaries of the call, which only heightened tensions. He added that even if they did have a good call, they would need to fix the broken process and come up with a new mechanism to reach a breakthrough.

I proposed setting up a channel of communication between his skillful foreign minister, Sheikh Mohammed bin Abdulrahman, and Saudi deputy defense minister Khalid bin Salman. I could work with the foreign minister of Kuwait, Dr. Ahmed, to mediate the discussions.

Tamim asked if the Saudis would agree to my proposal.

Putting aside all of the formality of being in the palace of an emir, I took a page out of my old commercial deal-making playbook: "Let me ask him. Do you have a conference room I can use to call MBS?"

His aide showed me to the conference room next door. Soon I was on the phone with MBS, briefing him on my discussion with Tamim. MBS assured me that if Tamim was sincere in his desire to resolve the dispute, he would meet him more than halfway.

"Would you be open to setting up a channel between Prince Khalid and Sheikh Mohammed, which I would personally mediate, to try to resolve the outstanding issues in the documents?" I asked.

MBS agreed immediately.

"Do you mind holding on for one minute?" I handed the phone to Avi, with whom MBS always enjoyed conversing, and walked down the short hallway to Tamim's office.

"MBS has agreed to the channel as a way to resolve the open issues," I said. "I have him on the phone. I think it would help build confidence for you to hear from each other. Would you be willing to talk to him?"

Tamim maintained his poker face while he weighed the consequences. Then he consented.

I went back to the conference room. "Hold one second. I'm going to put you on with Tamim. He's ready to talk."

I walked into Tamim's office and put the phone on speaker. Tamim greeted MBS in Arabic, and the two leaders spoke for about ten minutes as everyone in the room listened. Not fluent in Arabic, I stood by nervously, trying to read the facial expressions of Tamim and his advisers, since I had no idea what they were saying. When Tamim hung up, he paused for a moment to look at the phone and then handed it over to me. The room was silent.

I broke the silence and asked, "Was that a good call or bad call?" Everyone erupted in nervous laughter.

Tamim thanked me and said that it was a great call and that he was open to resuming talks if they could make progress. He expressed sadness that the feud had led to so much bitterness between the citizens of their countries.

We discussed next steps, and Tamim gave me his full support for us to try to resolve the conflict. It felt like we were on the cusp of another breakthrough.

Resolving the rift was critical for advancing American interests in the region. The blockade had forced flights in and out of Qatar to traverse Iranian airspace, which not only enriched Iran, but also endangered travelers, including Americans, and hampered economic partnerships in the region. Perhaps most importantly of all, ending the rift would create an opening for more countries to join the Abraham Accords. As long as Saudi Arabia, Qatar, and the rest of the Gulf Cooperation Council countries were divided, they would be less likely to create formal ties with Israel. But if these countries were united, they would be free to bridge relations with Israel.

On the drive back to the hotel, Brian Hook asked me a question: "Did you plan to do that?"

"No," I said. "But I read his reactions and decided to try. The worst thing that could have happened is that he would have said no."

"I've been around Washington for twenty years," Hook said. "I've worked with the best diplomats. No diplomat would have ever done that. You just broke every rule of diplomacy, and it worked."

54

The Abraham Accords

The first flight between Israel and the UAE was the beginning of a new and mutual appreciation between the Israelis and the Emiratis. Shortly after, Bibi confirmed his attendance at a White House event we had planned for September 15, and MBZ committed to send his brother, foreign minister Abdullah bin Zayed, as his representative.

With these key details locked in, we focused on finalizing the normalization agreement with Bahrain. I discreetly previewed Bahrain's interest to the Israelis and Emiratis, and both countries were eager to include the affluent Gulf country in the September signing. Adding a second Arab nation would serve as a force multiplier in shifting the regional paradigm.

Avi, General Correa, and the rest of our team worked tirelessly over the next several weeks to finalize the details of the agreement, which included sharing foreign policy resources, expanding trade, helping with oil and gas development, and deepening our already strong military relationship.

Through his thirty-year career in the US Army, Correa had earned a reputation for being a trustworthy and fair operator in the Middle East. He had been stationed in Abu Dhabi in 2017 to serve as a defense attaché to the US embassy, but his strong relationships with the Emiratis quickly created resentment among career State Department officials, and he was forced to leave the post. As we negotiated the deals with the UAE and

Bahrain, Correa's experience, perspective, and trust in the region proved invaluable in advancing America's interests.

On September 10, the Bahraini government approved our proposal. The next day, on September 11, Trump commenced his second phone call to make peace—this time with Bibi and King Hamad. It was not lost on the team that we were marking this historic breakthrough for peace on the anniversary of the September 11 terrorist attacks.

I directed my team to draft the Abraham Accords Declaration, an overarching document that included all three parties and the United States.[60] I envisioned a framework that wouldn't interfere with the specific and sensitive material in the individual country agreements. Its broader principles would allow for additional signatories to join later, as we continued to change the paradigm across the Middle East.

While the Abraham Accords Declaration was the shortest of the three documents, it was by far the most delicate to write. We worked to avoid areas of discord and to make it acceptable to any supporter of peace between Jews, Muslims, and Christians. After I outlined what I wanted it to say, Avi, Brian Hook, and Scott Leith drafted and negotiated every word, uniting the three parties around a meaningful and lasting document. They ironed out the final details just hours before the signing, and then sent the text to the translators, who used our original English version to write final documents in Arabic and Hebrew.

The day before the ceremony, I received a call from Richard Moore, the longtime British diplomat, who had recently been appointed head of the MI6 intelligence bureau. He had worked with my team as a valuable partner since 2017, when he had joined our meetings with Boris Johnson, at that time the foreign secretary. He congratulated me on the Abraham Accords and expressed astonishment that we had kept both deals a secret until we announced them.

"It's the Kushner doctrine, nothing leaks," he said.

That night I called the president to discuss the plan for the signing. Between campaign events, the ongoing COVID-19 response, and other responsibilities, his focus was divided in multiple ways. I wanted to make sure he was ready, but Trump dispelled any concern.

"Do you have a great speech for me?" he asked. "I want it to be great." I took copious notes as he walked me through several key points he wanted to address.

Trump brimmed with energy during his one-on-one meetings with the visiting leaders the morning of Tuesday, September 15. Abdullah bin Zayed, the Emirati foreign minister, impressed Trump with his eloquent, heartfelt remarks about the significance of the day. In a meeting with the Bahraini foreign minister, Abdullatif bin Rashid al-Zayani, Trump joked that the best wristwatch he ever owned was a gift he had received decades earlier from the emir of Bahrain.

"This watch was beautiful, and it worked for twenty-five years," he said. "Some old watches just stop ticking after a while—like Joe Biden."

When Trump met with Bibi, he whipped out his signature gift—an oversize bronze "key to the White House" in a wooden box carved with the presidential seal. Trump had designed the key himself to give to special guests.

"This is the first key I'm giving to anyone," he said. "Even when I'm not president anymore, you can walk up to the front gate of the White House and present it, and they will let you in."

Avi and I tried to keep from laughing. We had heard the line before, and Trump had delivered it a little too earnestly. Yet Bibi beamed. He and Trump were proud of what they had achieved.

Just before 1:00 p.m., the four leaders gathered in the Oval Office and then walked over to the Blue Room. Waiting for them on the South Lawn were over seven hundred guests, including foreign dignitaries, cabinet members, lawmakers, business leaders, and foreign policy experts. Secretary Pompeo, Avi, Robert O'Brien, David Friedman, and the rest of my team took seats in the front row. Most importantly for me, Ivanka was there, along with my parents and two sisters, Dara and Nikki, who came to help me celebrate the milestone.

"We're here this afternoon to change the course of history," Trump declared from the South Portico. "After decades of division and conflict, we mark the dawn of a new Middle East. Thanks to the great courage of the leaders of these three countries, we take a major stride toward a future in which people of all faiths and backgrounds live together in peace and prosperity. In a few moments, these visionary leaders will sign the first two peace deals between Israel and [an] Arab state in more than a quarter century. In Israel's entire history, there have previously been only two such agreements. Now we have achieved two in a single month, and there are more to follow."

Bibi spoke next. Unlike those at his previous White House event, his remarks showed true statesmanship: "For thousands of years, the Jewish people have prayed for peace. For decades, the Jewish state has prayed for peace. And this is why, today, we're filled with such profound gratitude."

When Bibi concluded, he handed the microphone to Emirati foreign minister Abdullah bin Zayed, who spoke in Arabic. "In our faith, we say 'O God, you are peace, and from you comes peace,'" he said. "The search for peace is an innate principle, yet principles are effectively realized when they are transformed into action. Today, we are ready—we are already witnessing a change in the heart of the Middle East, a change that will send hope around the world."

Bahraini foreign minister al-Zayani anchored the remarks with a forward-looking expression: "What was only dreamed of a few years ago is now achievable, and we can see before us a golden opportunity for peace, security, and prosperity for our region."

"Beautiful," Trump said, as he motioned for the leaders to follow him down the stairs to a platform on the South Lawn, where we had arranged a signing table.

The four leaders began to execute the documents that we had prepared for them. We provided each leader with copies of their signing documents in Arabic, Hebrew, and English. In the flurry of activity to prepare for the event, no one had clearly marked the signature lines so that the leaders would know where to sign on the documents that were not in their native language. The leaders looked for their aides, to no avail. In the lead-up to the event, everyone was angling to be in the historic photos, so I designed the event to keep all staffers away from the leaders and out of camera shot. The leaders deserved to be the focal point of the event. They were the ones who had created the conditions—and taken the risks—to make peace. Soon the leaders began helping each other figure out where to sign, and photographers captured their interactions with a series of memorable images that highlighted their distinct personalities. As Trump brought the ceremony to a conclusion, we all stood and cheered.

At the celebratory lunch that followed, Ivanka and I sat with the president and the other leaders. I was exhausted but profoundly happy. The magnitude of the moment and what it represented for the world finally started to sink in. After a long and hard journey, we had accomplished the unthinkable: we had made peace in the Middle East.

In the State Dining Room, I tried to soak up the moment. I watched Bibi share a meal and interact gregariously with the foreign ministers of Bahrain and the UAE. These former adversaries were beginning to form what I prayed would be a deep and lasting friendship. I hoped and believed that this day marked the beginning of an enduring change that would improve millions of lives.

Later that day, as the president prepared to depart the White House for an event in Philadelphia, he spoke with the press corps. It was one of his classic "chopper talks," with the engine of Marine One thundering in the background. He wanted to talk about what we had accomplished— and he surprised me with a comment that forced me to start thinking ahead.

"We have many other countries going to be joining us, and they're going to be joining us soon," he said. "We'll have, I think, seven or eight or nine."

This was classic Trump: even in his finest moments of achievement, he was raising the bar and pushing for more.

55

From Walter Reed to Election Night

I vanka's voice woke me up around 2:00 a.m. on Friday, October 2.

"Dad and Melania have COVID," she said.

"I was just with him a few hours ago, that can't be," I said, shaking off my sleepiness as I reached for my phone. When the screen lit up, the first notification I saw was the president's tweet from 12:54 a.m.: "Tonight, @FLOTUS and I tested positive for COVID-19. We will begin our quarantine and recovery process immediately. We will get through this TOGETHER!"

We were both shocked and worried by the news. On a personal level, I was concerned for my father-in-law, who contracted a virus that had proven to be fatal for many people over seventy. Ivanka and I love and admire him, and we were deeply worried about his wellbeing. We said a quick prayer asking for God to keep him safe and healthy for many years to come. On a professional level, I wondered what his diagnosis would mean for his presidency and for our country.

Around 10:30 a.m., I met with White House chief of staff Mark Meadows. His bleary eyes revealed his exhaustion. He had stayed with the president all night.

"I'm really nervous," he said, adding that Dr. Sean Conley recommended that the president go to Walter Reed National Military Medical Center as a precautionary measure.

Meadows and I suited up in full PPE—surgical gowns, masks, gloves, and goggles—and went to the residence. We looked like actors in a movie about a biohazard crisis. When we arrived in the president's bedroom, Trump was sitting up and reviewing documents. After asking how he was feeling, I revealed the purpose of our visit: "We strongly recommend that you go to Walter Reed."

"I'm already feeling better," Trump said. "Just give me some time to rest up, and then we can make a determination later. I don't like how it looks to our adversaries to have the leader of America in the hospital."

"I'm advising you as both a family member and a senior member of your staff," I said. "Even if you don't want to go for you, this is about the office of the presidency. You have an obligation to go to the place where they can give you the very best care and monitor you perfectly. Even if the care is one percent better, it's worth it."

Trump didn't think it was necessary, but he agreed to go. Before exiting his room, I made one more request: "I know this is the last thing you want to do, but people are really nervous. They want to know that you're okay. Would you be willing to shoot a quick video right before we depart, letting people know that you are okay and thanking them for the well wishes? It will go a long way."

Trump agreed, and he recorded it in one take before walking out to the helicopter on the South Lawn. "I want to thank everybody for the tremendous support. I'm going to Walter Reed Hospital," he said, wearing his usual suit and tie. "I think I'm doing very well, but we're going to make sure that things work out. The First Lady is doing very well. So thank you very much. I appreciate it. I will never forget it. Thank you."

The next morning, a Saturday, I went directly to Walter Reed. Upon arriving, a military doctor gave me protective gear and escorted me to the presidential suite, which included a full medical unit, a conference room, a dining room, kitchen, several sitting rooms, and additional space for staff. When I walked in, the president was already up, dressed in khakis and a button-down dress shirt, and working at a table. Mark Meadows and Dan Scavino were also in the room in full protective gear. Trump was feeling strong and wanted to discuss his campaign.

Trump couldn't have come down with COVID-19 at a worse time. Just three days earlier, on September 29, he had traveled to Cleveland, Ohio, for the first presidential debate of the 2020 general election cycle. The debate occurred the day after Yom Kippur, the holiest day on the Jewish calendar, so between that and Rosh Hashanah the week before, I had missed the debate prep sessions. Yet I had been in the room when the president spoke with American historian and conservative commentator Victor Davis Hanson.

"It's going to be a lot tougher to debate against a guy like Joe than you think," said Hanson. "He's potentially senile so he will say he's always been for a position—and you will cut him off, point out that it's not true, and he will say 'That's not how I remember it.' And that will be a true statement because his memory is gone. You're going to have to work extra hard to not come across as committing elder abuse."

Hanson's warning was prophetic. Trump viewed the debate as a rare chance to draw a stark contrast with his Democratic rival. He was also frustrated that the media had refused to ask Biden any tough questions or scrutinize his controversial policy positions. He came out swinging hard and put Biden on the ropes when Biden refused to say whether he supported defunding the police, and again when Trump asked why Biden's son Hunter received tens of millions of dollars from Chinese and Russian sources. Both times, however, debate moderator Chris Wallace cut off the conversation before Trump could land a knockout blow. It was like watching a biased referee unfairly separate boxers in the middle of a round.

As the president recovered at Walter Reed, we all recognized that the campaign would have to wait until Trump was both physically strong and medically cleared to return to the trail. In the meantime, he spoke directly to Americans through social media to update them on his recovery. In his first video from the hospital, Trump said, "I came here, wasn't feeling so well. I feel much better now." He also explained why he took the risk of continuing to attend events during the pandemic: "This is America. This is the United States. This is the greatest country in the world. This is the most powerful country in the world. I can't be locked up in a room upstairs totally safe. . . . As a leader, you have to confront problems." I admired my father-in-law's spirit and determination. I knew he was feeling better when he requested one of his favorite meals: a Mc-Donald's Big Mac, Filet-o-Fish, fries, and a vanilla shake.

Meanwhile, America's best scientists were on the cusp of delivering a vaccine. Both Pfizer and Moderna were nearing the completion of their third and final phase of clinical trials. Albert Bourla, the CEO of Pfizer, went on both the *Today Show* and *Face the Nation* to announce that the vaccine would be ready by the end of October. The president's investment in Operation Warp Speed was paying off, and we were on course to have a vaccine even sooner than our ambitious timelines projected. This was excellent news for America and the world, but it was so unexpected that while it should have been welcomed as good news, it prompted top Democrats to accuse us of rushing the process. Many claimed that they would be reluctant to take a vaccine approved by the FDA under the Trump administration. Among them was Biden: "I trust vaccines. I trust scientists, but I don't trust Donald Trump, and at this moment, the American people can't, either."

Sadly, the Democratic pressure campaign worked. Just as Pfizer prepared to announce the completion of its phase three trial, the FDA changed the guidelines for approval. On October 6, the FDA regulators modified the safety standards they had released in June, forcing companies to wait an additional eighteen days before seeking FDA approval of their vaccines upon completion of clinical trials. This last-minute revision meant that Pfizer could not submit its application for approval until after the election. When Adam Boehler and Brad Smith asked FDA commissioner Stephen Hahn about the decision, he seemed to suggest that the FDA made the change to avoid the perception that the vaccine had been approved for political reasons.

The FDA's decision delayed the vaccine approval by at least two weeks, just as a new wave of cases was slamming the country. During this period, the United States averaged thousands of coronavirus deaths per day, and many Americans lost an opportunity to receive a vaccine that was more than 90 percent effective.

By October 10, Trump's symptoms were nearly gone. The doctors confirmed that he was no longer contagious and cleared him to resume public events. That same week, the Commission on Presidential Debates announced that the next debate on the schedule would be virtual "in order to protect the health and safety of all involved." This decision made no sense, and Trump felt it was politically motivated: fewer Americans would watch a virtual debate, which played into Biden's strategy of running a low-profile campaign that avoided talking about what he stood for. The president refused to participate, and his campaign released a statement proposing that the next two debates be moved back a week so that both could still be held in person, as planned. Biden seized the opening to pull out of the second debate altogether, and the media applauded him for doing so. On the same day that second debate was originally scheduled to occur, both candidates safely participated in town halls, proving that there was no actual risk. Former Republican nominee for president and longtime Kansas senator Bob Dole called me, sharp as ever at ninety-seven years old. He thought the decision revealed an anti-Trump bias among the Republican members of the debate commission. I asked if he would put out a statement, which he later tweeted: "The Commission on Presidential Debates is supposedly bipartisan w/ an equal number of Rs and Ds. I know all of the Republicans and most are friends of mine. I am concerned that none of them support @realDonaldTrump. A biased Debate Commission is unfair."

The final debate took place in Nashville on October 22. Trump can masterfully adjust when the moment calls for it. He knew what he needed to do, and he nailed it: he answered questions with substance, responded with good humor, and allowed Biden to ramble before forcefully pushing back on false claims in exactly the right places.

In the final three weeks of the campaign, Trump hit his stride, holding rally after rally in battleground states. Just as he had done in 2016, he stayed on message, drew big crowds, and gave everything he had. He spoke at three, four, and even five events a day. Campaign manager Bill Stepien, RNC chairwoman Ronna McDaniel, and RNC political director Chris Carr were directing our get-out-the-vote operation, which was one of the best in the history of presidential campaigns, and it was clear that Trump's voters were energized. Our internal polling showed Trump gaining momentum by the day—and even surpassing Biden. The public polling, however, forecasted a Biden victory. The RealClearPolitics unweighted national average showed Biden up 7.8 points, and FiveThirty-Eight predicted that Biden had an 89 percent chance of winning.

We knew from 2016 that public polling heavily favored the Democratic candidate, causing misperceptions about the true state of the race. Yet in 2020, we contended with additional challenges. Many Democratic states had altered their voting rules during the pandemic. This introduced two new variables that made predictions even more difficult: the amount of early mail-in voting and the level of voter turnout on Election Day. We knew that an unprecedented number of people were casting ballots early, and that many of these voters were Democrats. What we didn't know was whether it was too late to turn the tide.

* * *

On the morning of Election Day—November 3, 2020—I knew the results would be tight. The energy, enthusiasm, and momentum we had felt in the closing sequence of rallies—seventeen in eight states over the final four days—convinced me that Trump had a shot to pull off another improbable, come-from-behind victory. That night I tried to temper my enthusiasm as I walked into the White House's Map Room, which the campaign had converted into a makeshift war room. Flatscreen televisions lined the walls. Computer monitors pumped out data from precincts in swing states. Bill Stepien, Mark Meadows, Justin Clark, Jason Miller, Gary Coby, and the campaign's data whiz Matt Oczkowski analyzed the latest results. Ivanka, Don Jr., Eric, Lara, Tiffany, and Kimberly Guilfoyle joined us to watch as the results came in. Upstairs on the first floor of the Executive Residence, hundreds of the president's closest friends, advisers, and campaign donors followed the coverage and sampled from a generous spread of food.

At 11:04 p.m., Fox News anchor Bret Baier flashed on-screen with a breaking news alert: "The Fox News decision desk can now project that President Donald Trump will win the state of Florida, twenty-nine electoral votes, and he will win it convincingly." Our best-case scenario had unfolded in the Sunshine State, with strong support from seniors and Hispanics, and we immediately interpreted it as a favorable sign for the rest of the country. Things were also looking good in Ohio, another state we had to win. Since 1964, every presidential candidate who had won Ohio had won the election.

Then, at 11:21 p.m., Fox News interrupted a panel discussion with an update: with just 73 percent of the votes counted in Arizona, the network called the state for Joe Biden. Republicans had carried the state in every presidential election since 1996. Trump had won it by 3.5 percent in 2016. We knew it would be harder to win in 2020, but we believed Arizona would remain red.

"That is a big get for the Biden campaign," Baier said. "Biden picking up Arizona changes the math."

The shocking projection brought our momentum to a screeching halt. It instantly changed the mood among our campaign's leaders, who were scrambling to understand the network's methodology. Many felt that the early call would embolden people who were looking to play dirty with the vote counting in the outstanding swing states.

Up to that moment, Trump was performing even better than our models had forecast in several key states that immediately reported the results. Voter turnout was far higher than predicted, showing that our expansive ground operation had worked. We had mobilized our base, which was always an important factor in elections. But losing Arizona would drastically narrow our path to victory.

I dialed Rupert Murdoch and asked why Fox News had made the Arizona call before hundreds of thousands of votes were tallied. Rupert said he would look into the issue, and minutes later he called back. "Sorry, Jared, there is nothing I can do," he said. "The Fox News data authority says the numbers are ironclad—he says it won't be close."

Our campaign had a different view: based on the remaining votes to be counted, we believed that Arizona's outstanding votes would favor

Trump and that it would be razor close. After Arizona, however, negative news came in from other swing states. Unlike in 2016, when it was clear how many outstanding votes each precinct needed to count and report within hours of the polls closing, 2020 was full of electoral anomalies. At 1:40 a.m., with 93 percent of the vote counted, Trump was hanging on by a thread in Georgia with 50.7 percent, down from his lead of 12.7 percentage points earlier in the night.

Trump addressed his guests in the East Room of the White House at 2:20 a.m.: "This is a fraud on the American public," he said. "This is an embarrassment to our country. We were getting ready to win this election. Frankly, we did win this election. So our goal now is to ensure the integrity for the good of this nation."

My phone rang a few minutes later. It was Karl Rove, the man who in 2000 had helped George W. Bush win the closest presidential election in US history.

"The president's rhetoric is all wrong," he said. "He's going to win. Statistically, there's no way the Democrats can catch up with you now."

"Call the president and tell him that," I said.

The next morning, I went over to the campaign's Arlington headquarters. Stepien and communications adviser Jason Miller walked me through the data. They believed Arizona was a true toss-up, given the number of outstanding ballots from likely Republican voters. Georgia would be close, but it looked like we were in a position to hold the state. Trump was still up by roughly 600,000 votes in Pennsylvania, but we kept getting different official numbers for how many votes were left to be counted. If those three states went our way, Trump would surpass the 270 electoral votes he needed to win reelection. Yet no one could predict precisely how the outstanding votes would break.

The results remained inconclusive for days, but discouraging numbers began to trickle in. The day after the election, the Associated Press called both Michigan and Wisconsin for Biden. In Arizona, Trump was inching forward as officials continued to tally votes, but he still trailed. By Friday, Georgia was still too close to call. With more than eight thousand votes remaining to be received, Biden led by about four thousand votes. On

Saturday morning, the AP declared Biden the victor in Pennsylvania, giving him more than enough electoral votes to win the presidency if the results held in the other states.

Ultimately, after more than nearly 158.4 million votes were tallied, the election came down to fewer than 42,918 votes in three states—20,682 in Wisconsin, 10,457 in Arizona, and 11,779 in Georgia.[62]

Trump earned more than seventy-four million votes—more votes than any other incumbent president in American history. He did so in the midst of the COVID-19 pandemic, arguably the greatest global crisis since World War II. Despite this challenge, Trump made incredible inroads with African American and Hispanic voters. He outperformed the predictions of nearly every major pollster. As Democratic political consultant David Shor wrote in his autopsy of the 2020 election, "When the polls turned out to be wrong—and Trump turned out to be much stronger than predicted—a lot of people concluded that turnout models must have been off. . . . Trump didn't exceed expectations by inspiring higher-than-anticipated Republican turnout. He exceeded them mostly through persuasion. A lot of voters changed their minds between 2016 and 2020."

In the days that followed the election, I participated in several discussions about how to investigate the many incoming allegations of election fraud. I was still trying to develop a comprehension of the issues when Rudy Giuliani asked the president to put him in charge of the effort. The president wasn't ready to make a decision at first, but Giuliani persisted. Citing his experience at the Justice Department, he claimed, "I know how to run these kinds of investigations. I will prove the fraud if you put me in charge."

Two days after the election, Mark Meadows tested positive for COVID. I had been in close contact with Meadows for an extended period of time and started to feel under the weather. When I began to lose my sense of taste, Ivanka and I quarantined in New Jersey. By the time I returned to the White House from my quarantine, the president had appointed Giuliani and his team of lawyers to lead the effort. I discussed the situation with Eric Herschmann, a talented trial lawyer who had left behind his partnership at a major law firm in 2020 to join the White House staff. I told him to keep an eye on the developments while I focused on my Middle East peace efforts and Operation Warp Speed. Like millions of Americans, I was disappointed by the outcome of the election. Yet I was proud of all that we had achieved over the past four years. Now, with precious time left on the clock, I was determined to make the best use of every remaining minute.

56

Landing Planes on an Aircraft Carrier

I f my time in Washington had taught me anything, it was that challenging circumstances can lead to unforeseen opportunities. I never would have guessed that the president's contentious relationship with big tech companies would pave the way to another peace agreement, but that's precisely what happened.

During the lame-duck session that followed the November 3 election, Congress prepared to pass an annual bill to authorize funding for the military. The National Defense Authorization Act (NDAA) was a sprawling $700 billion package. Passing it was usually an uncontroversial and bipartisan affair. This had been the case for the first three years of Trump's term, as Congress responded favorably to Trump's requests to rebuild the military and establish a new US Space Force, among other priorities. As the 2020 version of the bill moved closer to his desk, however, Trump decided to use it as leverage to fight for a change that he believed would safeguard our democracy.

Ever since Twitter and Facebook had taken the unprecedented step to censor conservatives, including the president, over the summer, Trump had threatened to take action against technology companies for violating the free speech of Americans. He believed that social media platforms played a central role in facilitating public discourse, and that they abused their power when they censored people who had done nothing more than espouse conservative or nonconformist political ideas. Yet a law passed back when people still used dial-up modems and floppy disks shielded these massive corporations from lawsuits. Trump questioned the law, section 230 of the Communications Decency Act, believing that social media censorship posed a "serious threat to our national security and election integrity." He insisted on including a provision to terminate section 230 in the NDAA.

When I asked Mark Meadows whether Congress would modify the law, he said it was unlikely: "Inhofe isn't budging." Jim Inhofe, the Republican chairman of the Senate Armed Services Committee, had enormous power to determine what provisions would become a part of the NDAA.

"Just do me a little favor," I said to Meadows. "Make sure the president knows that Inhofe is holding this up, and he's the reason we don't have a peace deal with Morocco."

Six months earlier, the president had discussed the Western Sahara issue with Inhofe, who had implored him not to change US policy. For the Moroccans, their generations-old claim on the Western Sahara was a matter of territorial sovereignty and national security. If Morocco obtained US recognition of the territory, it would be much more plausible for the Arab country to reach beyond its borders and normalize relations with Israel. Inhofe was an instrumental ally in the Senate and worked with our administration on many national security priorities. Yet he had long held the position that the United States should support the Polisario Front's desire for a referendum on self-determination in the Western Sahara. Although Trump appreciated what a breakthrough could mean for Israeli-Arab relations, he had previously told Inhofe that he would not move forward with the recognition. Now that the senator was blocking the section 230 provision, however, Trump was less concerned about the senator's opposition. This created an unexpected opening for us to revisit the issue with the president.

I asked Avi to call the foreign minister of Morocco, Nasser Bourita, to see if his country would still honor the terms of the peace deal we had discussed six months prior. A skilled diplomat, Bourita possessed a deep reservoir of knowledge on the issues, which he paired with his vast intellect and a creative mind. He always gave us honest feedback. Avi told Bourita that it was a long shot, but we wanted to know if the Moroccans were ready and willing to move quickly. After checking with the king, Bourita confirmed that they were on board.

We had little margin for error. One misstep or poorly timed comment would sink an eleventh-hour deal. If word leaked that we were on the brink of an agreement, the enemies of normalization might rally and defeat our initiative. To mitigate the risk—and avoid getting ahead of the president—we didn't tell any Israeli officials about the potential deal. Unlike the agreement with the UAE, Israel wouldn't need to make any concessions. All Bibi would have to do is accept the offer, which was clearly in Israel's national interest. One million Jews are of Moroccan descent, and normalization would make it easier for Israeli families to reconnect with relatives and visit ancestral sites.

In early December, the NDAA negotiations dragged on, but without resolution on the section 230 issue. As the president's chief negotiator on the bill, Meadows urged Inhofe to include Trump's request. Each morning, Avi tiptoed into Meadows's office to see if there was an update. It got to the point where anytime Meadows crossed paths with Avi, he would chuckle and say: "I don't have an update yet, but I'll let you know as soon as I do!"

When the House and Senate negotiators released their final version of the NDAA on December 3, it did not include the section 230 provision, which deeply disappointed the president. Shortly thereafter, Meadows brought Trump a presidential proclamation we had drafted to recognize Morocco's sovereignty over the Western Sahara. After confirming that the details were in line with our previous discussions, Trump signed the document, and we set up a call for him to speak with King Mohammed on December 10, the following day.

That night, at the annual White House Hanukkah reception, Avi pulled Israeli ambassador Ron Dermer aside and gave him a heads-up.

"We have another surprise," he started. "Tomorrow, the president will recognize Morocco's sovereignty over the Western Sahara, and the kingdom will announce its readiness to normalize with Israel."

Amazed, Dermer commented that getting a deal like this done in the lame-duck period might have been even more impressive than the previous agreements.

The next morning, after Trump spoke with King Mohammed, he announced Morocco's decision to fully normalize with Israel in a series of tweets:

"Today, I signed a proclamation recognizing Moroccan sovereignty over the Western Sahara. Morocco's serious, credible, and realistic autonomy proposal is the ONLY basis for a just and lasting solution for enduring peace and prosperity!"

"Another HISTORIC breakthrough today! Our two GREAT friends Israel and the Kingdom of Morocco have agreed to full diplomatic relations—a massive breakthrough for peace in the Middle East!"

"Morocco recognized the United States in 1777. It is thus fitting we recognize their sovereignty over the Western Sahara."

The news reverberated throughout the Middle East. "This step, a sovereign move, contributes to strengthening our common quest for stability, prosperity, and just and lasting peace in the region," tweeted MBZ of the UAE. President Abdel Fattah el-Sisi of Egypt praised the announcement as an "important step towards more stability and regional cooperation."

I almost couldn't believe that we had secured another peace agreement. Getting this deal done was like trying to land a plane on an aircraft carrier in the middle of a storm: we had to navigate through many uncontrollable variables, fly at just the right speed, and hope that we'd hit the tarmac at exactly the right moment. Almost miraculously, we managed to make the runway.

But several other planes were still in the air, and we needed to land them in rapid succession before our time expired. In November, after we had decided to sell the F-35 stealth fighter jet to the UAE, a problem surfaced: Republican senator Rand Paul and Democratic senators Bob Menendez and Chris Murphy introduced legislation to block the arms sale. Paul had a history of objecting to US foreign military sales, but Menendez and Murphy had a different reason. They claimed that we had committed a process foul by not informally clearing the deal with the foreign relations committee before announcing it.

While the president would veto any congressional resolution blocking the sale—and the Senate would not have the two-thirds majority needed to override a veto—the public display of opposition would embarrass the Emiratis and prompt concerns about their relationship with Democratic leaders just before Biden assumed the presidency. It was an unwanted development, and one that could even jeopardize the Abraham Accords in their infancy. Avi and I worked with Pompeo, UAE ambassador Yousef Al Otaiba, and Israeli ambassador Ron Dermer to call nearly every senator, explain the importance of the military sale, and answer their questions. Dermer told reporters that Israel was "very comfortable" with the sale and called the UAE an "ally in confronting Iran." Ultimately, most senators decided that the sale would tilt the regional balance of power against Iran without compromising Israel's security. They also understood that in the absence of our deal, the UAE would likely buy weapons from China or Russia. It was clearly in our interests to keep the Emiratis in America's orbit.

After intense engagement, the Senate rejected the legislation. With the exception of Paul, every Republican present voted with us. After the vote, Yousef called to express his thanks and noted that Ambassador Dermer was very talented, and that working with him was a much different experience when he was an ally.

Around the same time, another outstanding issue emerged. Back in August, a government minister in Sudan had said in a tweet that his country should normalize relations with Israel. Unfortunately, the minister had deleted his tweet and was fired.[63]

We saw the incident as encouraging—or at least worthy of pursuit. Secretary Pompeo made a special trip to Sudan, a predominately Arab country in North Africa. Meeting with leaders from Sudan's governing factions, he confirmed the possibility that Sudan would be open to joining the Abraham Accords. First, however, the Sudanese wanted to resolve several issues. Their most urgent request was to be removed from America's State Sponsors of Terrorism list. Being on that list barred Sudan from receiving aid from the United States and put it in a category with bad actors such as Iran, North Korea, and Syria. Sudan had earned its place on the list for supporting Hamas and for providing a safe haven for Osama bin Laden and his fellow al-Qaeda terrorists, who had operated from within Sudan to coordinate the deadly bombings of US embassies in Kenya and Tanzania in 1998 and the USS *Cole* in 2000. In 2019, however, the country overthrew its brutal dictator Omar al-Bashir, who had ruled for more than three decades and had committed atrocities against the Sudanese people. A transitional government was inching toward democracy. In exchange for removal from the list, Sudan agreed to pay a $335 million court judgment for the victims of the 1998 and 2000 bombings. It also agreed to normalize with Israel.

We were under no illusions about the tumultuous state of affairs in Sudan, but we saw the country's interest as a way for the United States to give it a chance to chart a new path. Too often in diplomacy, we allow sins from the past to prevent opportunities for change. Getting Sudan to join the Abraham Accords also carried symbolic value. In 1967, following Israel's victory in the Six-Day War, the Arab League convened in Sudan's capital city and passed its infamous Khartoum Resolution. This hateful document had proclaimed "The Three Nos": no peace with Israel, no recognition of Israel, and no negotiations with Israel. Now Sudan was finally willing to redeem its past.

After an intense diplomatic effort, the United States, Israel, and Sudan released a joint statement in October: "The leaders agreed to the normalization of relations between Sudan and Israel and to end the state of belligerence between their nations." The statement noted that the two countries would begin economic relations and would meet in the coming weeks to negotiate potential areas of cooperation.

In December, however, another issue arose. Sudan wanted the United States to grant their country sovereign immunity, indemnifying its new leadership from legal liability for actions committed under the former dictator Omar al-Bashir. For this, we needed legislative approval. Congress granted sovereign immunity in the year-end spending bill, which Trump signed on December 27. This sealed Sudan's participation in the Abraham Accords and continued the positive shift in the Middle East.[64]

Diplomacy is a fragile business. Everything done can suddenly be undone. The three issues we tackled after the election—the Western Sahara recognition, the F-35 sale to the UAE, and sovereign immunity for Sudan—may have seemed like relatively minor sticking points. Yet peace is not a piece of parchment. It's a process that requires constant attention and ongoing trust, which is most fragile in the beginning. This was the moment to prove that the United States was a reliable partner, and the Abraham Accords were an ironclad commitment.

57

Pardons, Pfizer, and Peace

A s my official duties started to wind down, Ivanka and I prepared for a personal transition. After four years of being on the clock every day, I was excited to make some adjustments to my life. At the top of my list was being more present for my three kids. I also wanted to go back to fully observing the Sabbath on Friday evenings at sundown—a weekly practice Ivanka and I had cherished before entering government service.

On Friday, December 18, I was hoping to make it home in time to light the Shabbat candles with my children—but the day was packed with activity. At ten o'clock that morning, I joined a weekly conference call with the Operation Warp Speed board. The FDA had authorized the use of Pfizer's COVID-19 vaccine the week before, and now it was approving the Moderna vaccine. Our program had delivered two safe and effective vaccines in ten months—a full year faster than many experts had predicted. The unprecedented vaccine effort was poised to save hundreds of thousands of lives, beginning immediately. Thanks to our meticulous planning and big investments in manufacturing, the government shipped millions of vaccine doses to all fifty states and every US territory within twenty-four hours of the FDA approval.

On the conference call that morning, the primary point of discussion was Pfizer. In July the government had purchased a hundred million doses of the Pfizer vaccine, pending the FDA's emergency approval, for $1.95 billion. Though the company had accepted the money, it rejected our offer to use the Defense Production Act's authorities to help accelerate production. Apparently Pfizer did not want to disclose how many doses it was selling to other countries, which was required under the DPA. Since then, however, the pharmaceutical giant had struggled to acquire the raw materials it needed and had fallen behind on its production schedule. It had promised its first twenty million doses by November and another twenty million doses in December. But the company blew through November without delivering a single dose, and it was on track to deliver only half of the promised doses by year's end. As a result, tensions had grown between Pfizer and the administration.

After the Pfizer vaccine received FDA authorization in December, Secretary of Health and Human Services Alex Azar launched negotiations with Pfizer CEO Albert Bourla to purchase an additional hundred million doses in 2021. The negotiations, however, were reaching a stalemate. I offered to call Bourla to resolve the open issues and to make sure that the agreement included a faster production schedule. I believed that, with the support of the federal government, Pfizer could produce the doses more quickly than the company projected—and when they did I wanted to ensure that the United States got the extra doses before other countries. For many Americans, this could mean the difference between life and death.

I was supposed to join the president at around 2:30 p.m. for a meeting about pardons. The meeting kept getting pushed back until it finally landed at 4:30 p.m.—twenty minutes before sundown. As had happened on so many Fridays, I set aside my religious observance to fulfill my government duties. I couldn't justify going home early to pray when I had a chance to advocate for people who would otherwise remain unjustly locked in prison.

The pardon is one of the most awesome powers afforded to the president, and when he exercised it, Trump took people who would have spent the rest of their lives in prison and gave them a second chance at life. The more Trump was persecuted through partisan investigations, the more he condemned the injustice of overzealous prosecutors and wanted to help others who had been treated unfairly. I loved watching the way he would immerse himself in the details of each case as if he had no other responsibilities in the world: he studied the facts, called lawyers and advocates to hear from them directly, and weighed all the variables.

As I walked into the Oval Office, White House counsel Pat Cipollone and his lead lawyer on pardons, Deirdre Eliot, were already seated. In the Trump White House, they served as the main line of communication to the Department of Justice. They collected the information on potential worthy pardons and presented it to the president so that he could make decisions informed by all the facts.

Soon after the discussion began, Molly Michael, the executive assistant of the president, walked into the Oval Office and passed me a note: Albert Bourla was on the line. I rarely stepped out of Oval Office meetings, but I made an exception and took the call from my cell phone as I paced around the Cabinet Room. After we exchanged niceties, Bourla explained that before we could even talk about speeding up delivery of vaccines, he had an issue with the termination clause in the contract.

"The clause is standard for government contracts," I said. "Normally I would give you my word that I would personally safeguard our agreement, but since I will no longer be a government employee when this matters, let me see what I can do."

After hanging up the phone, I ping-ponged between the Oval and the Cabinet Room, making calls about vaccines and trying to push for pardons. I was bouncing between two life-or-death issues.

By the time I departed the West Wing, it was past 8:30 p.m. When I got home, our youngest son, Theo, was already asleep, but Arabella and Joseph were still up, reading books with Ivanka. She had lit candles and fed the kids earlier, and the four of us sang the two customary Shabbat songs and said the blessing over the wine and challah. Ivanka and I tucked the kids into bed and then sat down for our Shabbat meal.

"I don't think we are going to get that wind-down period we had hoped for," I said. "It's been a wild five years, but in thirty days, we'll have a lot less responsibility and we will get our lives back. I'm ready. We just have to keep going hard for thirty more days."

As soon as I said the words, I thought about my marathon training in high school and how my father would always push me to find the strength I didn't know I had to pick up the pace in the final stretch of the race. I knew I wouldn't let myself do anything less than press forward until the end.

That Saturday, Ivanka and I went for a run through Rock Creek Park. It was a cold December day, but we enjoyed the chance to jog through Washington for one of the final times before our service ended. When I got back to the house, I spoke to Bourla, who like me was out for a walk to clear his head. We agreed on a compromise to resolve the outstanding legal issues, and he pledged to review his manufacturing plan and see how to expedite our next hundred million doses.

The following Monday, December 21, I departed for Israel with Avi. Ambassador David Friedman, Adam Boehler, and General Miguel Correa joined us. It was our last trip to the Jewish state before the end of our term. Following the playbook we used for the UAE and the subsequent flight to Bahrain, we scheduled the first-ever commercial flight from Israel to Morocco. Though Morocco had announced that it would normalize relations with Israel on December 10, the two countries still needed to sign an agreement. I learned from my experiences with the UAE and Bahrain that if we didn't take the initiative, the signing might not happen for months, if at all. With less than thirty days remaining in Trump's term, we couldn't afford to drag out the process, so Avi and I worked to organize and introduce both parties, resolve the final issues, and schedule the first flight before the month's end.

Upon our arrival in Israel, we were escorted to the Grove of Nations for an event that had popped onto my schedule at the last minute as a surprise addition. Located in the Jerusalem Forest, the grove is home to dozens of olive trees planted by heads of state as a symbol of the promise of peace in the Middle East. Traditionally, Israel invites visiting leaders to plant a single olive tree. But Bibi had decided to inaugurate the Kushner Garden of Peace with eighteen olive trees to commemorate the unique and unprecedented transformation we had brought to the region. When we arrived in the piney forest in the Judean Hills, we entered a white tent that the Israelis had erected for the event, and Bibi took the makeshift stage:

"It is fitting that we choose to honor Jared Kushner in this way because, Jared, you played a critical role in the inception and the implementation of the Abraham Accords. . . . In planting the Kushner Garden of Peace as a permanent presence in this Grove of Nations, we will ensure that future generations will know what your contribution has been. And I personally want to express my deep affection and my appreciation for the fact that the young teenager who I met many years ago, in fact in your house, in your room, has grown to be a man of stature who has helped change the history of our region and the history of Israel."

As Bibi and I shoveled dirt over the roots of the first sapling, I whispered a joke to the prime minister: "Only the Israelis can get someone to do free landscaping work while giving them an honor." Bibi chuckled. I was not used to being the center of attention, but I was moved by Bibi's magnanimity and grateful for the public recognition of a garden that would stand as a living testimony to the budding peace in the region.

From there, we drove to the American embassy in Jerusalem, where Ambassador Friedman had received special permission to dedicate the courtyard in my honor. "Unlike a lot of my other initiatives, this one is fully aboveboard and sanctioned," said Friedman in good humor. "I got all the sign-offs, including from State Department lawyers and Secretary Pompeo." Hanging in the courtyard was a bronze plaque that read: "Kushner Courtyard: Dedicated in honor of Jared Kushner and inspired by his relentless pursuit of peace." He told me that this was one of only a few times in State Department history that a US government official had received such an honor.

After a brief celebration, we returned to our usual business of negotiations. One of the final points of disagreement between Israel and Morocco involved an embassy. Israel wanted Morocco to open one, and Morocco wanted to start the new relationship with liaison offices. Foreign minister Nasser Bourita of Morocco had become so frustrated over the spat that he threatened to call off the deal entirely. I promised him that we'd get the Israelis to the right place.

That evening, I sat with Bibi in his study. I began by thanking him again for the beautiful ceremony earlier that day in the Grove of Nations.

Although he had honored me, I wanted him to know how grateful I was for his partnership on the Abraham Accords. Bibi had spent years laying the groundwork with the Arab world to create the conditions for peace. When the Obama administration proposed the Iran deal, he traveled to Washington to forcefully oppose the bill in Congress. Bibi knew this diplomatic foray was doomed from the start: Obama was going to sign the deal no matter what. But his public lobbying, which culminated in a nationally televised address to Congress, drew the anger of the Obama administration and damaged Israel's relationship with the United States, its most important ally. His advocacy was a watershed moment, however, in Israel's relationship with the Arab states in the Gulf. It revealed common ground on their top priority, and it showed that Israel could be more valuable as a friend than a foe. Like Trump, Bibi was fearless. This could sometimes be polarizing, but it also made him a powerful catalyst for change.

We didn't have long to reflect. We still needed to finalize the terms of the impending peace deal with Morocco. Bibi raised his disappointment with the liaison offices and said that we should push harder for a better deal.

This echoed sentiments we had heard from Ambassador Friedman, who had been lobbying against the deal, conveying his strong reservations to Avi.

"Please trust my judgment on this one," I urged. "The king is a very deliberate and instinctive person. We have worked through the embassy issue, and this is all we are going to get at this point. The smart move is to show them trust and take less now. I promise that if you give them trust, ultimately they will give you much more than you bargained for."

By the end of our meeting, less than twelve hours before the historic first flight to Morocco, Bibi signed off on the final terms of the declaration.

At the airport the next morning, we were greeted by the Israeli delegation, led by Meir Ben-Shabbat, the Israeli national security adviser, whose parents were born in Morocco. Upon landing in Rabat, we were immediately escorted to the Mausoleum of Mohammed V, where we signed a guest book and laid a wreath on the graves of the late Moroccan sovereigns Mohammed V and Hassan II, who defended the Jewish people against persecution.

That evening we went to the palace, where we were escorted into the king's office, a large wood-paneled room that smelled of incense and was adorned with stunning damask fabrics. Two neat rows of chairs faced one another—one side for the Moroccan officials and the other for me, Avi, Boehler, and Meir Ben-Shabbat. King Mohammed VI sat at the head of the room, splitting the rows, in front of a massive mural depicting his family tree, which dated back to the Prophet Muhammad. Known for his impeccable taste, the king was dressed in a well-tailored black suit. Seated directly next to the king was his son, Moulay Hassan, the high-school-age crown prince who had impressed me at our dinner back in 2019. The king greeted me as warmly as he could while adhering to strict COVID-19 protocols.

As the cameras captured the moment for the world to see, we signed the joint declaration between Israel, Morocco, and the United States.[65] The document restored full diplomatic relations between Israel and Morocco. It granted authorization for direct flights between the countries, opened liaison offices in Rabat and Tel Aviv, and promoted economic collaboration on trade, investment, technology, visas, tourism, water, food security, and more.

I paused a few seconds before applying my signature as the representative of the United States. I had signed lots of documents in my business life. The action was the same—pressing the pen to paper to complete a deal—but the difference in significance couldn't have been more dramatic. This deal would lead to connections and activities that would make the world more peaceful and prosperous. In business deals, parties change ownership; in peace deals, people change minds.

Afterward, I handed the king a present: the US State Department's official new map of Morocco, which included the Western Sahara within the country's territory. The king was jubilant for the recognition as well as his country's newly established ties with Israel.

That evening, as people celebrated in the streets to mark the momentous agreement, the Moroccans prepared a kosher meal for us in the sprawling guest palace.

During dinner, I felt the all-too-familiar buzz of my phone. It was Albert Bourla of Pfizer. After much deliberation, he had decided to go forward with the contract, and he was willing to accept the federal government's assistance in acquiring supplies to expedite production. This was a win-win partnership for both parties, but we had one condition: American-made vaccines would go to Americans first.

"We will get you the supplies you need," I said. "I just want to be very clear, we need your first hundred million doses in the second quarter. We will not let those doses leave the country."

"Why are you playing God?" Bourla shot back. "Why do you get to determine whether an American gets a dose of the vaccine versus someone from Japan or Israel?"

"Because I represent America," I said. "That's the country I work for. My job is to get as many doses for the American people as possible, and you are an American company. If you ramp up your production to the levels we anticipate, it will be in part because of the help of the US government. What we ask in return is that you prioritize saving American lives."

The next day, Pfizer announced that it would supply the United States with an additional hundred million doses by July, securing a total of two hundred million Pfizer vaccines for Americans by the first half of 2021. It was another critical step to ensuring that every American who wanted a vaccine could get one. This time, Pfizer delivered on its promise.

I landed back in Washington in the afternoon on December 23. As I made my way to the White House, I got an unexpected call from the president.

"Jared, I just signed a full pardon for your dad," he said. "A few days ago, I called your father and asked if he wanted a pardon, and he said no. I know his case well, and I believe he got screwed. Because of his unfortunate experience, we enacted major criminal justice reforms that have helped tens of thousands of people. I hope he won't be mad at me, but I'm very proud to be able to do this. Your dad is a great guy."

I was so overwhelmed I didn't know what to say. I asked if he had called my dad to tell him, but he said he was still working on a pile of cases and would try him later. He told me to feel free to call him in the meantime.

As soon as we hung up, I called Ivanka. Together, we conferenced in my dad and mom and shared the news. I could hear my dad's voice crack. "When Donald asked me about this, I really told him that I didn't need one," he said. "I am at peace with what happened and have rebuilt my life in a way where I have all of the right priorities and am comfortable with who I am. I didn't want to cause Donald any controversy. But truthfully, hearing this news makes me realize how much I really did want one but was too proud to ask. This brings me closure to a very hard period of my life."

I was overwhelmed with joy and relief for our family, and even more so because I knew that thousands of families had experienced the same joy and relief due to the reforms we enacted nationwide.

Fifteen years earlier, when I was visiting my father each week in prison, I never dreamed I would be having this conversation. I certainly never imagined that the president of the United States would grant my own father a pardon. In that moment, I felt that only God's hand could have written this real-life script, and that His plans are always bigger than ours.

58

Reconciliation

I was getting ready to leave the White House for Joint Base Andrews and a flight to Saudi Arabia on the morning of January 3, 2021, when I received a call from the foreign minister of Qatar, Sheikh Mohammed. He was calling on behalf of his boss—the ruler of Qatar, Emir Tamim bin Hamad.

The deal was off, he said. He thanked me for working so hard to resolve the dispute, but told me that the decision was final and nothing more could be done at this point.

I was supposed to join Tamim and MBS at a signing agreement to end the three-year blockade of Qatar by Saudi Arabia, Bahrain, Egypt, and the UAE. Ever since I facilitated the call between Tamim and MBS back in September, my team and I had helped the two sides work through their differences. In December I had traveled back to the region to complete the agreement. After two seven-hour negotiating sessions, I thought we had resolved all of the open issues, and MBS and Tamim were planning to meet at the Gulf Cooperation Council (GCC) summit on January 5. They invited me to come to Saudi Arabia for the event to witness the agreement, which was both an honor and an opportunity to help ensure the negotiations crossed the finish line.

Over the previous three months, Sheikh Mohammed had masterfully negotiated each delicate issue he encountered, so I could sense his palpable disappointment through the phone. Until his call, I thought we were on track to sign the deal, but he explained that the Saudis had not yet agreed to lift their airspace restrictions in advance of the summit. Since our first discussion, Tamim had made clear that he was willing to travel to Saudi Arabia to sign the agreement, but only if the country opened the airspace beforehand: if his citizens couldn't fly, then he didn't want to fly.

During the negotiations, I had communicated Tamim's position to the Saudis, who assured me that they would remove the restrictions in time. Now, just two days before the summit, the flight restrictions had yet to be lifted.

When I heard the news, I bypassed the Saudi negotiating team and called MBS directly. "We have a big problem," I said.

To my surprise, MBS's reaction revealed that this was the first he'd heard of Tamim's request. He said that Qatar's ask was a "re-trade," using a commercial term for renegotiating the price after the parties had come to an initial agreement. He interpreted the request as a sign that the Qataris weren't sincere in wanting to resolve the dispute.

I pushed back emphatically: "In four years, I have never lied to you. I promise you that Tamim has made this a condition from my very first meeting on the topic. Your team knew about this request. If you want to be upset about this, be upset at your team, be upset at me, but don't think Tamim is playing games here."

MBS assured me that he understood the stakes and would talk to his team and see whether they could resolve the issue.

All the while, our military plane was waiting on the tarmac at Joint Base Andrews. Once pilots go on the clock, federal regulations say they have fourteen hours before they are required to break. Because a direct flight to Saudi Arabia is more than twelve hours, we kept asking the pilots to push back our start time so that we wouldn't have to stop halfway through the trip to spend the night, and then arrive late to the summit.

As I paced around my house, I considered canceling the trip and letting the Saudis and Qataris figure it out for themselves, but I knew that the president wanted the rift resolved. An agreement would advance American interests by strengthening America's position in the region, unifying two of our important partners, and eliminating a constant point of contention that obstructed potential peace agreements. If we failed to strike this agreement, Iran would have an opening to further exploit the rift.

I moved my flight time to the latest possible window—8:00 a.m. the next morning. Before going to bed, I told Sheikh Mohammed that MBS was prepared to open Saudi airspace, and he said that he would take the message back to Tamim. This put the deal back in play, but as I waited for word from Qatar, I wondered if it was too late. I woke up at 1:00 a.m. to check for an update, but all was quiet. I felt like the deal was slipping away.

Early the next morning, we headed to Joint Base Andrews, but we still had not heard from the Qataris. The clock was ticking. If we didn't take off before 9:00 a.m., we wouldn't make the summit in time.

I called Sheikh Mohammed and asked him to relay a message to Tamim: "I'm boarding the plane now and heading to the summit. I would strongly suggest that the emir come and take advantage of a rare opportunity to resolve this issue. Tell him that while I know there is little trust right now, I will be there personally to ensure that he is treated with the utmost respect. If you don't come now, I believe the ice will get thicker, not thinner. Both sides are rightfully skeptical of each other, and the Saudis will interpret the last-minute cancellation as a sign of bad intent. It's unlikely that you will find another US government official who will bridge the two countries, and you could remain in the blockade for the next twenty years." Sheikh Mohammed agreed and promised me that he would do what he could.

We lifted off for Saudi Arabia without knowing what the Qataris would do. After three hours in the air, I received a message from MBS: Tamim had called to say he appreciated the Saudis' flexibility, and he had decided to come. The deal was back on. Excited and relieved, I told my team that we were closing in on another critical peace agreement—this one between Arab neighbors who had been locked in a years-long conflict.

As our plane descended into Al-Ula, an ancient city in northwestern Saudi Arabia, all I could see was an endless landscape of sand and rock formations. After a short drive on the ground, we arrived at a newly built compound—a collection of modular units covered by tent roofs, giving visitors the experience of camping in the desert. As I spoke with the Arab royalty assembled there, I received a call from Sheikh Mohammed.

"We're turning our plane around," he said.

I nearly shouted: "What do you mean you're not coming?"

There had been a last-minute dispute about the execution of the agreement.

I was standing with Dr. Ahmed Nasser Al-Mohammed Al-Sabah, the foreign minister of Kuwait, who had been my partner in negotiating this deal. With Sheikh Mohammed on the line, we walked over to MBS, pulled him away from a discussion, and described the problem. MBS took my phone and walked away.

Several minutes later, the crown prince returned. "Problem solved," he said. MBS had given his word that he would deliver, and the Qataris decided to proceed on his honor.

Minutes later, as the cameras rolled, Tamim walked down the stairs of his plane and was greeted by MBS. Disregarding pandemic protocols, the two leaders hugged. The embrace between the former rivals was broadcast on television screens throughout the world. Much like the first flight between Israel and the UAE, it was a powerful image that reflected the burgeoning change in the Middle East. It signaled to people across the region that they could move on from past tensions and seek a better future.

The end of the blockade on Qatar dominated global headlines by the next morning. "Saudis, Qatar to Settle Feud, Aiding U.S. Efforts on Iran," read the *Wall Street Journal* headline. "Saudi Arabia and Allies to Restore Full Ties with Qatar, Says Foreign Minister," proclaimed Reuters. "Qatar Crisis: Saudi Arabia and Allies Restore Diplomatic Ties with Emirate," reported the BBC.

On the morning of January 6, 2021, we departed the Middle East for the final time during our government service. I had grown accustomed to using the long return flights to debrief with my team, reflect on our meetings with foreign leaders, and plan our next moves. With this final deal closed, there were no next moves. We were done.

"In the history of American diplomacy, no one has achieved more peace deals than this team," said Brian Hook in an impromptu speech. "Looking back to when we first entered office, we were dealt a terrible hand. It's clear just how ripe the region was for new thinking and approaches. That could only come from someone like you who was outside the think tank industry, which has been using the same talking points from the 1970s. You didn't have the baggage of what passes for 'expertise.'"

I thanked Hook for his kind words and for the crucial role he had played. He was an essential member of the team who believed wholeheartedly in Trump's policies and had been instrumental in achieving some of the president's greatest successes. We all continued to share stories about our favorite moments, cultural snafus, and the unforgettable people we had met. As we laughed and swapped stories, I felt like a lead weight was being lifted off my back. On so many of our trips, we had spent the flight home digesting the knowledge we had gained and planning our next steps in pursuit of what felt like an ever-elusive breakthrough. This trip was different. We were leaving office having brokered six peace deals: the agreements between Israel and the UAE, Bahrain, Morocco, Sudan, and Kosovo, plus a reconciliation between Saudi Arabia and Qatar.

In my four years in government, that plane ride was a high point. I reflected on the many challenges we had faced. So many of them had felt like existential threats at the time, but now seemed like footnotes. A tinge of nostalgia swelled up in my chest. But more than anything, I was content. Our quest for peace was coming to an end. I had played the game until the final whistle and always tried to do what was right rather than what was easy. Now I was ready to pass on the immense responsibility, return to a quieter life, spend more time with my family, and have some adventures of my own. Maybe I'll even be able to take my kids sightseeing before we leave town, I thought.

My momentary reflection was interrupted by a phone call from Eric Herschmann.

"Where are you?" he asked.

"I'm in the air, heading back from Saudi Arabia," I responded. "What's going on?"

"Rioters have broken into the Capitol," he said. "I'll give you an update when you land."

We touched down in the midafternoon on the all-too-familiar grounds of Joint Base Andrews. As I climbed into my SUV, the Secret Service warned me that there were large crowds on the National Mall and around the Capitol and recommended that we head straight home to Kalorama. On the drive, I called Ivanka to check in. As I spoke to her, I detected a strain in her voice that only a husband can truly understand. She encouraged me to head home to see the children and told me that she would see me a bit later.

When I arrived home, exhausted from our thirteen-hour flight, I went to our room and turned on the shower. But before I could get in, I received a call from Kevin McCarthy, the Republican leader in the House of Representatives, asking me if I could help the situation. He sounded nervous, so I took the call seriously and told him that I would see what I could do. I shut off the shower, put on a clean suit, and went to the White House. By the time I arrived, the president had already released a video statement addressing the riot.

That night, after I learned more about what happened at the Capitol earlier that day, Ivanka and I started working with the team on a proposed speech for the president to deliver the next day. In the afternoon of January 7, Trump delivered remarks expressing our sentiment, and that of millions of his supporters: "The demonstrators who infiltrated the Capitol have defiled the seat of American democracy. To those who engaged in the acts of violence and destruction, you do not represent our country." He committed to a "smooth, orderly, and seamless transition of power." As he concluded, he said, "This moment calls for healing and reconciliation. . . . We must revitalize the sacred bonds of love and loyalty that bind us together as one national family." Ivanka and I stood nearby as he read the statement, which we had drafted with a few others.

The violent storming of the Capitol was wrong and unlawful. It did not represent the hundreds of thousands of peaceful protesters, or the tens of millions of Trump voters, who were good, decent, and law-abiding citizens. What is clear to me is that no one at the White House expected violence that day. I'm confident that if my colleagues or the president had anticipated violence, they would have prevented it from happening. After more than six hundred peaceful Trump rallies, these rioters gave Trump's critics the fodder they had wanted for more than five years. It allowed them to say that Trump's supporters were crazed and violent thugs. The claim was as false as the narrative that the violent Antifa rioters who desecrated American cities that summer were representative of the millions of peaceful demonstrators who had marched for equality under the law. In the aftermath of January 6, the morale in the White House sank to an all-time low. Some staff members resigned. Others came to my office prepared to offer their resignation. I encouraged them to stay.

"You took an oath to the country," I said. "This is a moment when we have to do what's right, not what's popular. If the country is better off with you here, then stay. If it doesn't matter, then do what you want."

During our remaining days in office, Ivanka and I continued to work on presidential pardons, but I reserved most of my attention for completing the presidential transition.

Back in December, I had begun periodic meetings with the Biden transition team to brief them on all the information and operational knowledge needed. I was especially focused on Operation Warp Speed and the COVID-19 response. I worked closely with Secretary Azar and his staff at the Department of Health and Human Services to prepare a wing of offices for the Biden team to use during the transition. On the day Biden's representatives were scheduled to arrive, Azar's team was surprised that no one showed up—apparently for fear of catching COVID. This demoralized the HHS staff, who for months had risked their personal health to work around the clock during the pandemic.

I invited Jeff Zients, who was slated to lead Biden's COVID task force, to come to the West Wing with his team. We had been communicating regularly. Brad Smith, Adam Boehler, Dr. Deborah Birx, Paul Mango, and I walked him through our administration's ongoing efforts to confront the pandemic. Over the previous ten months, we distributed tens of millions of masks and other PPE and had rebuilt the Strategic National Stockpile. In January 2020, the stockpile was down to 13 million N95 masks, 5 million gowns, and 16 million gloves.[66] The United States had completed 250 million COVID tests, and we had created the capacity to complete 1.3 billion tests in the first half of 2021. By January 2021, it had 237 million N95 masks, 52 million gowns, and 159 million gloves. And through Operation Warp Speed, we had delivered close to 40 million vaccine doses to communities across America, with an additional 100 million doses expected to be delivered by the end of March. By June of 2021, every American who wanted a vaccine would be able to get one. We were surging resources into therapeutics, and on January 12 we announced a $2.63 billion purchase for 1.25 million doses of Regeneron's monoclonal antibody treatment, which was proven to reduce mortality. At the end of the meeting, I thanked Zients for his willingness to serve in government, adding that we were all available to him 24/7, both then and after Biden assumed office. I knew he had a tough job ahead, and I wished him the best.

I also met with Jake Sullivan, Biden's incoming national security adviser, to brief him on our peace deals and review the countries that we believed were close to normalizing with Israel. He stressed that the Biden administration's top priorities would be the three Cs: COVID-19, climate change, and China. I urged him to take a fresh look at the Middle East, as a lot had changed in the four years since he had been in government. I detailed my ongoing discussions and predicted with confidence that with six months of focused execution, the United States could build on the momentum and achieve between four and six additional peace deals. I didn't care who got the credit. This was about keeping Americans safe and improving the lives of millions.

As we entered our final week, pardon requests were stacking up and awaiting the president's final decision. Some of the best clemency recommendations came from Ivanka, who had volunteered to help identify deserving individuals and work with the White House Counsel's Office to vet them. When we met with the president, he liked the candidates that Ivanka presented. She was advocating for people who didn't typically have a champion in Washington. They weren't celebrities or well-connected individuals. They were men and women who had come from difficult circumstances, made mistakes they regretted, and had reformed their lives while in prison. Local nonprofit organizations like #cut50 and advocates like Alice Johnson brought their cases to the White House.

"Bring me more like these," he said. "I want the Ivanka cases."

One evening, with just a few days left in office, the president called me. "What do you think I should do with Bannon?" he asked. "He's been lobbying hard for a pardon." Bannon had gotten himself into legal trouble and was being charged with fraud.

"I haven't reviewed his case, but I don't oppose him getting a pardon based on our past," I said. "You know me. I'm a softy. I err on the side of mercy."

"Seriously?" replied Trump. "You would really be for that? After everything he did to you?"

"I don't forget, but I do forgive," I said. "If you think it's a good idea, I'm okay with it. Steve was incredibly destructive to your first year in office, but he was there for you on the first campaign when few were."

Bannon single-handedly caused more problems for me than anyone else in my time in Washington. He probably leaked and lied about me more than everyone else combined. He played dirty and dragged me into the mud of the Russia investigation. But now that he was in trouble, I felt like helping him was the right thing to do.

I hadn't forgotten the lesson I learned from my father's situation. Nothing is achieved from harboring resentment. It's better to forgive and let God be the judge of the rest.

59

Hourglass

There was an unfamiliar stillness in the West Wing as the clock slid toward midnight on our last full day in office. The lights were off, the desks were cleared, and the hallways were eerily empty. Staff had said their goodbyes and gone home, save for a handful of us who remained: Ivanka and me, Mark Meadows, Dan Scavino, White House counsel Pat Cipollone, a few members of our staff, and the president.

Cipollone and his legal team had worked around the clock to finalize the legal documents for the few remaining pardons the president had approved. Shortly before midnight, Trump granted clemency to an additional 143 individuals. Ivanka began calling the families whose loved ones had just received a pardon. It was late, but she knew that families would not want to sleep through one more night waiting to find out if their loved one was coming home.

As Ivanka made calls, I headed back to my office and wrote a note to Mike Donilon, who would move into my office as Biden's senior adviser. I wished him luck and told him that amazing things could happen from that small, unassuming office, and that I was rooting for him to accomplish a lot for our country. Though we worked for presidents from different parties, ultimately, we were all on the same team. Along with the note, I left a few items in the top drawer that would come in handy for any job conducted from that office: Extra Strength Tylenol, Purell, and a bottle of Macallan scotch.

Then I walked the few feet down the hall to the Oval Office. I was a bit surprised to see the president still sitting at his desk. He was finishing his letter to incoming president Joe Biden. He handed it to me. I read it and was genuinely moved. It was a beautiful letter, gracious and from the heart—a presidential tribute to the country he loved.

As I was closing this chapter of my life, I wanted to remember this day. That morning, my first order of business had been a visit to the Navy Mess, where I thanked the dedicated service members posted there. They had kept me standing for the past four years. Each day they'd made me the same lunch: a special chopped salad topped with sliced avocado and grilled kosher chicken. In keeping with my New Jersey roots, where we frequented diners and had eggs three times a day, the Navy Mess staff also made an exception to their strict "no breakfast after 9:00 a.m." rule, and they would fry an omelet with American cheese for dinner whenever I asked.

Later that day, Ivanka had arrived with the kids. Arabella, Joseph, and Theo raced into the Oval Office to greet their grandpa. He gave them a big hug and, as usual, opened his desk drawer and pulled out boxes of presidential M&Ms. The kids handed personalized cards to our Secret Service detail. Arabella's card summed up our gratitude best: "Thank you for keeping me safe . . . you have been so kind to me. For example: you go fishing with Joseph, you go on golf cart rides with Theo, and you listen to my terrible jokes. Thank you!!! You guys and gals are my best friends." Next to a picture of an American flag she had drawn with markers, she added, "Yes, there are 50 stars."

As I watched our kids gallivant through the West Wing, handing homemade cookies to the Secret Service agents and the custodial staff, I couldn't believe how much they'd grown during the past four years. Theo hadn't even had his first birthday when we came to Washington. He had crawled for the first time on the weekend after inauguration, in the White House State Dining Room, no less. Now, at four, he was strutting around in his loosened tie, unaware that anyone had won or lost an election. Arabella had grown about a foot and was nine going on nineteen, with the charm and sass of her mom. Joseph, who was just three when we moved and had the hardest time adjusting to our more demanding work schedules, had discovered a love for fishing. I promised him that when we got to Florida, where Ivanka and I had decided to settle, I would replace his Secret Service agents as his fishing partner.

I had planned to pack everything up in an orderly fashion, but by the final day I had barely started, so I asked Charlton and Cassidy to help me pile my stuff into a few boxes. "We came in a storm, and we left in a storm," I said half jokingly to Avi, Cassidy, and Charlton as we parted ways that evening.

As I prepared to head home, I dropped by Meadows's office for a final time. I found him with cell phones in both ears, sitting in his familiar spot on the couch with documents spread across the coffee table in front of him. The fireplace was burning, and he smiled and nodded at me as he wound down his calls. I thanked him for stepping into the role and for all of the incredible things he had accomplished and problems he'd helped avoid, which history would likely never know or appreciate.

Before making the final walk down the creaky, narrow stairs of the West Wing, I paused and silently said goodbye to my cave of an office, where I had spent most of my waking hours over the past four years. The walls were blank, stripped of the photos, presidential proclamations, and recognitions I had collected. The narrow room looked small, dark, and lifeless—almost exactly as it did when I entered it in 2017. Few would ever know all the heated conversations, agonizing decisions, and sweet moments of victory that occurred within these walls. I picked up the last two items I had left until the very end: the mezuzah on my doorway and an hourglass that Chris Liddell had given me. As I prepared to leave the White House for the final time, I thought about what Liddell had said: "Every day here is sand through an hourglass, and we have to make it count."

I knew I had lived by those words. I never forgot that my office wasn't really my office. I was just the current inhabitant. From the day we arrived, I never stopped working. My responsibility was to give every ounce of energy I had to help the president advance his vision for the American people. Even when I was at home, I thought about the job. I could never predict when I would receive an urgent phone call with an unexpected request. I could never shake the sense that if I convened one more meeting, maybe I would find a solution to an impossible problem or help improve one more person's life. My duty to serve the president of the United States came first, even before family. Ivanka and I were a unique case: we were senior White House staffers who were also family members, adding another level of stress and scrutiny. There was never a moment of true calm in the White House, never a moment of pure enjoyment. There was always action, always a crisis, always high velocity.

Now, as our time in office drew to a close, I was at peace. I had given my all and was proud of what we had achieved. While many throw up their hands and say "Washington is broken," I came to view it differently. I learned that the system is complex, but that it can work if people think with creativity and put in the effort that the job demands. What we accomplished on four seemingly unsolvable problems—trade, criminal justice reform, Operation Warp Speed, and Middle East peace—was proof of this concept.

After decades of outdated trade deals that sent American jobs overseas, we replaced NAFTA with the US-Mexico-Canada Agreement, the largest trade agreement in history. We had also taken the first significant steps to confront China's unfair trade practices and protect American farmers and workers. Against opposition from both Democrats and Republicans, we found common ground, gained the president's support, and enacted the most significant criminal justice reform in a generation. In the midst of a devastating pandemic, we delivered a COVID-19 vaccine in record time. And through unconventional diplomacy and relentless resolve, we overcame a history of stalemate in the Middle East and forged the Abraham Accords.

No one could take those accomplishments away. They were real. Most importantly, these bold policies changed lives for the better. I thought of the countless former inmates who were now reunited with their families and were determined to make the most of their second chance in life. I thought of the manufacturing workers and farmers who greeted us with gratitude when we visited their communities and thanked the president for bringing back their jobs and restoring their pride. I thought of the grandparents and other vulnerable citizens who would now be able to safely reunite with their families. And I thought of the millions of people in the Middle East who would now be able to travel between Israel and the UAE, Bahrain, Morocco, and eventually Sudan. The deals not only linked people to their geographical neighbors but also opened new economic opportunities and established cultural ties that transcend religion and race. We had shown that the conflicts that held back generations of the past no longer had to constrain the generations of the future. Working in government was a grind. It put enormous pressure on me and my family. But I didn't regret a single minute of my 1,461 days on the government clock. The White House is the most daunting, thrilling, exhausting, and meaningful place to work in the world. The responsibility is difficult to comprehend, and so too is the potential for impact.

I turned off the lights to my office and walked down the narrow stairs that led to the ground-floor corridor. As I passed the West Wing lobby, I said a final goodbye to the uniformed Secret Service agent at the desk before exiting the double doors to West Executive Avenue. I climbed into my SUV and didn't look back.

Our rented Kalorama home was bare, save for a few piles of boxes. Over the previous weeks, we had sold, donated, or shipped most of our belongings. Ivanka was still making her way through her list of calls to the families whose loved ones had just received pardons. I joined her, and we finished together. By the time we made the last call, it was 3:00 a.m. Exhausted but grateful and proud of what we had helped achieve, we turned off the lights and gave each other a kiss good night.

These four years had brought Ivanka and me closer together. Her deep involvement in the pardons perfectly exemplified her tenure in Washington. She was happiest behind the scenes, using her influence to help others in ways that most people would never know. While many speculated about her motivations, she never had any political aspirations. She tolerated the politics to drive the impact. She wanted to use her unique position to give back to a country she loves. In Washington, that made her an anomaly.

Ivanka didn't have to work in government, but she chose to close her successful businesses to serve. She advanced reforms to uplift families across the country, especially those who were most forgotten. She spearheaded the effort to double the child tax credit, allowing hardworking American families to keep more of their tax dollars to provide for their children. She created a workforce training initiative that helped countless Americans hone their craft, progress in their careers, and work in jobs they love. She helped pass historic legislation to stop the heinous crime of human trafficking. And when Americans were at their most vulnerable during the pandemic, she launched the Farmers to Families Food Box Program, which fed tens of millions of people. Throughout our time in Washington, she managed to find a way to give our kids the love and attention they needed. She was a loyal and loving daughter to her father, and a constant source of strength. There's no way I could have survived the four years in Washington without Ivanka, my best friend and partner. Her constant encouragement, companionship, support, and insight sustained me throughout our journey.

Early in the morning on January 20, we packed up the final boxes, piled into an SUV with our kids, and left our house for the last time.

When we arrived at Joint Base Andrews, Ivanka and I found Eric and Lara Trump, Don Jr. and Kimberly Guilfoyle, and Tiffany Trump and her fiancé Michael Boulos. We reminisced for a few moments about our experiences as a family the past four years. Don Jr. and I had been absurdly accused of treason. Eric must have broken a Guinness World Record for congressional subpoenas. Lara and Kimberly campaigned across the country. Tiffany made it through law school in an era of outrage, and did so with elegance. We had all taken this unexpected journey together. Not only had we survived, we had grown closer.

As we stood on the tarmac on the cold, crisp morning, we heard the familiar noise of Marine One's rotor blades. The forest-green helicopter descended, and the president and Melania stepped off to the sound of applause, with several hundred staff members cheering them on. In a few hours Trump would no longer be commander in chief. He would be an American civilian who had served as the forty-fifth president of the United States.

Trump built one of the strongest economies our country had ever seen and advanced policies that benefited all Americans. Before the pandemic, unemployment had reached a fifty-year low, wages had hit a record high, and middle-class income had increased an average of $6,000. Trump's economic policies created seven million new jobs and made America the number one producer of oil and natural gas. Through his foreign policy of peace through strength, Trump prevented new wars, and America regained its military might. Our enemies feared us, our partners respected us, and our allies could once again count on us. American troops were coming home, and peace was burgeoning in the Middle East.

When COVID-19 struck, the president mobilized all of America to respond. The United States acquired, delivered, and ramped up the production of masks, PPE, ventilators, testing supplies, and other lifesaving materials. The economy rebounded faster than experts had predicted, with the GDP growing at a rate of 33 percent in the third quarter of 2020. And because Trump took a calculated risk and invested billions of dollars in Operation Warp Speed, America delivered lifesaving therapeutics and a safe and effective vaccine in less than a year, far faster than anyone thought possible. Operation Warp Speed succeeded only because Trump believed in the ingenuity of America's private sector and the ability of America's military to save hundreds of thousands of lives.

Donald Trump arguably accomplished more than any other president in my lifetime. I was proud to serve in his White House, and I was grateful that he gave me the chance to help him deliver on his promises to the American people.

"It is my greatest honor and privilege to have been your president," he said in his final public remarks as president. "I will always fight for you. I will be watching, I will be listening. And I will tell you that the future of this country has never been better. So just, a goodbye, we love you, we will be back in some form."

As Trump departed on Air Force One, a familiar tune began to play.

"I traveled each and every highway
And more, much more than this
I did it, I did it my way . . ."

Frank Sinatra's "My Way" was one of Trump's favorites. It captured the moment the way only a great American song can.

Exactly four years earlier, Ivanka and I had arrived on this same tarmac with president-elect Trump. We left our lives in New York and moved our three young kids to Washington for the journey of a lifetime. We got to know Americans from all walks of life who were making tremendous sacrifices to provide for their families and give their children the very best. We traveled the globe and met the most powerful leaders in the world. We navigated through a controversial time, compounded by West Wing infighting, vicious investigations, media attacks, partisan divides, geopolitical conflicts, and an unexpected cancer scare.

We weathered ups and downs together and learned a great deal about politics, human nature, and ourselves.

I learned that to make it in Washington I needed to have a spine of steel. I learned to stay away from petty fights and power struggles, to make fewer enemies and more friends, and to talk less and do more. As hard as it was to hear people spread lies about me and my family, I tried to ignore the noise and focus on improving the lives of others. The personal cost was a small price to pay for the opportunity to change the world. Instead of relying on conventional wisdom, I viewed issues from a fresh perspective, put myself in the shoes of others, and found common ground. Despite countless setbacks and criticism that threatened to derail our efforts, I reached breakthroughs that benefited our country and the world. Through it all, I stayed true to my core conviction: life is too short to remain stuck in the past. It's up to us to make the most of the lives we're given, help others, and create the future we want for our children and grandchildren.

I squeezed Ivanka's hand as Air Force One disappeared into the clear blue sky. As quickly as the journey had begun, it jolted to an end. Our time was up. Our duty was done.

As the roar of the 747 engine faded into the distance, I thought of the words that had guided me since I was a young man wrestling with my father's prison sentence and wondering what God could possibly have in store:

Don't look back, look forward.

Acknowledgments

I couldn't have completed this project without the encouragement and support of many people I am blessed to have in my life.

From the first day I started writing down a few stories—having no idea whether I would ever share them beyond the confines of our living room—Ivanka has joined me on this journey. She patiently listened to me recount anecdotes she already knew and added details from her own memory. She read and reread drafts, providing keen edits and suggestions that only a wife can make. Ivanka—you are my constant source of strength and the best friend and life partner I could ask for. I love you and thank you. And to our three incredible kids—Arabella, Joseph, and Theodore—thank you for keeping me grounded and reminding me what's worth fighting for. Growing up with your parents working in the White House was your unique contribution to the country, and I hope that one day you will read this book and better understand why your father was not as present as he would have liked to be. Ivanka, Arabella, Joseph, and Theodore: You are my life.

President Donald J. Trump gave me the chance to make the two best decisions of my life: asking for Ivanka's hand in marriage and accepting an opportunity to serve in the White House. Donald—thank you for having the courage to shake up Washington and fight for the forgotten men and women of this country. Your presidency achieved historic results, and I am forever grateful for all that I have learned from your leadership.

My parents, Charles and Seryl Kushner, have shaped me into the person I am. They prepared me for my time in government in ways we could have never known at the time. They taught me to value hard work, honor my word, ignore critics, cherish our family, hold on to my faith, and always remember that life is a gift and it's up to each of us to make the most of it. I am grateful for their steadfast love and their constant encouragement throughout this journey. Mom and Dad—your love for each other and sacrifice for our family inspire me every day.

To my siblings and their spouses, Dara and David, Nikki and Tuvi, and Josh and Karlie—we have been through many highs and lows together, and each challenge has brought us closer. Thanks to each of you for your support and wisdom throughout this process. I am blessed beyond measure to call you family.

Eric and Lara and Don Jr. and Kimberly were instrumental in Trump's campaigns, they hung tough under intense pressure during his presidency, and I'm grateful for their friendship. Barron and Tiffany continued their studies with devotion and strength in the face of criticism, and their love and support lifted us all. And Tiffany and Michael—we are so excited for you as you begin this new chapter together. To each of you—I deeply respect how you are always there for each other. To Melania—I am grateful that you welcomed me into the Trump family with open arms from the very beginning.

Robert Thompson has been a friend for almost two decades. He has always brought perspective, wisdom, and wit whenever I sought his advice, and I am grateful for his brilliant suggestions and candid thoughts throughout this process.

Jonathan Burnham put the full weight of HarperCollins behind this project. He bet on this book when it was merely an idea. He offered superb advice throughout the process and shepherded a complex and sprawling manuscript into a tight, coherent narrative.

Eric Nelson saw the potential of this project for Broadside Books. His editorial expertise, savvy suggestions, good humor, brutal red pen, and exceeding patience helped a first-time author transform an idea into reality.

Brittany Baldwin is one of the most positive people I have ever worked with and also was one of the best presidential speechwriters at the White

House. Days after our government service came to an end, I called her and she accepted another unexpected assignment. When we commenced this project, I had no idea where it would go or what shape it would take, and through her listening ear, astute insights, and bold vision—and many late nights behind the keyboard—we have brought to life this true adventure story. She poured her heart into this monumental undertaking and helped me put words to the experiences that have defined my life.

Nick Butterfield served as deputy assistant to the president and deputy White House policy coordinator. Prior to that appointment, he was an invaluable member of the Staff Secretary's office, helping to ensure that every speech, proclamation, executive order, and document that reached the president was perfect. After our time in the White House concluded, I asked Nick to come to Florida for a month to help me organize a potential book project. I soon discovered that in addition to his exceptional understanding of complex policy issues, he's an excellent writer and superb editor. His devotion, drive, and relentless pursuit of perfection have helped us create this primary account of history.

Avi Berkowitz was the first person I hired at the White House, and he was instrumental in achieving groundbreaking peace deals in the Middle East. We experienced the daily highs and lows of this unique journey often not more than a few feet apart. I am grateful for his wise counsel, friendship, and devotion to our initiatives. He added many details from his own memory and offered careful edits and insightful feedback throughout the drafting process.

Ken Kurson was one of the few Republican friends I had when my father-in-law decided to run for office. His perspective helped me better understand what was happening in the country, and he also helped bring experienced operatives and creative ideas to an upstart campaign. My time in politics deepened our two-decade friendship. From the inception of this endeavor, Ken's brutally honest feedback and inventive suggestions have made this a better book.

Cassidy Luna played a pivotal role over the last five years in helping me manage my responsibilities. When people asked how we accomplished so much with so few people, I would say that she was my army of one. She sat through hours of interviews for this project and added her own recollections. She offered skillful suggestions, helped verify important details, oversaw the photo selection process, and has kept our team organized every step of the way.

Charlton Boyd was a constant team member in the White House and continues to work with me today. His careful transcription of hours of interviews made this book possible.

John J. Miller edited the manuscript with swiftness, artistry, and sterling professionalism. He offered an invaluable outside perspective, trimmed fat in the manuscript, and helped us ensure that each chapter opened with verve.

Jennifer Montazzoli researched and verified thousands of facts that are woven into nearly every page of this book. Her steadfast work and careful research have helped make this book a highly detailed account of history. In the final phase of the project, Robert Gabriel helped oversee the fact-check process, and I am grateful for his swift and thorough assist.

My agents David Vigliano and Tom Flannery offered sound advice and fresh perspective from start to finish. Tom made excellent edits to keep this an engaging and fast-paced read.

I want to thank the entire team at HarperCollins who made this book possible, as well the numerous friends who graciously read the manuscript and offered invaluable suggestions. I especially want to thank Hope Hicks, Dan Scavino, Brian Hook, Bob Lighthizer, Bobby Marilyn Stadtmauer, Loretta, and Robin for their insightful feedback. I would also like to thank the many colleagues and counterparts—including Luis Videgaray Caso, Yousef Al Otaiba, Dina Powell, Jason Greenblatt, David Friedman, Dan Brouillette, Adam Boeller, Brad Smith, Paul Mango, Kimberly Breier, Matt Pottinger, Ric Grenell, Brooke Rollins, and Ja'Ron Smith—who sat for interviews and helped with our extensive fact-checking. Thanks as well to Eric Herschmann, Abbe Lowell, Chad Mizelle, and Ben Wizner for their counsel.

The experience and achievements described in these pages would never have happened without the tireless efforts of dozens of people quietly working behind the scenes. Many friends and acquaintances stepped up to help our country when needed and these acts of selfless altruism were appreciated and made a big impact. There are far too many to name, but I am forever grateful to the many devoted men and women in the White House, the Secret Service, and across the federal government who served—and in some cases, continue to serve—our country with the highest standards of professionalism.

Notes

The citations below document sources such as books, government reports, and statistics. In general, the citations do not include news articles, quotations from public speeches, and other information readily available through public transcripts and press reports on the internet. As stated in the preface, many of the quotes in the book are drawn from published records, such as transcripts, but others come from private conversations. In these cases, I've relied on my memory and extensive interviews with colleagues and counterparts. In some cases, dialogue has been recreated. Quotes from foreign leaders are based on my memory and therefore are not direct quotes.*

1. Bernard Schoenburg, "Donald Trump Packs Prairie Capital Convention Center in Springfield," *Peoria Journal Star*, November 9, 2015, https://www.pjstar.com/story /news/2015/11/10/donald-trump-packs-prairie-capital/33140750007/.
2. Charles Murray, *Coming Apart: The State of White America, 1960–2010* (New York: Cox and Murray, 2012).
3. Sarah Frier, "Trump Campaign Said It Was Better at Facebook. Facebook Agrees," *Bloomberg*, April 3, 2018, https://www.bloomberg.com/news/articles/2018-04-03 /trump-s-campaign-said-it-was-better-at-facebook-facebook-agrees.
4. Andrew Bosworth, "The NYT Recently Obtained a Copy of a Post I Made to the Wall of My Internal Profile," Facebook, January 7, 2020, 1:51 p.m., https://www.facebook .com/boz/posts/10111288357877121.
5. Amity Shales, *The Forgotten Man: A New History of the Great Depression* (New York: HarperCollins, 2007).
6. "Chapo Guzman's Rumored $100M Bounty on Donald Trump Is 'Reason for Concern,'" Fox News, January 10, 2017, https://www.foxnews.com/world/chapo-guzmans -rumored-100m-bounty-on-donald-trump-is-reason-for-concern.
7. Jonathan Allen and Amie Parnes, *Shattered: Inside Hillary Clinton's Doomed Campaign* (New York: Crown, 2018).
8. Peter Navarro, *Death by China: Confronting the Dragon—A Global Call to Action* (Upper Saddle River, NJ: Pearson, 2011).
9. H. R. McMaster, *Dereliction of Duty: Lyndon Johnson, Robert McNamara, the Joint Chiefs of Staff, and the Lies That Led to Vietnam* (New York: HarperCollins, 1998).
10. Bob Woodward, *Fear: Trump in the White House* (New York: Simon & Schuster, 2019).
11. "Jared Kushner's Statement on Russia to Congressional Committees," CNN, July 24, 2017, https://www.cnn.com/2017/07/24/politics/jared-kushner-statement-russia-2016 -election/index.html.
12. Chris Whipple, *The Gatekeepers: How the White House Chiefs of Staff Define Every Presidency* (New York: Broadway Books, 2017).
13. Sun Tzu, *The Art of War* (London: Harper Press, 2011).
14. Henry Kissinger, *Diplomacy* (New York: Simon & Schuster, 1994).
15. Bernard Gwertzman, "Kissinger Urges Israelis to Map a New Truce Line," *New York Times*, May 6, 1974, https://www.nytimes.com/1974/05/06/archives/kissinger-urges-is raelis-to-map-a-new-truce-line.html.
16. Document 88, *Foreign Relations of the United States, 1969–1976,* vol. 26, *Arab-Israeli Dispute, 1974–1976,* ed. Adam M. Howard (Washington, DC: Government Printing

Office, 1974), https://history.state.gov/historicaldocuments/frus1969–76v26/d88.

17. Dore Gold, *The Fight for Jerusalem: Radical Islam, the West, and the Future of the Holy City* (Washington, DC: Regnery, 2009).

18. Jimmy Carter, *Palestine: Peace Not Apartheid* (New York: Simon & Schuster, 2006).

19. "Arab Peace Initiative: Full Text," *Guardian*, March 28, 2002, https://www.theguardian.com/world/2002/mar/28/israel7.

20. David Pollock, "Half of Jerusalem's Palestinians Would Prefer Israeli to Palestinian Citizenship," Washington Institute for Near East Policy, August 21, 2015, https://www.washingtoninstitute.org/policy-analysis/half-jerusalems-palestinians-would-prefer-israeli-palestinian-citizenship.

21. Charles Dickens, *Great Expectations*, ed. Edgar Rosenberg (New York: W. W. Norton, 1999).

22. *Annual Operational Report 2018*, United Nations Relief and Works Agency, 2019, https://www.unrwa.org/sites/default/files/content/resources/2019_annual_operation al_report_2018_-_final_july_20_2019.pdf.

23. Robert E. Scott, "Growth in U.S.-China Trade Deficit between 2001 and 2015 Cost 3.4 Million Jobs," Economic Policy Institute, January 31, 2017, https://www.epi.org/publication/growth-in-u-s-china-trade-deficit-between-2001-and-2015-cost-3-4-million-jobs-heres-how-to-rebalance-trade-and-rebuild-american-manufacturing; "Trade in Goods with China," United States Census Bureau, https://www.census.gov/foreign-trade/balance/c5700.html.

24. "Billion-Dollar Secrets Stolen," Federal Bureau of Investigation, May 27, 2020, https://www.fbi.gov/news/stories/scientist-sentenced-for-theft-of-trade-secrets-0527 20#:~:text=Tan's%20theft%20of%20a%20trade,of%20U.S.%20companies%20 and%20facilities; "Treasury Designates China as a Currency Manipulator," U.S. Department of the Treasury, August 5, 2019, https://home.treasury.gov/news/press-releases /sm751#:~:text=Under%20Section%203004%20of%20the,trade.%E2%80%9D%20 Secretary%20Mnuchin%2C%20under; Robert E. Scott, "Growth in U.S.–China Trade Deficit," Economic Policy Institute, January 31, 2017; *Findings of the Investigation into China's Acts, Policies, and Practices Related to Technology Transfer, Intellectual Property, and Innovation under Section 301 of the Trade Act of 1974*, Office of the United States Trade Representative, March 22, 2018, https://ustr.gov/sites/default/files/Sec tion%20301%20FINAL.PDF.

25. Michael Pillsbury, *The Hundred-Year Marathon: China's Secret Strategy to Replace America as the Global Superpower* (New York: Henry Holt, 2015).

26. "Did the Trump Administration's Child Tax Credit 'Put over $2,000 into the Pockets of 40 Million American Families'?," CNN Facts First, https://edition.cnn.com/factsfirst /politics/factcheck_5f7e8ab2-5389-4690-9f24-86a149a749dd.

27. Dov Lieber, "Bitter Abbas to Trump: We reject your peace 'deal of the century,'" *Times of Israel*, January 14, 2018, https://www.timesofisrael.com/bitter-abbas-to-trump -we-reject-your-peace-deal-of-the-century/.

28. Peter Wagner and Wanda Bertram, "'What Percent of the U.S. Is Incarcerated?' (and Other Ways to Measure Mass Incarceration)," Prison Policy Initiative, January 6, 2020, https://www.prisonpolicy.org/blog/2020/01/16/percent-incarcerated/.

29. "Federal Sentencing Reform," American Bar Association, accessed February 28, 2022, https://www.americanbar.org/advocacy/governmental_legislative_work/priorities_poli cy/criminal_justice_system_improvements/federalsentencingreform/.

30. Timothy Hughes and Doris J. Wilson, *Reentry Trends in the United States*, U.S. Department of Justice Bureau of Justice Statistics, revised April 14, 2004, https://bjs.ojp.gov /content/pub/pdf/reentry.pdf.

31. Anita Kumar, "Trump White House Departs from Recent Security Clearance Norms,"

McClatchy Washington Bureau, February 20, 2018, https://www.mcclatchydc.com/news/politics-government/white-house/article199997364.html.

32. Hakeem Jeffries to "Colleague," Washington, DC, May 18, 2018, https://www.politico.com/f/?id=00000163-73c9-d627-a5e3-7fcfc3110001.

33. Andrew Chatzky, James McBride, and Mohammed Aly Sergie, "NAFTA and the USMCA: Weighing the Impact of North American Trade," Council on Foreign Relations, updated July 1, 2020, https://www.cfr.org/backgrounder/naftas-economic-impact.

34. "U.S. Exports to Mexico by 5-Digit End-Use Code, 2011–2020," Foreign Trade, United States Census Bureau, accessed February 28, 2020, https://www.census.gov/foreign-trade/statistics/product/enduse/exports/c2010.html.

35. Ana Swanson and Jim Tankersley, "Trump Just Signed the U.S.M.C.A. Here's What's in the New NAFTA," *New York Times*, January 29, 2020, https://www.nytimes.com/2020/01/29/business/economy/usmca-deal.html.

36. Matthew Kassel, "Ultrafiltered Milk Sparks a U.S.–Canada Trade Battle," *Wall Street Journal*, May 14, 2017, https://www.wsj.com/articles/ultrafiltered-milk-sparks-a-u-s-canada-trade-battle-1494813601.

37. Caitlin Dewey, "Trump's Sudden Preoccupation with Canadian Milk, Explained," *Washington Post*, April 25, 2017, https://www.washingtonpost.com/news/wonk/wp/2017/04/25/president-trumps-sudden-preoccupation-with-milk-explained.

38. Chris Christie, *Let Me Finish: Trump, the Kushners, Bannon, New Jersey, and the Power of In-Your-Face Politics* (New York: Hachette, 2019).

39. Christopher Mann, *FY2020 Defense Reprogrammings for Wall Funding: Backgrounder*, Congressional Research Service, March 24, 2020, https://crsreports.congress.gov/prod/uct/pdf/IN/IN11274.

40. The original words of this phrase are "Never shall I forget the small faces of the children whose bodies I saw transformed into smoke under a silent sky." Elie Wiesel, "Never Shall I Forget," in *Night*, transl. Marion Wiesel (New York: Hill and Wang, 2006).

41. *Vision 2030*, Kingdom of Saudi Arabia, https://www.vision2030.gov.sa/media/rc0b5oy1/saudi_vision203.pdf.

42. "Economic Framework," Trump White House Archives, January 20, 2020, https://trumpwhitehouse.archives.gov/peacetoprosperity/economic/.

43. "West Bank and Gaza," Human Rights Report Excerpt, Bureau of Democracy, Human Rights, and Labor, U.S. Department of State, accessed April 1, 2022, https://www.state.gov/report/custom/05a7423036/.

44. Memorandum from Kirstjen M. Nielsen, Secretary, U.S. Department of Homeland Security, to L. Francis Cissna, Director, U.S. Citizenship and Immigration Services, et al., subject "Policy Guidance for Implementation of the Migrant Protection Protocols," Washington, DC, January 25, 2019, https://www.dhs.gov/sites/default/files/publications/19_0129_OPA_migrant-protection-protocols-policy-guidance.pdf.

45. Paul Sonne, Michael Kranish, and Matt Viser, "The Gas Tycoon and the Vice President's Son: The Story of Hunter Biden's Foray into Ukraine," *Washington Post*, September 28, 2019, https://www.washingtonpost.com/world/national-security/the-gas-tycoon-and-the-vice-presidents-son-the-story-of-hunter-bidens-foray-in-ukraine/2019/09/28/1aadff70-dfd9-11e9-8fd3-d943b4ed57e0_story.html; Julia Ainsley, "Hunter Biden Tax Inquiry Focused on Chinese Business Dealings," NBC News, December 10, 2020, https://www.nbcnews.com/politics/justice-department/hunter-biden-tax-probe-focused-chinese-business-dealings-n1250772; Colleen McCain Nelson and Julian E. Barnes, "Biden's Son Hunter Discharged from Navy Reserve after Failing Cocaine Test," *Wall Street Journal*, October 16, 2014, https://www.wsj.com/articles/bidens-son-hunter-discharged-from-navy-reserve-after-failing-cocaine-test-1413499657.

46. Adam Taylor, "Hunter Biden's New Job at a Ukrainian Gas Company Is a Problem for

U.S. Soft Power," *Washington Post*, May 14, 2014, https://www.washingtonpost.com /news/worldviews/wp/2014/05/14/hunter-bidens-new-job-at-a-ukrainian-gas -company-is-a-problem-for-u-s-soft-power/; Brie Stimson, "Hunter Biden Got $83G per Month for Ukraine 'Ceremonial' Gig: Report," Fox News, October 19, 2019, https://www.foxnews.com/politics/hunter-biden-paid-80g-per-month-while-on-board -of-ukranian-gas-company-report.

47. Courtney Subramanian, "Explainer: Biden, Allies Pushed Out Ukrainian Prosecutor Because He Didn't Pursue Corruption Cases," *USA Today*, October 3, 2019, https:// www.usatoday.com/story/news/politics/2019/10/03/what-really-happened-when -biden-forced-out-ukraines-top-prosecutor/3785620002/.

48. "Read Trump's Phone Conversation with Volodymyr Zelensky," CNN, September 24, 2019, https://www.cnn.com/2019/09/25/politics/donald-trump-ukraine-transcript -call/index.html.

49. Steve Herman (@W7VOA), "Purported Soleimani photo evidence tweeted by an Iranian journalist," Twitter, January 2, 2020, 9:08 p.m., https://twitter.com/w7voa/status /1212918590774632448?lang=en.

50. Lawrence Wright, *Thirteen Days in September* (New York: Penguin Random House, 2014).

51. "Trade in Goods with European Union," United States Census Bureau, https://www .census.gov/foreign-trade/balance/c0003.html.

52. Hannah Ritchie et al., "Coronavirus (COVID-19) Testing," Our World in Data, https://ourworldindata.org/coronavirus-testing.

53. Ivan Watson et al., "How This South Korean Company Created Coronavirus Test Kits in Three Weeks," CNN, March 12, 2020, https://www.cnn.com/2020/03/12/asia /coronavirus-south-korea-testing-intl-hnk/index.html.

54. Google Communications (@Google_Comms), "Statement from Verily: 'We are developing a tool to help triage individuals for Covid-19 testing. Verily is in the early stages of development, and planning to roll testing out in the Bay Area, with the hope of expanding more broadly over time,'" Twitter, March 13, 2020, 5:16 p.m., https://twitter .com/Google_Comms/status/1238574670686928906.

55. "15 Days to Slow the Spread," Trump White House Archives, March 16, 2020, https:// trumpwhitehouse.archives.gov/articles/15-days-slow-spread/.

56. Amy Stillman, Javier Blas, Grant Smith, and Salma El Wardany, "Mexico Reaches Deal with U.S. to Cut Oil Production Allowing for OPEC+ Output Cuts," *Time*, April 9, 2020, https://time.com/5818938/opec-oil-deal-coronavirus-mexico.

57. *2020 Annual Report*, Commission on Accreditation for Law Enforcement Agencies, Inc., https://calea.org/annual-reports/CALEA-AR_2020_final.html.

58. "Joint Statement of the United States, the State of Israel, and the United Arab Emirates," U.S. Embassy in Israel, August 13, 2020, https://il.usembassy.gov/joint-statement-of -the-united-states-the-state-of-israel-and-the-united-arab-emirates/.

59. Aya Batrawy and Isabel Debre, "Saudi Arabia to lift Qatar embargo, easing the Gulf crisis," Associated Press, January 4, 2021, https://apnews.com/article/qatar-saudi-arabia -united-arab-emirates-kuwait-dubai-0a09c370d8430b93d32e403aaf4d3f2a.

60. "The Abraham Accords Declaration," Bureau of Near Eastern Affairs, U.S. Department of State, https://www.state.gov/the-abraham-accords/.

61. *Abraham Accords: Declaration of Peace, Cooperation, and Constructive Diplomatic and Friendly Relations*, U.S. Department of State, September 15, 2020, https://www.state .gov/wp-content/uploads/2020/09/Bahrain_Israel-Agreement-signed-FINAL-15 -Sept-2020–508.pdf.

62. *Official 2020 Presidential General Election Results*, Federal Election Commission, January 28, 2021, https://www.fec.gov/resources/cms-content/documents/2020presgere sults.pdf.

63. *Times of Israel* Staff, "Sudan Fires Spokesman Who Confirmed Peace Talks with Israel," *Times of Israel,* August 19, 2020, https://www.timesofisrael.com/sudan-fires-spokesman-who-confirmed-peace-talks-with-israel/.

64. *The Abraham Accords Declaration, for the Republic of Sudan,* U.S. Department of State, https://www.state.gov/wp-content/uploads/2021/01/Sudan-AA.pdf.

65. *Agreement between the United States of America, Morocco and Israel,* U.S. Department of State, December 22, 2020, https://www.state.gov/wp-content/uploads/2021/05/20-1222-Morocco-Israel-Joint-Declaration.pdf.

66. The facts in this paragraph originated from "The COVID-19 Response: A Whole America Effort," White House Internal Backgrounder.

Index

About the Author

JARED KUSHNER is the founder of Affinity Partners, a global investment firm. Previously, he served as senior adviser to President Donald J. Trump, and before that, as CEO of Kushner Companies. He also cofounded two technology companies, Cadre and WiredScore. In 2015, he was named to *Fortune*'s 40 under 40, and in 2017 was named one of *Time*'s 100 Most Influential People. In 2018, Jared received the Aztec Eagle Award, Mexico's highest honor, for his work on the USMCA trade agreement. He was given a presidential citation for helping to architect Operation Warp Speed, which produced COVID-19 vaccines in record time. In recognition of his success negotiating the Abraham Accords, Israel planted the Kushner Garden of Peace in the Grove of Nations outside Jerusalem. He also was awarded the National Security Medal, the Department of Defense Medal for Distinguished Public Service, and the Grand Cordon of the Order of Ouissam Alaouite from King Mohammed VI of Morocco. He lives in Florida with his wife, Ivanka, and their three children.